MW00843407

Springer Series in Statistics

Springer
New York
Berlin
Heidelberg
Barcelona
Hong Kong
London
Milan
Paris
Singapore
Tokyo

Springer Series in Statistics

Andersen/Borgan/Gill/Keiding: Statistical Models Based on Counting Processes.
Atkinson/Riani: Robust Diagnotstic Regression Analysis.
Berger: Statistical Decision Theory and Bayesian Analysis, 2nd edition.
Bolfarine/Zacks: Prediction Theory for Finite Populations.
Borg/Groenen: Modern Multidimensional Scaling: Theory and Applications
Brockwell/Davis: Time Series: Theory and Methods, 2nd edition.
Chan/Tong: Chaos: A Statistical Perspective.
Chen/Shao/Ibrahim: Monte Carlo Methods in Bayesian Computation.
David/Edwards: Annotated Readings in the History of Statistics.
Devroye/Lugosi: Combinatorial Methods in Density Estimation.
Efromovich: Nonparametric Curve Estimation: Methods, Theory, and Applications.
Eggermont/LaRiccia: Maximum Penalized Likelihood Estimation, Volume I: Density Estimation.
Fahrmeir/Tutz: Multivariate Statistical Modelling Based on Generalized Linear Models, 2nd edition.
Farebrother: Fitting Linear Relationships: A History of the Calculus of Observations 1750-1900.
Federer: Statistical Design and Analysis for Intercropping Experiments, Volume I: Two Crops.
Federer: Statistical Design and Analysis for Intercropping Experiments, Volume II: Three or More Crops.
Fienberg/Hoaglin/Kruskal/Tanur (Eds.): A Statistical Model: Frederick Mosteller's Contributions to Statistics, Science and Public Policy.
Fisher/Sen: The Collected Works of Wassily Hoeffding.
Glaz/Naus/Wallenstein: Scan Statistics.
Good: Permutation Tests: A Practical Guide to Resampling Methods for Testing Hypotheses, 2nd edition.
Gouriéroux: ARCH Models and Financial Applications.
Grandell: Aspects of Risk Theory.
Gu: Smoothing Spline ANOVA Models.
Haberman: Advanced Statistics, Volume I: Description of Populations.
Hall: The Bootstrap and Edgeworth Expansion.
Härdle: Smoothing Techniques: With Implementation in S.
Harrell: Regression Modeling Strategies: With Applications to Linear Models, Logistic Regression, and Survival Analysis
Hart: Nonparametric Smoothing and Lack-of-Fit Tests.
Hartigan: Bayes Theory.
Hastie/Tibshirani/Friedman: The Elements of Statistical Learning: Data Mining, Inference, and Prediction
Hedayat/Sloane/Stufken: Orthogonal Arrays: Theory and Applications.
Heyde: Quasi-Likelihood and its Application: A General Approach to Optimal Parameter Estimation.
Huet/Bouvier/Gruet/Jolivet: Statistical Tools for Nonlinear Regression: A Practical Guide with S-PLUS Examples.

(continued after index)

Chong Gu

Smoothing Spline ANOVA Models

With 42 Illustrations

Chong Gu
Department of Statistics
Purdue University
West Lafayette, IN 47907
USA

Library of Congress Cataloging-in-Publication Data
Gu, Chong.
Smoothing spline ANOVA models / Chong Gu
p. cm. — (Springer series in statistics)
Includes bibliographical references and indexes.
ISBN 0-387-95353-1 (alk. paper)
1. Analysis of variance. 2. Spline theory. 3. Smoothing (Statistics) I. Title.
II. Series
QA279.G8 2002
519.5′38—dc21 2001053052

Printed on acid-free paper.

Production managed by Lesley Poliner; manufacturing supervised by Jeffrey Taub.
Camera-ready copy provided by the author from his LaTeX files.
Printed and bound by Maple-Vail Book Manufacturing Group, York, PA.
Printed in the United States of America.

9 8 7 6 5 4 3 2 1

ISBN 0-387-95353-1 SPIN 10851021

Springer-Verlag New York Berlin Heidelberg
A member of BertelsmannSpringer Science+Business Media GmbH

To my father
For the books and the bookcases

Preface

Thirty years have passed since the pioneering work of Kimeldorf and Wahba (1970a, 1970b, 1971) and Good and Gaskins (1971), and during this time, a rich body of literature has been developed on smoothing methods with roughness penalties. There have been two books solely devoted to the subject prior to this one, of which Wahba (1990) compiled an excellent synthesis for work up to that date, and Green and Silverman (1994) provided a mathematically gentler introduction to the field through regression models that are largely univariate.

Much has happened in the past decade, and more has been done with the penalty method than just regression. In this book, I have tried to assemble a comprehensive treatment of penalty smoothing under a unified framework. Treated are (i) regression with Gaussian and non-Gaussian responses as well as with censored lifetime data, (ii) density and conditional density estimation under a variety of sampling schemes, and (iii) hazard rate estimation with censored lifetime data and covariates. The unifying themes are the general penalized likelihood method and the construction of multivariate models with certain ANOVA decompositions built in. Extensive discussions are devoted to model (penalty) construction, smoothing parameter selection, computation, and asymptotic convergence. There are, however, many omissions, and the selection and treatment of topics solely reflect my personal preferences and views. Most of the materials have appeared in the literature, but a few items are new, as noted in the bibliographic notes at the end of the chapters.

An adequate treatment of model construction in the context requires some elementary knowledge of reproducing kernel Hilbert spaces, of which a self-contained introduction is included early in the book; the materials should be accessible to a second-year graduate student with a good training in calculus and linear algebra. Also assumed is a working knowledge of basic statistical inference such as linear models, maximum likelihood estimates, etc. To better understand materials on hazard estimation, prior knowledge of basic survival analysis would also help.

Most of the computational and data analytical tools discussed in the book are implemented in R, an open-source clone of the popular S/Splus language. Code for regression is reasonably polished and user-friendly and has been distributed in the R package `gss` available through CRAN, the Comprehensive R Archive Network, with the master site at

`http://cran.r-project.org`

The use of `gss` facilities is illustrated in the book through simulated and real-data examples.

Remaining on my wish list are (i) polished, user-friendly software tools for density estimation and hazard estimation, (ii) fast computation via approximate solutions of penalized likelihood problems, and (iii) handling of parametric random effects such as those appearing in longitudinal models and hazard models with frailty. All of the above are under active development and could be addressed in a later edition of the book or, sooner than that, in later releases of `gss`.

The book was conceived in Spring 1996 when I was on leave at Department of Statistics, University of Michigan, which offered me the opportunity to teach a course on the subject. Work on the book has been on and off since then, with much of the progress being made in the 1997–1998 academic year during my visit at National Institute of Statistical Sciences, and in Fall 2000 when I was teaching a course on the subject at Purdue.

I am indebted to Grace Wahba, who taught me smoothing splines, and to Doug Bates, who taught me statistical computing. Bill Studden carefully read various drafts of Chapters 1, 2, and 4; his questions alerted me to numerous accounts of mathematical sloppiness in the text and his suggestions led to much improved presentations. Detailed comments and suggestions by Nancy Heckman on a late draft helped me to fix numerous problems throughout the first five chapters and to shape the final organization of the book (e.g., the inclusion of §1.4). For various ways in which they helped, I would also like to thank Mary Ellen Bock, Jerry Davis, Nels Grevstad, Wensheng Guo, Alan Karr, Youngju Kim, Ping Ma, Jerry Sacks, Jingyuan Wang, Yuedong Wang, Jeff Wu, Dong Xiang, Liqing Yan, and the classes at Michigan and Purdue. Last but not least, I would like to thank the R Core Team, for creating a most enjoyable platform for statistical computing.

Chong Gu
West Lafayette, Indiana
July 2001

Contents

Preface ... vii

1 Introduction ... 1
1.1 Estimation Problem and Method ... 2
1.1.1 Cubic Smoothing Spline ... 2
1.1.2 Penalized Likelihood Method ... 4
1.2 Notation ... 5
1.3 Decomposition of Multivariate Functions ... 6
1.3.1 ANOVA and Averaging Operator ... 6
1.3.2 Multiway ANOVA Decomposition ... 7
1.3.3 Multivariate Statistical Models ... 10
1.4 Case Studies ... 12
1.4.1 Water Acidity in Lakes ... 12
1.4.2 AIDS Incubation ... 14
1.4.3 Survival After Heart Transplant ... 15
1.5 Scope ... 17
1.6 Bibliographic Notes ... 18
1.7 Problems ... 19

2 Model Construction ... 21
2.1 Reproducing Kernel Hilbert Spaces ... 22
2.1.1 Hilbert Spaces and Linear Subspaces ... 22
2.1.2 Riesz Representation Theorem ... 27

2.1.3 Reproducing Kernel and Non-negative Definite Function . . . 27
2.2 Smoothing Splines on $\{1, \ldots, K\}$ . . . 30
2.3 Polynomial Smoothing Splines on $[0, 1]$ . . . 32
2.3.1 A Reproducing Kernel in $\mathcal{C}^{(m)}[0, 1]$ . . . 32
2.3.2 Computation of Polynomial Smoothing Splines . . . 34
2.3.3 Another Reproducing Kernel in $\mathcal{C}^{(m)}[0, 1]$ . . . 35
2.4 Smoothing Splines on Product Domains . . . 38
2.4.1 Tensor Product Reproducing Kernel Hilbert Spaces . . . 38
2.4.2 Reproducing Kernel Hilbert Spaces on $\{1, \ldots, K\}^2$ . . . 39
2.4.3 Reproducing Kernel Hilbert Spaces on $[0, 1]^2$ . . . 40
2.4.4 Reproducing Kernel Hilbert Spaces on $\{1, \ldots, K\} \times [0, 1]$ . . . 43
2.4.5 Multiple-Term Reproducing Kernel Hilbert Spaces . . . 43
2.5 Bayes Model . . . 46
2.5.1 Shrinkage Estimates as Bayes Estimates . . . 46
2.5.2 Polynomial Splines as Bayes Estimates . . . 47
2.5.3 Smoothing Splines as Bayes Estimates . . . 49
2.6 Minimization of Penalized Functional . . . 50
2.6.1 Existence of Minimizer . . . 50
2.6.2 Penalized and Constrained Optimization . . . 52
2.7 Bibliographic Notes . . . 53
2.8 Problems . . . 55

3 Regression with Gaussian-Type Responses **59**
3.1 Preliminaries . . . 60
3.2 Smoothing Parameter Selection . . . 62
3.2.1 Unbiased Estimate of Relative Loss . . . 63
3.2.2 Generalized Cross-Validation . . . 65
3.2.3 Restricted Maximum Likelihood . . . 68
3.2.4 Weighted and Replicated Data . . . 69
3.2.5 Empirical Performance . . . 70
3.3 Bayesian Confidence Intervals . . . 72
3.3.1 Posterior Distribution . . . 72
3.3.2 Confidence Intervals on Sampling Points . . . 74
3.3.3 Across-the-Function Coverage . . . 75
3.4 Computation: Generic Algorithms . . . 76
3.4.1 Algorithm for Fixed Smoothing Parameters . . . 76
3.4.2 Algorithm for Single Smoothing Parameter . . . 77
3.4.3 Algorithm for Multiple Smoothing Parameters . . . 79
3.4.4 Calculation of Posterior Variances . . . 80
3.5 Software . . . 81

3.5.1 RKPACK 81
3.5.2 R Package gss 82
3.6 Model Checking Tools 86
3.6.1 Cosine Diagnostics 87
3.6.2 Examples 87
3.6.3 Concepts and Heuristics 91
3.7 Case Studies 93
3.7.1 Nitrogen Oxides in Engine Exhaust 93
3.7.2 Ozone Concentration in Los Angeles Basin 94
3.8 Computation: Special Algorithms 99
3.8.1 Fast Algorithm for Polynomial Splines 100
3.8.2 Monte Carlo Cross-Validation 101
3.9 Bibliographic Notes 103
3.10 Problems 105

4 More Splines **111**
4.1 Partial Splines 112
4.2 Splines on the Circle 113
4.2.1 Periodic Reproducing Kernel Hilbert Spaces 114
4.2.2 Splines as Low-Pass Filters 114
4.2.3 More on Asymptotics of §3.2 116
4.3 L-Splines 119
4.3.1 Trigonometric Splines 120
4.3.2 Chebyshev Splines 122
4.3.3 General Construction 126
4.3.4 Case Study: Weight Loss of Obese Patient 130
4.3.5 Fast Algorithm 134
4.4 Thin-Plate Splines 135
4.4.1 Semi-Kernels for Thin-Plate Splines 136
4.4.2 Reproducing Kernels for Thin-Plate Splines 138
4.4.3 Tensor Product Thin-Plate Splines 140
4.4.4 Case Study: Water Acidity in Lakes 141
4.5 Bibliographic Notes 143
4.6 Problems 144

5 Regression with Exponential Families **149**
5.1 Preliminaries 150
5.2 Smoothing Parameter Selection 151
5.2.1 Performance-Oriented Iteration 152
5.2.2 Direct Cross-Validation 155
5.2.3 Empirical Performance 158
5.3 Approximate Bayesian Confidence Intervals 159
5.4 Software: R Package gss 161
5.4.1 Binomial Family 162
5.4.2 Poisson Family 163

5.4.3 Gamma Family 164
5.4.4 Inverse Gaussian Family 165
5.4.5 Negative Binomial Family 166
5.5 Case Studies 168
5.5.1 Eruption Time of Old Faithful 168
5.5.2 Spectrum of Yearly Sunspots 170
5.5.3 Progression of Diabetic Retinopathy 171
5.6 Bibliographic Notes 173
5.7 Problems 175

6 Probability Density Estimation **177**
6.1 Preliminaries 178
6.2 Poisson Intensity 182
6.3 Smoothing Parameter Selection 183
6.3.1 Kullback-Leibler Loss and Cross-Validation 183
6.3.2 Modifications of Cross-Validation Score 185
6.3.3 Empirical Performance 187
6.4 Computation 187
6.5 Case Studies 190
6.5.1 Buffalo Snowfall 190
6.5.2 Eruption Time of Old Faithful 191
6.5.3 AIDS Incubation 191
6.6 Biased Sampling and Random Truncation 192
6.6.1 Biased and Truncated Samples 193
6.6.2 Penalized Likelihood Estimation 194
6.6.3 Empirical Performance 196
6.6.4 Case Study: AIDS Incubation 198
6.7 Conditional Densities 198
6.7.1 Penalized Likelihood Estimation 199
6.7.2 Case Study: Penny Thickness 200
6.7.3 Logistic Regression 202
6.8 Response-Based Sampling 204
6.8.1 Response-Based Samples 204
6.8.2 Penalized Likelihood Estimation 206
6.9 Bibliographic Notes 206
6.10 Problems 209

7 Hazard Rate Estimation **211**
7.1 Preliminaries 212
7.2 Smoothing Parameter Selection 214
7.2.1 Kullback-Leibler Loss and Cross-Validation 215
7.2.2 Empirical Performance 217
7.3 Case Studies 218
7.3.1 Treatments of Gastric Cancer 218
7.3.2 Survival After Heart Transplant 219

7.4 Penalized Partial Likelihood 221
7.4.1 Partial Likelihood and Biased Sampling 221
7.4.2 Case Study: Survival After Heart Transplant 221
7.5 Models Parametric in Time 222
7.5.1 Accelerated Life Models 222
7.5.2 Weibull Family 224
7.5.3 Log Normal Family 225
7.5.4 Log Logistic Family 226
7.5.5 Case Study: Survival After Heart Transplant 227
7.6 Bibliographic Notes 228
7.7 Problems 230

8 Asymptotic Convergence **231**
8.1 Preliminaries 231
8.2 Rates for Density Estimates 234
8.2.1 Linear Approximation 235
8.2.2 Approximation Error and Main Results 237
8.2.3 Semiparametric Approximation 239
8.2.4 Convergence Under Incorrect Model 242
8.2.5 Estimation Under Biased Sampling 243
8.2.6 Estimation of Conditional Density 244
8.2.7 Estimation Under Response-Based Sampling 245
8.3 Rates for Hazard Estimates 245
8.3.1 Martingale Structure 246
8.3.2 Linear Approximation 247
8.3.3 Approximation Error and Main Result 248
8.3.4 Semiparametric Approximation 250
8.3.5 Convergence Under Incorrect Model 253
8.4 Rates for Regression Estimates 253
8.4.1 General Formulation 254
8.4.2 Linear Approximation 254
8.4.3 Approximation Error and Main Result 255
8.4.4 Convergence Under Incorrect Model 257
8.5 Bibliographic Notes 258
8.6 Problems 259

References **261**

Author Index **273**

Subject Index **277**

1 Introduction

Data and models are two sources of information in a statistical analysis. Data carry noise but are "unbiased," whereas models, effectively a set of constraints, help to reduce noise but are responsible for "biases." Representing the two extremes on the spectrum of "bias-variance" trade-off are standard parametric models and constraint-free nonparametric "models" such as the empirical distribution for a probability density. In between the two extremes, there exist scores of nonparametric or semiparametric models, of which most are also known as smoothing methods. A family of such nonparametric models in a variety of stochastic settings can be derived through the penalized likelihood method, forming the subject of this book.

The general penalized likelihood method can be readily abstracted from the cubic smoothing spline as the solution to a minimization problem, and its applications in regression, density estimation, and hazard estimation set out the subject of study (§1.1). Some general notation is set in §1.2. Multivariate statistical models can often be characterized through function decompositions similar to the classical analysis of variance (ANOVA) decomposition, which we discuss in §1.3. To illustrate the potential applications of the methodology, previews of selected case studies are presented in §1.4. Brief summaries of the chapters to follow are given in §1.5.

1.1 Estimation Problem and Method

The problem to be addressed in this book is flexible function estimation based on stochastic data. To allow for flexibility in the estimation of η, say, soft constraints of the form $J(\eta) \leq \rho$ are used in lieu of the rigid constraints of parametric models, where $J(\eta)$ quantifies the roughness of η and ρ sets the allowance; an example of $J(\eta)$ for η on $[0,1]$ is $\int_0^1 (d^2\eta/dx^2)^2 dx$. Solving the constrained maximum likelihood problem by the Lagrange method, one is led to the penalized likelihood method.

In what follows, a brief discussion of the cubic smoothing spline helps to motivate the idea, and a simple simulation illustrates the role of ρ through the Lagrange multiplier, better known as the smoothing parameter in the context. Following a straightforward abstraction, the penalized likelihood method is exemplified in regression, density estimation, and hazard estimation.

1.1.1 Cubic Smoothing Spline

Consider a regression problem $Y_i = \eta(x_i) + \epsilon_i$, $i = 1, \ldots, n$, where $x_i \in [0,1]$ and $\epsilon_i \sim N(0, \sigma^2)$. In a classical parametric regression analysis, η is assumed to be of the form $\eta(x, \boldsymbol{\beta})$, known up to the parameters $\boldsymbol{\beta}$, which are to be estimated from the data. When $\eta(x, \boldsymbol{\beta})$ is linear in $\boldsymbol{\beta}$, one has a standard linear model. A parametric model characterizes a set of rigid constraints on η. The dimension of the model space (i.e., the number of unknown parameters) is typically much smaller than the sample size n.

To avoid possible model misspecification in a parametric analysis, otherwise known as bias, an alternative approach to estimation is to allow η to vary in a high-dimensional (possibly infinite) function space, leading to various nonparametric or semiparametric estimation methods. A popular approach to the nonparametric estimation of η is via the minimization of a penalized least squares score,

$$\frac{1}{n}\sum_{i=1}^{n}(Y_i - \eta(x_i))^2 + \lambda \int_0^1 \ddot{\eta}^2 dx, \tag{1.1}$$

with $\ddot{\eta} = d^2\eta/dx^2$, where the first term discourages the lack of fit of η to the data, the second term penalizes the roughness of η, and the smoothing parameter λ controls the trade-off between the two conflicting goals. The minimization of (1.1) is implicitly over functions with square integrable second derivatives. The minimizer η_λ of (1.1) is called a cubic smoothing spline. As $\lambda \to 0$, η_λ approaches the minimum curvature interpolant. As $\lambda \to \infty$, η_λ approaches the simple linear regression line. Note that the linear polynomials $\{f : f = \beta_0 + \beta_1 x\}$ form the so-called null space of the roughness penalty $\int_0^1 \ddot{f}^2 dx$, $\{f : \int_0^1 \ddot{f}^2 dx = 0\}$.

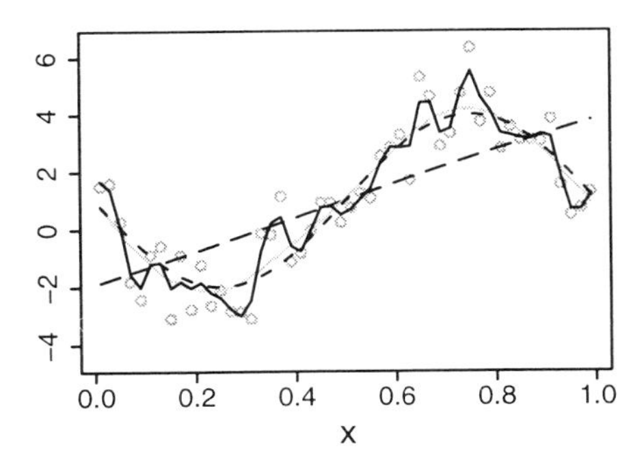

FIGURE 1.1. Cubic Smoothing Splines. The test function is indicated by the dashed line and the estimates are indicated by the solid, faded solid, and long-dashed lines. The data are superimposed as circles.

To illustrate, consider a simple simulation with $x_i = (i - 0.5)/50$, $i = 1, \ldots, 50$, $\eta(x) = 1 + 3\sin(2\pi x - \pi)$, and $\sigma^2 = 1$. The estimate η_λ was calculated at $\log_{10} n\lambda = 0, -3, -6$. Plotted in Figure 1.1 are the test function (dashed line), the estimates (solid, faded solid, and long-dashed lines), and the data (circles). The rough fit corresponds to $\log_{10} n\lambda = -6$, the near straight line to $\log_{10} n\lambda = 0$, and the close fit to $\log_{10} n\lambda = -3$.

An alternative derivation of the cubic smoothing spline is through a constrained least squares problem, which solves

$$\min \frac{1}{n} \sum_{i=1}^{n} (Y_i - \eta(x_i))^2, \quad \text{subject to} \quad \int_0^1 \ddot{\eta}^2 dx \leq \rho, \tag{1.2}$$

for some $\rho \geq 0$. The solution to (1.2) usually falls on the boundary of the permissible region, $\int_0^1 \ddot{\eta}^2 dx = \rho$, and by the Lagrange method, it can be calculated as the minimizer of (1.1) with an appropriate Lagrange multiplier λ. Thus, up to the choices of λ and ρ, a penalized least squares problem with a penalty proportional to $\int_0^1 \ddot{\eta}^2 dx$ is equivalent to a constrained least squares problem subject to a soft constraint of the form $\int_0^1 \ddot{\eta}^2 dx \leq \rho$; see, e.g., Schoenberg (1964). See also §2.6.2.

The minimizer η_λ of (1.1) is called a cubic spline because it is a piecewise cubic polynomial. It is three times differentiable, with the third derivative jumping at the knots $\xi_1 < \xi_2 < \cdots < \xi_q$, the ordered distinctive sampling points x_i, and it is linear beyond the first knot ξ_1 and the last knot ξ_q. In the numerical analysis literature, a smoothing spline is also called a natural spline. See Schumaker (1981, Chapter 8) for a comprehensive treatment of smoothing splines from a numerical analytical perspective. See also de Boor (1978).

1.1.2 Penalized Likelihood Method

The cubic smoothing spline of (1.1) is a specialization of the general penalized likelihood method in univariate Gaussian regression. To estimate a function of interest η on a generic domain $\mathcal{X}$ using stochastic data, one may use the minimizer of

$$L(\eta|\text{data}) + \frac{\lambda}{2}J(\eta), \tag{1.3}$$

where $L(\eta|\text{data})$ is usually taken as the minus log likelihood of the data and $J(f)$ is a quadratic roughness functional with a null space $\mathcal{N}_J = \{f : J(f) = 0\}$ of low dimension; see §2.1.1 for the definition of quadratic functional. The solution of (1.3) is the maximum likelihood estimate in a model space $\mathcal{M}_\rho = \{f : J(f) = \rho\}$ for some $\rho \geq 0$, and the smoothing parameter λ in (1.3) is the Lagrange multiplier. See §2.6.2 for a detailed discussion of the role of λ as a Lagrange multiplier.

A few examples of penalized likelihood estimation follow.

Example 1.1 (Response data regression)
Assume

$$Y|x \sim \exp\{(y\eta(x) - b(\eta(x)))/a(\phi) + c(y,\phi)\},$$

an exponential family density with a modeling parameter η and a possibly unknown nuisance parameter ϕ. Observing independent data (x_i, Y_i), $i = 1, \dots, n$, the method estimates η via minimizing

$$-\frac{1}{n}\sum_{i=1}^{n}\{Y_i\eta(x_i) - b(\eta(x_i))\} + \frac{\lambda}{2}J(\eta). \tag{1.4}$$

When the density is Gaussian, (1.4) reduces to a penalized least squares problem; see Problem 1.1. Penalized least squares regression for Gaussian-type responses is the subject of Chapter 3. Penalized likelihood regression for non-Gaussian responses will be studied in Chapter 5. □

Example 1.2 (Density estimation)
Observing independent and identically distributed samples X_i, $i = 1, \dots, n$, from a probability density $f(x)$ supported on a bounded domain $\mathcal{X}$, the method estimates f by $e^\eta / \int_\mathcal{X} e^\eta dx$, where η minimizes

$$-\frac{1}{n}\sum_{i=1}^{n}\left\{\eta(X_i) - \log\int_\mathcal{X} e^\eta dx\right\} + \frac{\lambda}{2}J(\eta). \tag{1.5}$$

A side condition, say $\int_\mathcal{X} \eta\, dx = 0$, shall be imposed on η for a one-to-one transform $f \leftrightarrow e^\eta / \int_\mathcal{X} e^\eta dx$. Penalized likelihood density estimation is the subject of Chapter 6. □

Example 1.3 (Hazard estimation)
Let T be the lifetime of an item with survival function $S(t|u) = P(T > t|u)$, possibly dependent on a covariate U. The hazard function is defined as $e^{\eta(t,u)} = -\partial \log S(t|u)/\partial t$. Let Z be the left-truncation time and C be the right-censoring time, independent of T and of each other. Observing $(U_i, Z_i, X_i, \delta_i)$, $i = 1, \ldots, n$, where $X = \min(T, C)$, $\delta = I_{[T \leq C]}$, and $Z < X$, the method estimates the log hazard η via minimizing

$$-\frac{1}{n}\sum_{i=1}^{n}\left\{\delta_i \eta(X_i, U_i) - \int_{Z_i}^{X_i} e^{\eta(t,U_i)}dt\right\} + \frac{\lambda}{2}J(\eta); \tag{1.6}$$

see Problem 1.2 for the derivation of the likelihood. Penalized likelihood hazard estimation will be studied in Chapter 7. □

The two basic components of a statistical model, the deterministic part and the stochastic part, are well separated in (1.3). The structure of the deterministic part is determined by the construction of $J(\eta)$ for η on the domain $\mathcal{X}$, of which a comprehensive treatment is presented in Chapter 2. The stochastic part is reflected in the likelihood $L(\eta|\text{data})$ and determines, among other things, the natural measures with which the performance of the estimate is to be assessed. The minimizer of (1.3) with a varying λ defines a family of estimates, and from the cubic spline simulation shown in Figure 1.1, we have seen how differently the family members may behave. Data-driven procedures for the proper selection of the smoothing parameter are crucial to the practicability of penalized likelihood estimation, to which extensive discussion will be devoted in the settings of regression, density estimation, and hazard estimation in their respective chapters.

1.2 Notation

Listed below is some general notation used in this book. Context-specific or subject-specific notation may differ from that listed here, in which case every effort will be made to avoid possible confusion.

Domains are usually denoted by $\mathcal{X}$, $\mathcal{Y}$, $\mathcal{Z}$, etc., or subscripted as $\mathcal{X}_1$, $\mathcal{X}_2$, etc. Points on domains are usually denoted by $x \in \mathcal{X}$, $y \in \mathcal{Y}$, or $x_1, x_2, y \in \mathcal{X}$. Points on product domains are denoted by $x_1, x_2, y \in \mathcal{X} = \mathcal{X}_1 \times \mathcal{X}_2$, with $x_{1\langle 1\rangle}, x_{2\langle 1\rangle}, y_{\langle 1\rangle} \in \mathcal{X}_1$ and $x_{1\langle 2\rangle}, x_{2\langle 2\rangle}, y_{\langle 2\rangle} \in \mathcal{X}_2$, or by $z = (x, y) \in \mathcal{Z} = \mathcal{X} \times \mathcal{Y}$, with $x \in \mathcal{X}$ and $y \in \mathcal{Y}$. Ordinary subscripts are used to denote multiple points on a domain, but *not* coordinates of a point on a product domain.

Function spaces are usually denoted by $\mathcal{H}$, $\mathcal{G}$, etc. Functions in function spaces are usually denoted by $f, g, h \in \mathcal{H}$, $\eta, \phi, \xi \in \mathcal{H}$, etc. Derivatives of a univariate function $f(x)$ are denoted by $\dot{f} = df/dx$, $\ddot{f} = d^2f/dx^2$, or by the general notation $f^{(m)} = d^m f/dx^m$. Derivatives of multivariate functions $f(x_{\langle 1 \rangle}, x_{\langle 2 \rangle})$ on $\mathcal{X}_1 \times \mathcal{X}_2$ or $g(x, y)$ on $\mathcal{X} \times \mathcal{Y}$ are denoted by $f^{(3)}_{\langle 112 \rangle} = \partial^3 f / \partial x^2_{\langle 1 \rangle} \partial x_{\langle 2 \rangle}$, $\ddot{g}_{\langle xy \rangle} = \partial^2 g / \partial x \partial y$, etc.

Matrices are denoted by the standard notation of uppercase letters. Vectors, however, are often *not* denoted by boldface letters in this book. For a point on a product domain $\mathcal{X} = \prod_{\gamma=1}^{\Gamma} \mathcal{X}_\gamma$, we write $x = (x_{\langle 1 \rangle}, \ldots, x_{\langle \Gamma \rangle})$. For a function on domain $\mathcal{X} = \{1, \ldots, K\}$, we write $f = (f(1), \ldots, f(K))^T$, which may be used as a vector in standard matrix arithmetic. Boldface vectors are used where confusion may result otherwise. For example, $\mathbf{1} = (1, \ldots, 1)^T$ is used to denote a vector of all one's, and $\mathbf{c} = (c_1, \ldots, c_n)^T$ is used to encapsulate subscripted coefficients. In formulas concerning matrix computation, vectors are always set in boldface.

The standard O_p, o_p notation is used in the asymptotic analyses of §3.2, §4.2.3, §5.2, and Chapter 8. We write $X = O_p(Y)$ if $P(|X| > KY) \to 0$ for some constant $K < \infty$ and write $X = o_p(Y)$ if $P(|X| > \epsilon Y) \to 0$, $\forall \epsilon > 0$.

1.3 Decomposition of Multivariate Functions

An important aspect of statistical modeling, which distinguishes it from mere function approximation, is the interpretability of the results. Of great utility are decomposition of multivariate functions similar to the classical analysis of variance (ANOVA) decomposition and the associated notions of main effect and interaction. Higher-order interactions are often excluded in practical estimation to control the model complexity; the exclusion of all interactions yields the popular additive models. The selective exclusion of certain interactions also characterizes many interesting statistical models in a variety of stochastic settings.

Casting the classical one-way ANOVA decomposition as the decomposition of functions on a discrete domain, a simple averaging operator is introduced to facilitate the generalization of the notion to arbitrary domains. Multiway ANOVA decomposition is then defined, with the identifiability of the terms assured by side conditions specified through the averaging operators. Examples are given and a proposition is proved concerning certain intrinsic structures that are independent of the side conditions. The utility and implication of selective term trimming in an ANOVA decomposition are then briefly discussed in the context of regression, density estimation, and hazard estimation.

1.3.1 ANOVA Decomposition and Averaging Operator

Consider a standard one-way ANOVA model, $Y_{ij} = \mu_i + \epsilon_{ij}$, where μ_i are the treatment means at treatment levels $i = 1, \ldots, K$, and ϵ_{ij} are independent normal errors. Writing $\mu_i = \mu + \alpha_i$, one has the "overall mean" μ and the treatment effect α_i. The identifiability of μ and α_i are assured through a side condition, of which common choices include $\alpha_1 = 0$ with level 1 treated as the control and $\sum_{i=1}^{K} \alpha_i = 0$ with all levels treated symmetrically.

The one-way ANOVA model can be recast as $Y_j = f(x_j) + \epsilon_j$, where $f(x)$ is defined on the discrete domain $\mathcal{X} = \{1, \ldots, K\}$; the treatment levels are now coded by x and the subscript j labels the observations. The ANOVA decomposition $\mu_i = \mu + \alpha_i$ in the standard ANOVA model notation can be written as

$$f(x) = Af + (I - A)f = f_\emptyset + f_x,$$

where A is an averaging operator that "averages out" the argument x to return a constant function and I is the identity operator. For example, with $Af = f(1)$, one has $f(x) = f(1) + \{f(x) - f(1)\}$, corresponding to $\alpha_1 = 0$. With $Af = \sum_{x=1}^{K} f(x)/K = \bar{f}$, one has $f(x) = \bar{f} + (f(x) - \bar{f})$, corresponding to $\sum_{i=1}^{K} \alpha_i = 0$. Note that applying A to a constant function returns that constant, hence the name "averaging." It follows that $A(Af) = Af$, $\forall f$, or, simply, $A^2 = A$. The constant term $f_\emptyset = Af$ is the "overall mean" and the term $f_x = (I - A)f$ is the treatment effect, or "contrast," that satisfies the side condition $Af_x = 0$.

On a continuous domain, say $\mathcal{X} = [a, b]$, one may similarly define an ANOVA decomposition $f(x) = Af + (I - A)f = f_\emptyset + f_x$ through an appropriately defined averaging operator A, where f_x satisfies the side condition $Af_x = 0$. For example, with $Af = f(a)$, one has $f(x) = f(a) + \{f(x) - f(a)\}$. Similarly, with $Af = \int_a^b f dx/(b-a)$, one has $f(x) = \int_a^b f dx/(b-a) + \{f(x) - \int_a^b f dx/(b-a)\}$.

1.3.2 Multiway ANOVA Decomposition

Now, consider a function $f(x) = f(x_{\langle 1 \rangle}, \ldots, x_{\langle \Gamma \rangle})$ on a product domain $\mathcal{X} = \prod_{\gamma=1}^{\Gamma} \mathcal{X}_\gamma$, where $x_{\langle \gamma \rangle} \in \mathcal{X}_\gamma$ denotes the γth coordinate of $x \in \mathcal{X}$. Let A_γ be an averaging operator on $\mathcal{X}_\gamma$ that averages out $x_{\langle \gamma \rangle}$ from the active argument list and satisfies $A_\gamma^2 = A_\gamma$; $A_\gamma f$ is constant on the $\mathcal{X}_\gamma$ axis but not necessarily an overall constant function. An ANOVA decomposition of f can be defined as

$$f = \Big\{ \prod_{\gamma=1}^{\Gamma} (I - A_\gamma + A_\gamma) \Big\} f = \sum_{\mathcal{S}} \Big\{ \prod_{\gamma \in \mathcal{S}} (I - A_\gamma) \prod_{\gamma \notin \mathcal{S}} A_\gamma \Big\} f = \sum_{\mathcal{S}} f_{\mathcal{S}}, \quad (1.7)$$

where $\mathcal{S} \subseteq \{1, \ldots, \Gamma\}$ enlists the active arguments in $f_{\mathcal{S}}$ and the summation is over all of the 2^Γ subsets of $\{1, \ldots, \Gamma\}$. The term $f_\emptyset = \prod A_\gamma f$ is a constant, the term $f_\gamma = f_{\{\gamma\}} = (I - A_\gamma) \prod_{\alpha \neq \gamma} A_\alpha f$ is the $x_{\langle \gamma \rangle}$ main effect, the term $f_{\gamma,\delta} = f_{\{\gamma,\delta\}} = (I - A_\gamma)(I - A_\delta) \prod_{\alpha \neq \gamma,\delta} A_\alpha f$ is the $x_{\langle \gamma \rangle}$-$x_{\langle \delta \rangle}$ interaction, and so forth. The terms of such a decomposition satisfy the side conditions $A_\gamma f_{\mathcal{S}} = 0$, $\forall \mathcal{S} \ni \gamma$. The choices of A_γ, or the side conditions on each axes, are open to specification.

The domains $\mathcal{X}_\gamma$ are generic in the above discussion; in particular, they can be product domains themselves. As a matter of fact, the ANOVA decomposition of (1.7) can also be defined recursively through a series of nested constructions with $\Gamma = 2$; see, e.g., Problem 1.3.

The ANOVA decomposition can be built into penalized likelihood estimation through the proper construction of the roughness functional $J(f)$; details are to be found in §2.4.

Example 1.4
When $\Gamma = 2$, $\mathcal{X}_1 = \{1, \ldots, K_1\}$, and $\mathcal{X}_2 = \{1, \ldots, K_2\}$, the decomposition reduces to a standard two-way ANOVA decomposition. With averaging operators $A_1 f = f(1, x_{\langle 2 \rangle})$ and $A_2 f = f(x_{\langle 1 \rangle}, 1)$, one has

$$\begin{aligned}
f_\emptyset &= A_1 A_2 f = f(1,1), \\
f_1 &= (I - A_1) A_2 f = f(x_{\langle 1 \rangle}, 1) - f(1,1), \\
f_2 &= A_1 (I - A_2) f = f(1, x_{\langle 2 \rangle}) - f(1,1), \\
f_{1,2} &= (I - A_1)(I - A_2) f \\
&= f(x_{\langle 1 \rangle}, x_{\langle 2 \rangle}) - f(x_{\langle 1 \rangle}, 1) - f(1, x_{\langle 2 \rangle}) + f(1,1).
\end{aligned}$$

With $A_\gamma f = \sum_{x_{\langle \gamma \rangle}=1}^{K_\gamma} f(x_{\langle 1 \rangle}, x_{\langle 2 \rangle})/K_\gamma$, $\gamma = 1, 2$, one similarly has

$$\begin{aligned}
f_\emptyset &= A_1 A_2 f = f_{\cdot\cdot}, \\
f_1 &= (I - A_1) A_2 f = f_{x_{\langle 1 \rangle}\cdot} - f_{\cdot\cdot}, \\
f_2 &= A_1 (I - A_2) f = f_{\cdot x_{\langle 2 \rangle}} - f_{\cdot\cdot}, \\
f_{1,2} &= (I - A_1)(I - A_2) f \\
&= f(x_{\langle 1 \rangle}, x_{\langle 2 \rangle}) - f_{x_{\langle 1 \rangle}\cdot} - f_{\cdot x_{\langle 2 \rangle}} + f_{\cdot\cdot},
\end{aligned}$$

where $f_{\cdot\cdot} = \sum_{x_{\langle 1 \rangle}, x_{\langle 2 \rangle}} f(x_{\langle 1 \rangle}, x_{\langle 2 \rangle})/K_1 K_2$, $f_{x_{\langle 1 \rangle}\cdot} = \sum_{x_{\langle 2 \rangle}} f(x_{\langle 1 \rangle}, x_{\langle 2 \rangle})/K_2$, and $f_{\cdot x_{\langle 2 \rangle}} = \sum_{x_{\langle 1 \rangle}} f(x_{\langle 1 \rangle}, x_{\langle 2 \rangle})/K_1$. One may also use different averaging operators on different axes; see Problem 1.4. □

Example 1.5
Consider $\Gamma = 2$ and $\mathcal{X}_1 = \mathcal{X}_2 = [0,1]$. With $A_1 f = f(0, x_{\langle 2 \rangle})$ and $A_2 f = f(x_{\langle 1 \rangle}, 0)$, one has

$$\begin{aligned}
f_\emptyset &= A_1 A_2 f = f(0,0), \\
f_1 &= (I - A_1) A_2 f = f(x_{\langle 1 \rangle}, 0) - f(0,0), \\
f_2 &= A_1 (I - A_2) f = f(0, x_{\langle 2 \rangle}) - f(0,0), \\
f_{1,2} &= (I - A_1)(I - A_2) f \\
&= f(x_{\langle 1 \rangle}, x_{\langle 2 \rangle}) - f(x_{\langle 1 \rangle}, 0) - f(0, x_{\langle 2 \rangle}) + f(0,0).
\end{aligned}$$

With $A_\gamma f = \int_0^1 f dx_{\langle\gamma\rangle}$, $\gamma = 1, 2$, one has

$$\begin{aligned}
f_\emptyset &= A_1 A_2 f = \textstyle\int_0^1 \int_0^1 f dx_{\langle 1\rangle} dx_{\langle 2\rangle},\\
f_1 &= (I - A_1) A_2 f = \textstyle\int_0^1 (f - \int_0^1 f dx_{\langle 1\rangle}) dx_{\langle 2\rangle},\\
f_2 &= A_1 (I - A_2) f = \textstyle\int_0^1 (f - \int_0^1 f dx_{\langle 2\rangle}) dx_{\langle 1\rangle},\\
f_{1,2} &= (I - A_1)(I - A_2) f\\
&= f - \textstyle\int_0^1 f dx_{\langle 2\rangle} - \int_0^1 f dx_{\langle 1\rangle} + \int_0^1 \int_0^1 f dx_{\langle 1\rangle} dx_{\langle 2\rangle}.
\end{aligned}$$

Similar results with different averaging operators on different axes are also straightforward; see Problem 1.5. □

In standard ANOVA models, higher-order terms are frequently eliminated, whereas main effects and lower-order interactions are estimated from the data. One learns not to drop the $x_{\langle 1\rangle}$ and $x_{\langle 2\rangle}$ main effects if the $x_{\langle 1\rangle}$-$x_{\langle 2\rangle}$ interaction is considered, however, and not to drop the $x_{\langle 1\rangle}$-$x_{\langle 2\rangle}$ interaction when the $x_{\langle 1\rangle}$-$x_{\langle 2\rangle}$-$x_{\langle 3\rangle}$ interaction is included. Although the ANOVA decomposition as defined in (1.7) obviously depends on the averaging operators A_γ, certain structures are independent of the particular choices of A_γ. Specifically, for any index set $\mathcal{I}$, if $f_\mathcal{S} = 0$, $\forall \mathcal{S} \supseteq \mathcal{I}$ with a particular set of A_γ, then the structure also holds for any other choices of A_γ, as the following proposition asserts.

Proposition 1.1
For any two sets of averaging operators A_γ and $\tilde{A}_\gamma$ satisfying $A_\gamma^2 = A_\gamma$ and $\tilde{A}_\gamma^2 = \tilde{A}_\gamma$, $\prod_{\gamma\in\mathcal{I}}(I - A_\gamma)f = 0$ if and only if $\prod_{\gamma\in\mathcal{I}}(I - \tilde{A}_\gamma)f = 0$, where $\mathcal{I}$ is any index set.

Note that the condition $\prod_{\gamma\in\mathcal{I}}(I - A_\gamma)f = 0$ means that $f_\mathcal{S} = 0$, $\forall \mathcal{S} \supseteq \mathcal{I}$. For example, $(I - A_1)f = 0$ implies that all terms involving $x_{\langle 1\rangle}$ vanish, and $(I - A_1)(I - A_2)f = 0$ means that all terms involving both $x_{\langle 1\rangle}$ and $x_{\langle 2\rangle}$ disappear. Model structures that can be characterized through constraints of the form $\prod_{\gamma\in\mathcal{I}}(I - A_\gamma)f = 0$ permit a term $f_\mathcal{S}$ only when all of its "subset terms," $f_{\mathcal{S}'}$ for $\mathcal{S}' \subset \mathcal{S}$, are permitted. A simple corollary of the proposition is the obvious fact that an additive model remains an additive model regardless of the side conditions.

Proof of Proposition 1.1: It is easy to see that $(I - \tilde{A}_\gamma)A_\gamma = 0$. Suppose $\prod_{\gamma\in\mathcal{I}}(I - A_\gamma)f = 0$ and define the ANOVA decomposition in (1.7) using A_γ. Now, for any nonzero term $f_\mathcal{S}$ in (1.7), one has $\mathcal{S} \not\supseteq \mathcal{I}$, so there exists $\gamma \in \mathcal{I}$ but $\gamma \notin \mathcal{S}$, hence $f_\mathcal{S} = [\cdots A_\gamma \cdots]f$. The corresponding $(I - \tilde{A}_\gamma)$ in $\prod_{\gamma\in\mathcal{I}}(I - \tilde{A}_\gamma)$ then annihilates the term. It follows that all nonzero ANOVA terms in (1.7) are annihilated by $\prod_{\gamma\in\mathcal{I}}(I - \tilde{A}_\gamma)$, so $\prod_{\gamma\in\mathcal{I}}(I - \tilde{A}_\gamma)f = 0$. The converse is true by symmetry. □

1.3.3 Multivariate Statistical Models

Many multivariate statistical models can be characterized by selective term elimination in an ANOVA decomposition. Some of such models are discussed below.

Curse of Dimensionality and Additive Models

Recall the classical ANOVA models with $\mathcal{X}_\gamma$ discrete. In practical data analysis, one usually includes only the main effects, with the possible addition of a few lower-order interactions. Higher-order interactions are less interpretable yet more difficult to estimate, as they usually consume many more degrees of freedom than the lower-order terms. Models with only main effects included are called additive models.

The difficulty associated with function estimation in high-dimensional spaces may be perceived through the sparsity of the space. Take $\mathcal{X}_\gamma = [0, 1]$, for example, a k-dimensional cube with each side of length 0.5 has volume 0.5^k. Assume a uniform distribution of the data and consider a piecewise constant function with jumps only possible at $x_{\langle\gamma\rangle} = 0.5$. To estimate such a function in 1 dimension with 2 pieces, one has information from 50% of the data per piece, in 2 dimensions with 4 pieces, 25% per piece, in 3 dimensions with 8 pieces, 12.5% per piece, etc. The lack of data due to the sparsity of high-dimensional space is often referred to as the curse of dimensionality. Alternatively, the curse of dimensionality may also be characterized by the explosive increase in the number of parameters, or the degrees of freedom, that one would need to approximate a function well in a high-dimensional space. To achieve the flexibility of a 5-piece piecewise polynomial in 1 dimension, for example, one would end up with 125 pieces in 3 dimensions by taking products of the pieces in 1 dimension.

To combat the curse of dimensionality in multivariate function estimation, one needs to eliminate higher-order interactions to control the model complexity. As for classical ANOVA models, additive models with the possible addition of second-order interactions are among the most popular models used in practice.

Conditional Independence and Graphical Models

To simplify the notation, the marginal domains will be denoted by $\mathcal{X}$, $\mathcal{Y}$, $\mathcal{Z}$, etc. in the rest of the section instead of the subscripted $\mathcal{X}$ used in (1.7).

Consider a probability density $f(x)$ of a random variable X on a domain $\mathcal{X}$. Write

$$f(x) = \frac{e^{\eta(x)}}{\int_{\mathcal{X}} e^{\eta(x)} dx}, \tag{1.8}$$

known as the logistic density transform, the log density $\eta(x)$ is free of the positivity and unity constraints, $f(x) > 0$ and $\int_{\mathcal{X}} f(x) = 1$, that $f(x)$ must satisfy. The transform is not one-to-one, though, as $e^{\eta(x)} / \int_{\mathcal{X}} e^{\eta(x)} dx =$

$e^{C+\eta(x)}/\int_{\mathcal{X}} e^{C+\eta(x)}dx$ for any constant C. The transform can be made one-to-one, however, by imposing a side condition $A_x\eta = 0$ for some averaging operator A_x on $\mathcal{X}$; this can be achieved by eliminating the constant term in a one-way ANOVA decomposition $\eta = A_x\eta + (I - A_x)\eta = \eta_\emptyset + \eta_x$.

For a joint density $f(x, y)$ of random variables (X, Y) on a product domain $\mathcal{X} \times \mathcal{Y}$, one may write

$$f(x,y) = \frac{e^{\eta(x,y)}}{\int_{\mathcal{X}} dx \int_{\mathcal{Y}} e^{\eta(x,y)}dy} = \frac{e^{\eta_x+\eta_y+\eta_{x,y}}}{\int_{\mathcal{X}} dx \int_{\mathcal{Y}} e^{\eta_x+\eta_y+\eta_{x,y}}dy},$$

where η_x, η_y, and $\eta_{x,y}$ are the main effects and interaction of $\eta(x, y)$ in an ANOVA decomposition; the constant is eliminated in the rightmost expression for a one-to-one transform. The conditional distribution of Y given X has a density

$$f(y|x) = \frac{e^{\eta(x,y)}}{\int_{\mathcal{Y}} e^{\eta(x,y)}dy} = \frac{e^{\eta_y+\eta_{x,y}}}{\int_{\mathcal{Y}} e^{\eta_y+\eta_{x,y}}dy}, \tag{1.9}$$

where the logistic conditional density transform is one-to-one only for the rightmost expression with the side conditions $A_y(\eta_y + \eta_{x,y}) = 0$, $\forall x \in \mathcal{X}$, where A_y is the averaging operator on $\mathcal{Y}$ that help to define the ANOVA decomposition. The independence of X and Y, denoted by $X \perp Y$, is characterized by $\eta_{x,y} = 0$, or $(I - A_x)(I - A_y)\eta = 0$.

The domains $\mathcal{X}$ and $\mathcal{Y}$ are generic in (1.9); in particular, they can be product domains themselves. Substituting (y, z) for y in (1.9), one has

$$f(y,z|x) = \frac{e^{\eta_y+\eta_z+\eta_{y,z}+\eta_{x,y}+\eta_{x,z}+\eta_{x,y,z}}}{\int_{\mathcal{Y}} dy \int_{\mathcal{Z}} e^{\eta_y+\eta_z+\eta_{y,z}+\eta_{x,y}+\eta_{x,z}+\eta_{x,y,z}}dz},$$

where $\eta_{(y,z)}$ is expanded out as $\eta_y + \eta_z + \eta_{y,z}$ and $\eta_{x,(y,z)}$ is expanded out as $\eta_{x,y} + \eta_{x,z} + \eta_{x,y,z}$; see Problem 1.3. The conditional independence of Y and Z given X, denoted by $(Y \perp Z)|X$, is characterized by $\eta_{y,z} + \eta_{x,y,z} = 0$, or $(I - A_y)(I - A_z)\eta = 0$.

Now, consider the joint density of four random variables (U, V, Y, Z), with $(U \perp V)|(Y, Z)$ and $(Y \perp Z)|(U, V)$. It can be shown that such a structure is characterized by $\eta_{u,v} + \eta_{y,z} + \eta_{u,v,y} + \eta_{u,v,z} + \eta_{u,y,z} + \eta_{v,y,z} + \eta_{u,v,y,z} = 0$ in an ANOVA decomposition, or $(I - A_u)(I - A_v)\eta = (I - A_y)(I - A_z)\eta = 0$; see Problem 1.7.

As noted above, the ANOVA decompositions in the log density η that characterize conditional independence structures are all of the type covered in Proposition 1.1. The elimination of lower-order terms in (1.8) and (1.9) for one-to-one transforms only serve to remove technical redundancies introduced by the "overparameterization" of $f(x)$ or $f(y|x)$ by the corresponding unrestricted η.

Conditional independence structures can be represented as graphs, and models for multivariate densities with specified conditional independence structures built in are called graphical models; see, e.g., Whittaker (1990) for some general discussion and for the parametric estimation of graphical models.

Proportional Hazard Models and Beyond

For $\eta(t,u)$ a log hazard on the product of a time domain $\mathcal{T}$ and a covariate domain $\mathcal{U}$, an additive model $\eta(t,u) = \eta_\emptyset + \eta_t + \eta_u$ characterizes a proportional hazard model, with $e^{\eta_\emptyset + \eta_t}$ being the base hazard and e^{η_u} being the relative risk. When the interaction $\eta_{t,u}$ is included in the model, one has something beyond the proportional hazard model. The covariate domain can be a product domain itself, on which nested ANOVA decompositions can be introduced.

1.4 Case Studies

To illustrate potential applications of the techniques to be developed in this book, we shall now present previews of a few selected case studies. Full accounts of these studies are to be found in later chapters.

1.4.1 Water Acidity in Lakes

From the Eastern Lake Survey of 1984 conducted by the United States Environmental Protection Agency (EPA), Douglas and Delampady (1990) derived a data set containing geographic information, water acidity measurements, and main ion concentrations in 1798 lakes in 3 regions, northeast, upper midwest, and southeast, in the eastern United States. Of interest is the dependence of the water acidity on the geographic locations and other information concerning the lakes.

Preliminary analysis and consultation with a water chemist suggest that a model for the surface pH in terms of the geographic location and the calcium concentration is appropriate. A model of the following form is considered:

$$\text{pH} = \eta_\emptyset + \eta_c(\text{calcium}) + \eta_g(\text{geography}) + \eta_{c,g}(\text{calcium}, \text{geography}) + \epsilon.$$

The model can be fitted to the data using the tensor product thin-plate spline technique, to be discussed in §4.4, with the geographic location treated in an isotropically invariant manner. The isotropic invariance is in the following sense: After converting the longitude and latitude of the geographic location to the x-y coordinates (in distance) with respect to a local origin, the fitting of the model is invariant to arbitrary shift and rotation of the x-y coordinates. The geographic location is mathematically two dimensional, but, conceptually, it makes little sense to talk about north-south effect or east-west effect, or any other directional decomposition of

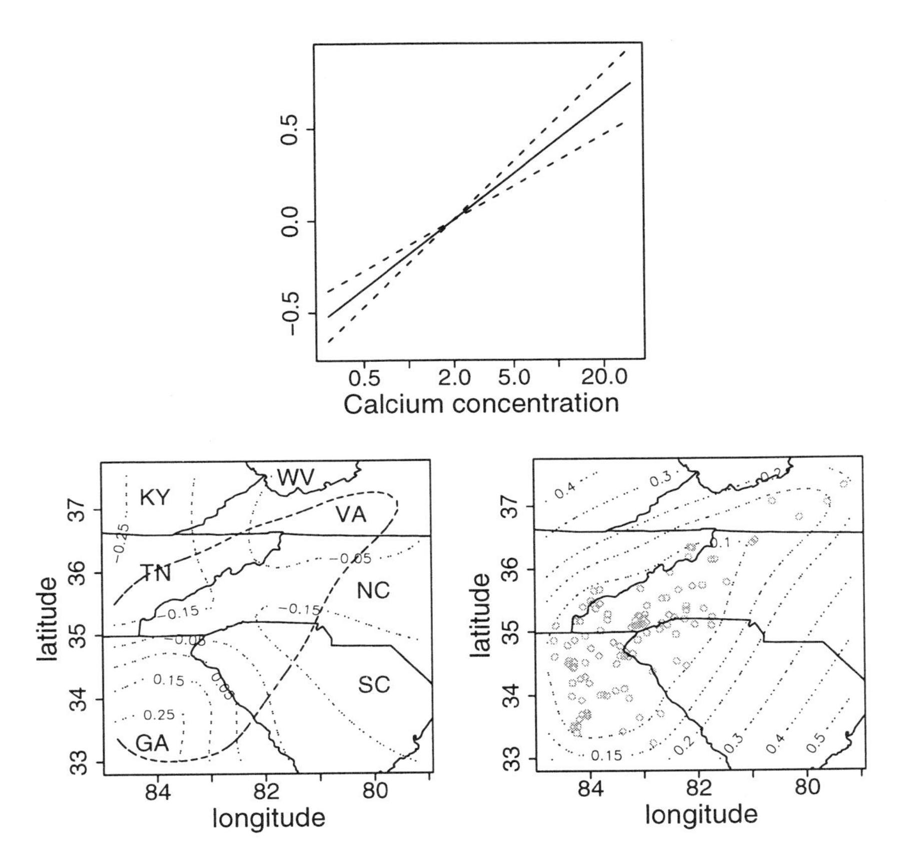

FIGURE 1.2. Water Acidity Fit for Lakes in the Blue Ridge. Top: Calcium effect with 95% confidence intervals. Left: Geography effect, with the 0.15 contour of standard errors superimposed as the dashed line. Right: Standard errors of geography effect with the lakes superimposed.

the geographic location, in the context. The isotropically invariant treatment preserves the integrity of the geographic location as an inseparable entity.

For illustration, consider the fitting of the model to 112 lakes in the Blue Ridge. As inputs to the fitting algorithm, the longitude and latitude were converted to x-y coordinates in distance, and a log transform was applied to the calcium concentration. The interaction $\eta_{c,g}$ was effectively eliminated by a data-driven model selection device to be discussed in §3.2.3, so an additive model was fitted. Plotted in Figure 1.2 are the fitted calcium effect with 95% confidence intervals, the estimated geography effect, and the standard errors of the estimated geography effect; see §3.3 for the definition and interpretation of the standard errors and confidence intervals. The 0.15

contour of the geography standard errors, which encloses all but one lake, is superimposed as the dashed line in the plot of the geography effect. The lakes are superimposed in the plot of geography standard errors. The fit has a R^2 of 0.55 and the "explained" variation in pH are roughly 66% by the calcium concentration and 34% by the geography.

A full account of the analysis is to be found in §4.4.4.

1.4.2 AIDS Incubation

To study the AIDS incubation time, a valuable source of information is in the records of patients who were infected with the HIV virus through blood transfusion, of which the date can be ascertained retrospectively. A data set collected by the Centers for Disease Control and Prevention (CDC) is listed in Wang (1989), which includes the time X from the transfusion to the diagnosis of AIDS, the time Y from the transfusion to the end of study (July 1986), both in months, and the age of the individual at the time of transfusion, for 295 individuals. It is clear that $X \leq Y$ (i.e., the data are truncated).

Assuming the independence of X and Y in the absence of truncation, and conditioning on the truncation mechanism, the density of (X, Y) is given by

$$f(x,y) = \frac{e^{\eta_x(x)+\eta_y(y)}}{\int_0^a dy \int_0^y e^{\eta_x(x)+\eta_y(y)} dx},$$

where $[0, a]$ is a finite interval covering the data. The penalized likelihood score (1.5) can be specified as

$$-\frac{1}{n}\sum_{i=1}^{n}\left\{\eta_x(X_i)+\eta_y(Y_i)-\log\int_0^a dy\int_0^y e^{\eta_x(x)+\eta_y(y)}dx\right\} + \frac{\lambda_x}{2}\int_0^a \ddot{\eta}_x^2 dx + \frac{\lambda_y}{2}\int_0^a \ddot{\eta}_y^2 dx, \quad (1.10)$$

where η_x and η_y satisfy certain side conditions such as $\int_0^a \eta_x dx = 0$ and $\int_0^a \eta_y dy = 0$.

Grouping the individuals by age, one has 141 "elderly patients" of age 60 or above. Estimating $f(x, y)$ for this age group through the minimization of (1.10), with $a = 100$ and λ_x and λ_y selected through a device introduced in §6.3, one obtains the estimate contoured in Figure 1.3, where the data are superimposed and the marginal densities $f(x) = e^{\eta_x} / \int_0^{100} e^{\eta_x} dx$ and $f(y) = e^{\eta_y} / \int_0^{100} e^{\eta_y} dy$ are plotted in the empty space on their respective axes.

A full account of the analysis of this data set will be presented in Chapter 6.

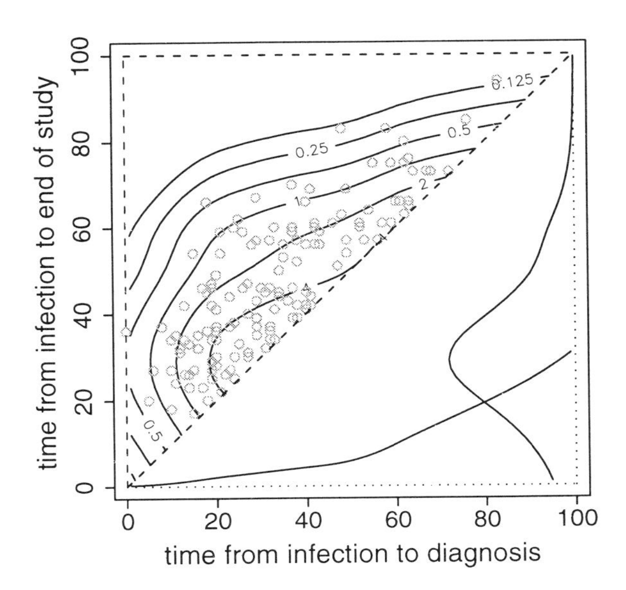

FIGURE 1.3. AIDS Incubation and HIV Infection of the Elderly. Contours are estimated density on the observable region surrounded by dashed lines. Circles are the observations. Curves over the dotted lines in the empty space are the estimated marginal densities.

1.4.3 Survival After Heart Transplant

One of the most demonstrated survival data is the Stanford heart transplant data. In this study, we consider the data listed in Miller and Halpern (1982). Recorded were survival or censoring times of 184 patients after the (first) heart transplant, in days, their ages at transplant, and a certain tissue-type mismatch score for 157 of the patients. There were 113 recorded deaths and 71 censorings. From the analysis by Miller and Halpern (1982) and others, the tissue-type mismatch score did not have significant impact on survival, so we will try to estimate the hazard as a function of time after transplant and the age of the patient at transplant.

In the notation of Example 1.3, $Z = 0$ and U is the age at transplant. With a proportional hazard model $\eta(t,u) = \eta_\emptyset + \eta_t + \eta_u$, the penalized likelihood score (1.6) can be specified as

$$-\frac{1}{n}\sum_{i=1}^{n}\left\{\delta_i(\eta_\emptyset + \eta_t(X_i) + \eta_u(U_i)) - e^{\eta_\emptyset+\eta_u(U_i)}\int_0^{X_i} e^{\eta_t(t)}dt\right\} + \frac{\lambda_t}{2}\int_0^{T^*}\ddot{\eta}_t^2 dt + \frac{\lambda_u}{2}\int_a^b \ddot{\eta}_u^2 du, \quad (1.11)$$

where $X_i \leq T^*$ and $U_i \in [a,b]$.

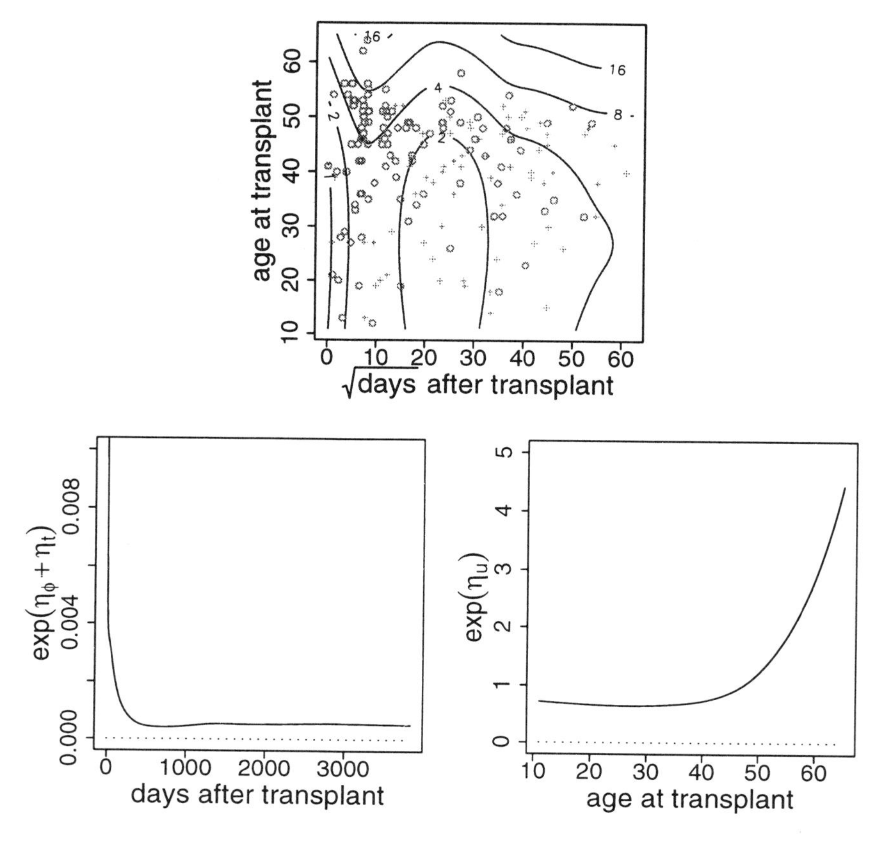

FIGURE 1.4. Hazard After Heart Transplant. Top: Contours of $100e^{\tilde{\eta}(t^*,u)}$, where $t^* = \sqrt{t}$, with deceased (circles) and censored (pluses) patients superimposed. Left: Base hazard $e^{\eta_\emptyset+\eta_t}$ on the original time scale. Right: Age effect e^{η_u}.

Before fitting the model to the data, the time axis was rescaled by a square root transform $t^* = \sqrt{t}$ to make X_i more evenly scattered. Once $e^{\tilde{\eta}(t^*,u)} = -d \log S(t^*, u)/dt^*$ is estimated, the hazard on the original time scale is simply

$$e^{\eta(t,u)} = e^{\tilde{\eta}(t^*,u)}(dt^*/dt) = e^{\tilde{\eta}(\sqrt{t},u)}/(2\sqrt{t}).$$

Fitting the proportional hazard model through the minimization of (1.11), with λ_t and λ_u selected via a device introduced in §7.2, one obtains the fit plotted in Figure 1.4: In the top frame, $e^{\tilde{\eta}(t^*,u)}$ is contoured with the data superimposed, and in the left and right frames, the base hazard $e^{\eta_\emptyset+\eta_t}$ (on the original time scale) and the age effect e^{η_u} are separately plotted.

A full account of the analysis of this data set will be presented in Chapter 7.

1.5 Scope

This book presents a systematic treatment of function estimation on generic domains using the penalized likelihood method. Main topics to be covered include model construction, smoothing parameter selection, computation, and asymptotic convergence.

Chapter 2 is devoted to the construction of $J(\eta)$ for use in (1.3) on generic domains; of particular interest is that on product domains with ANOVA decompositions built in. Among examples used to illustrate the construction are shrinkage estimates, polynomial smoothing splines, and their tensor products. Other issues that do not involve the stochastic structure of $L(\eta|\text{data})$ are also discussed in the chapter, which include the empirical Bayes model associated with (1.3) and the existence of the minimizer of (1.3).

Chapter 3 discusses penalized least squares regression with Gaussian-type responses. Effective methods for smoothing parameter selection and generic algorithms for computation are the main focus of the discussion. Open-source software packages are introduced and illustrated using examples. Also covered are data analytical tools, such as interval estimates and diagnostics for model redundancy, and fast algorithms for settings with certain special structures.

Chapter 4 enlists some generalizations and variations of the polynomial smoothing splines. Among subjects under discussion are the partial splines, the periodic splines, the L-splines, and the thin-plate splines. Conceptually, these are simply further examples of the general construction presented in Chapter 2, but some of the mathematical details are more involved.

Chapter 5 studies penalized likelihood regression with non-Gaussian responses. The central issue is, again, the effective selection of smoothing parameters and the related computation. Computational and data analytical tools developed in Chapter 3 are extended to non-Gaussian regression. Software tools are illustrated through examples.

Chapter 6 deals with penalized likelihood density estimation under a variety of sampling schemes. Beside the standard method of Example 1.2 for independent and identically distributed data, variation is also discussed for data subject to biased sampling and random truncation. Further variations include conditional density estimation and density estimation with data from response-based sampling. Methods for effective smoothing parameter selection are developed and the related computation is outlined.

Chapter 7 handles penalized likelihood hazard estimation. Under discussion are (i) the method of Example 1.3, (ii) the estimation of relative risk in a proportional hazard model via penalized partial likelihood, and (iii) the estimation of models parametric in time. The numerical structure of Example 1.3 parallels that of Example 1.2, and the partial likelihood is isomorphic to the likelihood under biased sampling, so the smoothing parameters in (i) and (ii) can be selected using the methods developed in

Chapter 6. For (iii), the smoothing parameters are selected by the methods developed in Chapter 5.

Chapter 8 investigates the asymptotic convergence of penalized likelihood estimates. Convergence rates are calculated in terms of problem-specific loss functions derived from the respective stochastic settings. Also noted are the mode and rates of convergence of the estimates when the models are incorrect.

1.6 Bibliographic Notes

Section 1.1

A discrete version of (1.1) for data smoothing dated back to Whittaker (1923). Early results on the modern theory of smoothing spline interpolation with exact data [i.e., with $\lambda = 0$ in (1.1) for $Y_i = f(x_i)$] can be found in, e.g., Schoenberg (1964) and de Boor and Lynch (1966), among others; see the Foreword of Wahba (1990) for further historical notes. A comprehensive treatment of smoothing splines from a numerical analytical perspective can be found in Schumaker (1981, Chapter 8). A popular reference on splines, especially on the popular B-splines, is de Boor (1978). B-splines, however, are *not* smoothing splines.

Pioneered by the work of Kimeldorf and Wahba (1970a, 1970b, 1971), the study of (1.1) and generalizations thereof in a statistical context has over the years produced a vast literature on penalized least squares regression. Historical breakthroughs can be found in Craven and Wahba (1979) and Wahba (1983), among others. Wahba (1990) compiled an excellent synthesis for work up to that date. See §3.9 for further notes on penalized least squares regression.

The penalized likelihood method was introduced by Good and Gaskins (1971) in the context of density estimation; the formulation of Example 1.2 by Gu and Qiu (1993) evolved from the work of Leonard (1978) and Silverman (1982). The penalized likelihood regression of Example 1.1 was formulated by O'Sullivan, Yandell, and Raynor (1986); see also Silverman (1978). The penalized likelihood hazard estimation of Example 1.3, which was formulated by Gu (1996), evolved from the work of Anderson and Senthilselvan (1980), O'Sullivan (1988a, 1988b), and Zucker and Karr (1990).

Section 1.3

Classical ANOVA models can be found in statistics textbooks of almost all levels. The definition (1.7) on generic domains can be found in Gu and Wahba (1991a, 1993b). The result of Proposition 1.1 on discrete domains can be found in standard graduate-level textbooks on linear models. See, e.g., Scheffe (1959, §4.1) and Seber (1977, page 277).

Additive models are routinely used in standard linear model analysis. Their use in nonparametric regression was popularized by the work of Stone (1985) and Hastie and Tibshirani (1986, 1990), among others. Graphical models have their roots in classical log linear models for categorical data; comprehensive modern treatments with a mixture of continuous and categorical data can be found in, e.g., Whittaker (1990) and Lauritzen (1996). The proportional hazard models, especially the so-called Cox models proposed by Cox (1972), are among standard tools found in most textbooks on survival analysis; see, e.g., Kalbfleisch and Prentice (1980) and Fleming and Harrington (1991).

Section 1.4

The EPA lake acidity data of §1.4.1 was used in Gu and Wahba (1993a) to illustrate tensor product thin-plate splines and in Gu and Wahba (1993b) to illustrate componentwise Bayesian confidence intervals.

The CDC blood transfusion data was used by Kalbfleisch and Lawless (1989) to motivate and illustrate methods for nonparametric (in the sense of empirical distribution) and parametric inference based on retrospective ascertainment. Wang (1989) analyzed the data using a semiparametric maximum likelihood method designed for truncated data. The analysis illustrated in §1.4.2 is similar to the one presented in Gu (1998b).

The Stanford heart transplant data has become a benchmark example for many researchers to showcase innovations in survival analysis. Early references on the analysis of the data include Turnbull, Brown, and Hu (1974), Miller (1976), and Crowley and Hu (1977). The analysis illustrated in §1.4.3 is similar to the one presented in Gu (1998b).

1.7 Problems

Section 1.1

1.1 Consider univariate regression on $\mathcal{X} = [0,1]$. Take $J(\eta) = \int \ddot{\eta}^2 dx$ in (1.4).

(a) For $Y|x \sim N(\mu(x), \sigma^2)$, verify that (1.4) with $\eta = \mu$ reduces to (1.1).

(b) For $Y|x \sim \text{Binomial}(1, p(x))$, specialize (1.4) with $\eta = \log\{p/(1-p)\}$ to obtain a score for penalized likelihood logistic regression.

(c) For $Y|x \sim \text{Poisson}(\lambda(x))$, specialize (1.4) with $\eta = \log \lambda$ to obtain a score for penalized likelihood Poisson regression.

1.2 Consider the hazard estimation problem in Example 1.3.

(a) Verify that $S(t|u) = \exp\{-\int_0^t e^{\eta(s,u)}ds\}$.

(b) The likelihood of exact lifetime T is simply its density $f(t)$ evaluated at T. The likelihood of right-censored lifetime $T > C$ is the survival probability $P(T > C) = S(C)$. Verify that the likelihood of (Z, X, δ) is $e^{\delta\eta(X)}S(X)/S(Z)$, where the dependence on the covariate U is suppressed from the notation.

(c) Verify that the first term in (1.6) is indeed the minus log likelihood of $(U_i, Z_i, X_i, \delta_i)$, $i = 1, \ldots, n$.

Section 1.3

1.3 For averaging operators A_γ on $\mathcal{X}_\gamma$, verify that

$$I - A_1A_2 = (I - A_1)A_2 + A_1(I - A_2) + (I - A_1)(I - A_2).$$

Use the result to construct the ANOVA decomposition of (1.7) with $\Gamma = 3$ through two nested constructions with $\Gamma = 2$.

1.4 For the discrete domains of Example 1.4, obtain $f_\emptyset$, f_1, f_2, and $f_{1,2}$ for $A_1f = f(1, x_{\langle 2\rangle})$ and $A_2f = \sum_{x_{\langle 2\rangle}=1}^{K_2} f(x_{\langle 1\rangle}, x_{\langle 2\rangle})/K_2$.

1.5 For the continuous domains of Example 1.5, obtain $f_\emptyset$, f_1, f_2, and $f_{1,2}$ for $A_1f = f(0, x_{\langle 2\rangle})$ and $A_2 = \int_0^1 f dx_{\langle 2\rangle}$.

1.6 The domains $\mathcal{X}_\gamma$ in (1.7) can be a mixture of different types. As a simple example, consider $\Gamma = 2$, $\mathcal{X}_1 = \{1, \ldots, K\}$, and $\mathcal{X}_2 = [0, 1]$, with $A_1f = \sum_{x_{\langle 1\rangle}=1}^{K} f(x_{\langle 1\rangle}, x_{\langle 2\rangle})/K$ and $A_2f = \int_0^1 f dx_{\langle 2\rangle}$. Obtain $f_\emptyset$, f_1, f_2, and $f_{1,2}$ in an ANOVA decomposition.

1.7 Prove that if the joint density of (U, V, Y, Z) has the expression

$$f(u, v, y, z) = \frac{e^{\eta_u+\eta_v+\eta_y+\eta_z+\eta_{u,y}+\eta_{u,z}+\eta_{v,y}+\eta_{v,z}}}{\int_{\mathcal{U}}\int_{\mathcal{V}}\int_{\mathcal{Y}}\int_{\mathcal{Z}} e^{\eta_u+\eta_v+\eta_y+\eta_z+\eta_{u,y}+\eta_{u,z}+\eta_{v,y}+\eta_{v,z}}},$$

then $(U \perp V)|(Y, Z)$ and $(Y \perp Z)|(U, V)$.

2
Model Construction

The two basic components of a statistical model, the deterministic part and the stochastic part, are well separated in the penalized likelihood score $L(f) + (\lambda/2)J(f)$ of (1.3). The deterministic part is specified via $J(f)$, which defines the notion of smoothness for functions on the domain $\mathcal{X}$. The stochastic part is characterized by $L(f)$, which reflects the sampling structure of the data.

In this chapter, we are mainly concerned with the construction of $J(f)$ for use in $L(f) + (\lambda/2)J(f)$. At the foundation of the construction is some elementary theory of reproducing kernel Hilbert spaces, of which a brief self-contained introduction is given in §2.1. Illustrations of the construction are presented on the domain $\{1, \ldots, K\}$ through shrinkage estimates (§2.2) and on the domain $[0, 1]$ through polynomial smoothing splines (§2.3); the discrete case also provides insights into the entities in a reproducing kernel Hilbert space through those in a standard vector space. The construction of models on product domains with the ANOVA structure of §1.3.2 built in is discussed in §2.4, with detailed examples on domains $\{1, \ldots, K_1\} \times \{1, \ldots, K_2\}$, $[0, 1]^2$, and $\{1, \ldots, K\} \times [0, 1]$.

Also included in this chapter are some general properties of the penalized likelihood score $L(f) + (\lambda/2)J(f)$ that are largely independent of $L(f)$. One such property is the fact that a quadratic functional $J(f)$ acts like the minus log likelihood of a Gaussian process prior for f, which leads to the Bayes model discussed in §2.5. Other important properties include the existence of the minimizer of $L(f) + (\lambda/2)J(f)$ and the equivalence of penalized minimization and constrained minimization (§2.6).

The definitions of numerous technical terms are embedded in the text. For convenient back reference, the terms are set in boldface at the point of definition.

Mathematically more sophisticated constructions, such as the L-splines and the thin-plate splines, are deferred to Chapter 4.

2.1 Reproducing Kernel Hilbert Spaces

By adding a roughness penalty $J(f)$ to the minus log likelihood $L(f)$, one considers only smooth functions in the space $\{f : J(f) < \infty\}$ or a subspace therein. To assist analysis and computation, one needs a metric and a geometry in the space, and the score $L(f) + (\lambda/2)J(f)$ to be continuous in f under the metric. The so-called reproducing kernel Hilbert space, of which a brief introduction is presented here, is adequately equipped for the purpose.

We start with the definition of Hilbert space and some of its elementary properties. The discussion is followed by the Riesz representation theorem, which provides the technical foundation for the notion of a reproducing kernel. The definition of reproducing kernel Hilbert space comes next and it is shown that a reproducing kernel Hilbert space is uniquely determined by its reproducing kernel, for which any non-negative definite function qualifies.

2.1.1 Hilbert Spaces and Linear Subspaces

As abstract generalizations of the familiar vector spaces, Hilbert spaces inherit many of the structures of the vector spaces. To provide insights into the technical concepts introduced here, abstract materials are followed by vector space examples set in italic.

For elements $f, g, h, \dots$, define the operation of **addition** satisfying the following properties: (i) $f + g = g + f$, (ii) $(f + g) + h = f + (g + h)$, and (iii) for any two elements f and g, there exists an element h such that $f + h = g$. The third property implies the existence of an element 0 satisfying $f+0 = f$. Further, define the operation of **scalar multiplication** satisfying $\alpha(f + g) = \alpha f + \alpha g$, $(\alpha + \beta)f = \alpha f + \beta f$, $1f = f$, and $0f = 0$, where α and β are real numbers. A set $\mathcal{L}$ of such elements form a **linear space** if $f, g \in \mathcal{L}$ implies that $f + g \in \mathcal{L}$ and $\alpha f \in \mathcal{L}$. A set of elements $f_i \in \mathcal{L}$ are said to be **linearly independent** if $\sum_i \alpha_i f_i = 0$ holds only for $\alpha_i = 0$, $\forall i$. The maximum number of elements in $\mathcal{L}$ that can be linearly independent defines its **dimension**.

Take real vectors of a given length as the elements; the standard vector addition and scalar-vector multiplication satisfy the conditions specified for the operations of addition and scalar multiplication. The notions of lin-

ear space, linear independence, and dimension reduce to those in standard vector spaces.

A **functional** in a linear space $\mathcal{L}$ operates on an element $f \in \mathcal{L}$ and returns a real number as its value. A **linear functional** L in $\mathcal{L}$ satisfies $L(f+g) = Lf + Lg$, $L(\alpha f) = \alpha Lf$, $f, g \in \mathcal{L}$, α real. A **bilinear form** $J(f,g)$ in a linear space $\mathcal{L}$ takes $f, g \in \mathcal{L}$ as arguments and returns a real value and satisfies $J(\alpha f + \beta g, h) = \alpha J(f,h) + \beta J(g,h)$, $J(f, \alpha g + \beta h) = \alpha J(f,g) + \beta J(f,h)$, $f, g, h \in \mathcal{L}$, α, β real. Fixing one argument in a bilinear form, one gets a linear functional in the other argument. A bilinear form $J(\cdot,\cdot)$ is **symmetric** if $J(f,g) = J(g,f)$. A symmetric bilinear form is **non-negative definite** if $J(f,f) \geq 0$, $\forall f \in \mathcal{L}$, and it is **positive definite** if the equality holds only for $f = 0$. For $J(\cdot,\cdot)$ non-negative definite, $J(f) = J(f,f)$ is called a **quadratic functional**.

Consider the linear space of all real vectors of a given length. A functional in such a space is simply a multivariate function with the coordinates of the vector as its arguments. A linear functional in such a space can be written as a dot product, $Lf = g_L^T f$, *where* g_L *is a vector "representing"* L. *A bilinear form can be written as* $J(f,g) = f^T B_J g$ *with* B_J *a square matrix, and* $J(f,g)$ *is symmetric, non-negative definite, or positive definite when* B_J *is symmetric, non-negative definite, or positive definite. A quadratic functional* $J(f) = f^T B_J f$ *is better known as a quadratic form in the classical linear model theory.*

A linear space is often equipped with an **inner product**, a positive definite bilinear form with a notation $(\cdot,\cdot)$. An inner product defines a **norm** in the linear space, $\|f\| = \sqrt{(f,f)}$, which induces a metric to measure the **distance** between elements in the space, $D[f,g] = \|f - g\|$. The Cauchy-Schwarz inequality,

$$|(f,g)| \leq \|f\| \, \|g\|, \tag{2.1}$$

with equality if and only if $f = \alpha g$, and the triangle inequality,

$$\|f + g\| \leq \|f\| + \|g\|, \tag{2.2}$$

with equality if and only if $f = \alpha g$ for some $\alpha > 0$, hold in such a linear space; see Problems 2.1 and 2.2.

Equip the linear space of all real vectors of a given length with an inner product $(f,g) = f^T g$; *one obtains the Euclidean space. The Euclidean norm* $\|f\| = \sqrt{f^T f}$ *induces the familiar Euclidean distance between vectors. The Cauchy-Schwarz inequality and the triangle inequality are familiar results in a Euclidean space.*

When $\lim_{n\to\infty} \|f_n - f\| = 0$ for a sequence of elements f_n, the sequence is said to **converge** to its **limit point** f, with a notation $\lim_{n\to\infty} f_n = f$ or $f_n \to f$. A functional L is **continuous** if $\lim_{n\to\infty} Lf_n = Lf$ whenever $\lim_{n\to\infty} f_n = f$. By the Cauchy-Schwarz inequality, (f,g) is continuous in f or g when the other argument is fixed.

In the Euclidean space, a functional is a multivariate function in the coordinates of the vector, and the definition of continuity reduces to the definition found in standard multivariate calculus.

A sequence satisfying $\lim_{n,m\to\infty} \|f_n - f_m\| = 0$ is called a **Cauchy sequence**. A linear space $\mathcal{L}$ is **complete** if every Cauchy sequence in $\mathcal{L}$ converges to an element in $\mathcal{L}$. An element is a **limit point of a set** A if it is the limit point of a sequence in A. A set A is **closed** if it contains all of its own limit points.

The Euclidean space is complete. In the two-dimensional Euclidean space, $(-\infty, \infty) \times \{0\}$ is a closed set, so is $[a_1, b_1] \times [a_2, b_2]$, where $-\infty < a_i \leq b_i < \infty$, $i = 1, 2$.

A **Hilbert space** $\mathcal{H}$ is a complete inner product linear space. A closed linear subspace of $\mathcal{H}$ is itself a Hilbert space. The **distance** between a point $f \in \mathcal{H}$ and a closed linear subspace $\mathcal{G} \subset \mathcal{H}$ is defined by $D[f, \mathcal{G}] = \inf_{g\in\mathcal{G}} \|f - g\|$. By the closedness of $\mathcal{G}$, there exists an $f_{\mathcal{G}} \in \mathcal{G}$, called the **projection** of f in $\mathcal{G}$, such that $\|f - f_{\mathcal{G}}\| = D[f, \mathcal{G}]$. Such an $f_{\mathcal{G}}$ is unique by the triangle inequality. See Problem 2.3.

In the two-dimensional Euclidean space, $\mathcal{G} = \{f : f = (a, 0)^T, a \text{ real}\}$ is a closed linear subspace. The distance between $f = (a_f, b_f)^T$ and $\mathcal{G}$ is $D[f, \mathcal{G}] = |b_f|$, and the projection of f in $\mathcal{G}$ is $f_{\mathcal{G}} = (a_f, 0)^T$.

Proposition 2.1
Let $f_{\mathcal{G}}$ be the projection of $f \in \mathcal{H}$ in a closed linear subspace $\mathcal{G} \subset \mathcal{H}$. Then, $(f - f_{\mathcal{G}}, g) = 0$, $\forall g \in \mathcal{G}$.

Proof: We prove by negation. Suppose $(f - f_{\mathcal{G}}, h) = \alpha \neq 0$, $h \in \mathcal{G}$. Write $\beta = (h, h)$ and take $g = f_{\mathcal{G}} + (\alpha/\beta)h \in \mathcal{G}$. It is easy to compute

$$\|f - g\|^2 = \|f - f_{\mathcal{G}}\|^2 - \alpha^2/\beta < \|f - f_{\mathcal{G}}\|^2,$$

a contradiction. □

The linear subspace $\mathcal{G}^c = \{f : (f, g) = 0, \forall g \in \mathcal{G}\}$ is called the **orthogonal complement** of $\mathcal{G}$. By the continuity of (f, g), $\mathcal{G}^c$ is closed. Using Proposition 2.1, it is easy to verify that

$$\begin{aligned}\|f - f_{\mathcal{G}} - f_{\mathcal{G}^c}\|^2 &= (f - f_{\mathcal{G}} - f_{\mathcal{G}^c}, f - f_{\mathcal{G}^c} - f_{\mathcal{G}}) \\ &= (f - f_{\mathcal{G}}, f - f_{\mathcal{G}^c}) - (f - f_{\mathcal{G}}, f_{\mathcal{G}}) \\ &\quad - (f_{\mathcal{G}^c}, f - f_{\mathcal{G}^c}) + (f_{\mathcal{G}^c}, f_{\mathcal{G}}) \\ &= 0,\end{aligned}$$

where $f_{\mathcal{G}} \in \mathcal{G}$ and $f_{\mathcal{G}^c} \in \mathcal{G}^c$ are the projections of f in $\mathcal{G}$ and $\mathcal{G}^c$, respectively. Hence, there exists a unique decomposition $f = f_{\mathcal{G}} + f_{\mathcal{G}^c}$ for every $f \in \mathcal{H}$. It is clear now that $(\mathcal{G}^c)^c = \mathcal{G}$. The decomposition $f = f_{\mathcal{G}} + f_{\mathcal{G}^c}$ is called a **tensor sum decomposition** and is denoted by $\mathcal{H} = \mathcal{G} \oplus \mathcal{G}^c$, $\mathcal{G}^c = \mathcal{H} \ominus \mathcal{G}$, or $\mathcal{G} = \mathcal{H} \ominus \mathcal{G}^c$. Multiple-term tensor sum decompositions can be defined recursively.

In the two-dimensional Euclidean space, the orthogonal complement of $\mathcal{G} = \{f : f = (a,0)^T, a \text{ real}\}$ *is* $\mathcal{G}^c = \{f : f = (0,b)^T, b \text{ real}\}$.

Consider linear subspaces $\mathcal{H}_0$ and $\mathcal{H}_1$ of a linear space $\mathcal{L}$, equipped with inner products $(\cdot,\cdot)_0$ and $(\cdot,\cdot)_1$, respectively. Assume the completeness of $\mathcal{H}_0$ and $\mathcal{H}_1$ so that they are Hilbert spaces. If $\mathcal{H}_0$ and $\mathcal{H}_1$ have only one common element 0, then one may define a tensor sum Hilbert space $\mathcal{H} = \mathcal{H}_0 \oplus \mathcal{H}_1$ with elements $f = f_0 + f_1$ and $g = g_0 + g_1$, where $f_0, g_0 \in \mathcal{H}_0$ and $f_1, g_1 \in \mathcal{H}_1$, and an inner product $(f,g) = (f_0,g_0)_0 + (f_1,g_1)_1$. It is easy to verify that such a bottom-up pasting is consistent with the aforementioned top-down decomposition; see Problem 2.4.

Consider the two-dimensional vector space. Equip the space $\mathcal{H}_0 = \{f : f = (a,0)^T, a \text{ real}\}$ *with the inner product* $(f,g)_0 = a_f a_g$, *where* $f = (a_f,0)^T$ *and* $g = (a_g,0)^T$, *and equip* $\mathcal{H}_1 = \{f : f = (0,b)^T, b \text{ real}\}$ *with the inner product* $(f,g)_1 = b_f b_g$, *where* $f = (0,b_f)^T$ *and* $g = (0,b_g)^T$. $\mathcal{H} = \mathcal{H}_0 \oplus \mathcal{H}_1$ *has elements of the form* $f = f_0 + f_1 = (a_f,0)^T + (0,b_f)^T = (a_f,b_f)^T$ *and* $g = (a_g,0)^T + (0,b_g)^T = (a_g,b_g)^T$, *and an inner product* $(f,g) = (f_0,g_0)_0 + (f_1,g_1)_1 = a_f a_g + b_f b_g$.

A non-negative definite bilinear form $J(f,g)$ in a linear space $\mathcal{H}$ defines a **semi-inner-product** in $\mathcal{H}$ which induces a square **seminorm** $J(f) = J(f,f)$. Unless $J(f,g)$ is positive definite, the **null space** $\mathcal{N}_J = \{f : J(f,f) = 0, f \in \mathcal{H}\}$ is a linear subspace of $\mathcal{H}$ containing more elements than just 0. With a nondegenerate $\mathcal{N}_J$, one typically can define another non-negative definite bilinear form $\tilde{J}(f,g)$ in $\mathcal{H}$ satisfying the following conditions: (i) it is positive definite when restricted to $\mathcal{N}_J$, so $\tilde{J}(f) = \tilde{J}(f,f)$ defines a square full norm in $\mathcal{N}_J$ and (ii) for every $f \in \mathcal{H}$, there exists $g \in \mathcal{N}_J$ such that $\tilde{J}(f-g) = 0$. With such an $\tilde{J}(f,g)$, it is easy to verify that $J(f,g)$ is positive definite in the linear subspace $\mathcal{N}_{\tilde{J}} = \{f : \tilde{J}(f,f) = 0, f \in \mathcal{H}\}$ and that $(J+\tilde{J})(f,g)$ is positive definite in $\mathcal{H}$. Hence, a semi-inner-product can be made a full inner product either via restriction to a subspace or via augmentation by an extra term, both through the definition of an inner product on its null space. If $\mathcal{H}$ is complete under the norm induced by $(J+\tilde{J})(f,g)$, then it is easy to see that $\mathcal{N}_J$ and $\mathcal{N}_{\tilde{J}}$ form a tensor sum decomposition of $\mathcal{H}$.

In the two-dimensional vector space $\mathcal{H}$ *with elements* $f = (a_f,b_f)^T$ *and* $g = (a_g,b_g)^T$, $J(f,g) = b_f b_g$ *defines a semi-inner-product with the null space* $\mathcal{N}_J = \{f : f = (a,0)^T, a \text{ real}\}$. *Define* $\tilde{J}(f,g) = a_f a_g$, *which satisfies the two conditions specified above. It follows that* $\mathcal{N}_{\tilde{J}} = \{f : f = (0,b)^T, b \text{ real}\}$, *in which* $J(f,g) = b_f b_g$ *is positive definite. Clearly,* $(J+\tilde{J})(f,g) = b_f b_g + a_f a_g$ *is positive definite in* $\mathcal{H}$.

Example 2.1 (L_2 space)
All square integrable functions on $[0,1]$ form a Hilbert space

$$\mathcal{L}_2[0,1] = \left\{f : \int_0^1 f^2 dx < \infty\right\}$$

with an inner product $(f,g)=\int_0^1 fg dx$. The space

$$\mathcal{G}=\{f: f=gI_{[x\leq .5]}, g\in\mathcal{L}_2[0,1]\}$$

is a closed linear subspace with an orthogonal complement

$$\mathcal{G}^c=\{f: f=gI_{[x\geq .5]}, g\in\mathcal{L}_2[0,1]\}.$$

Note that elements in $\mathcal{L}_2[0,1]$ are defined not by individual functions but by equivalent classes.

The bilinear form $J(f,g)=\int_0^{.5} fg dx$ defines a semi-inner-product in $\mathcal{L}_2[0,1]$, with a null space

$$\mathcal{N}_J=\mathcal{G}^c=\{f: f=gI_{[x\geq .5]}, g\in\mathcal{L}_2[0,1]\}.$$

Define $\tilde{J}(f,g)=C\int_{.5}^1 fg dx$, with $C>0$ a constant; one has an inner product $(f,g)=(J+\tilde{J})(f,g)=\int_0^{.5} fg + C\int_{.5}^1 fg$ on $\mathcal{L}_2[0,1]$. On $\mathcal{G}=\mathcal{L}_2\ominus\mathcal{N}_J$, $J(f,g)$ is a full inner product. □

Example 2.2 (Euclidean space)
Functions on $\{1,\ldots,K\}$ are vectors of length K. Consider the Euclidean K-space with an inner product

$$(f,g)=\sum_{x=1}^{K} f(x)g(x)=f^T g.$$

The space $\mathcal{G}=\{f: f(1)=\cdots=f(K)\}$ is a closed linear subspace with an orthogonal complement $\mathcal{G}^c=\{f:\sum_{x=1}^K f(x)=0\}$.

Write $\bar{f}=\sum_{x=1}^K f(x)/K$. The bilinear form

$$J(f,g)=\sum_{x=1}^{K}(f(x)-\bar{f})(g(x)-\bar{g})=f^T\left(I-\frac{1}{K}\mathbf{1}\mathbf{1}^T\right)g$$

defines a semi-inner-product in the vector space with a null space

$$\mathcal{N}_J=\mathcal{G}=\{f: f(1)=\cdots=f(K)\}.$$

Define $\tilde{J}(f,g)=C\bar{f}\bar{g}=Cf^T(\mathbf{1}\mathbf{1}^T/K)g$, with $C>0$ a constant; one has an inner product in the vector space,

$$(f,g)=(J+\tilde{J})(f,g)=f^T\left(I+\frac{C-1}{K}\mathbf{1}\mathbf{1}^T\right)g,$$

which reduces to the Euclidean inner product when $C=1$. On $\mathcal{G}^c=\{f:\sum_{x=1}^K f(x)=0\}$, $J(f,g)$ is a full inner product. □

2.1.2 Riesz Representation Theorem

For every g in a Hilbert space $\mathcal{H}$, $L_g f = (g, f)$ defines a continuous linear functional L_g. Conversely, every continuous linear functional L in $\mathcal{H}$ has a representation $Lf = (g_L, f)$ for some $g_L \in \mathcal{H}$, called the **representer** of L, as the following theorem asserts.

Theorem 2.2 (Riesz representation)
For every continuous linear functional L in a Hilbert space $\mathcal{H}$, there exists a unique $g_L \in \mathcal{H}$ such that $Lf = (g_L, f)$, $\forall f \in \mathcal{H}$.

Proof: Let $\mathcal{N}_L = \{f : Lf = 0\}$ be the null space of L. Since L is continuous, $\mathcal{N}_L$ is a closed linear subspace. If $\mathcal{N}_L = \mathcal{H}$, take $g_L = 0$. When $\mathcal{N}_L \subset \mathcal{H}$, there exists a nonzero element $g_0 \in \mathcal{H} \ominus \mathcal{N}_L$. Since $(Lf)g_0 - (Lg_0)f \in \mathcal{N}_L$, $((Lf)g_0 - (Lg_0)f, g_0) = 0$. Some algebra yields

$$Lf = \left(\frac{Lg_0}{(g_0, g_0)} g_0, f \right).$$

Hence, one can take $g_L = (Lg_0)g_0/(g_0, g_0)$. The uniqueness is trivial. □

The continuity of L is necessary for the theorem to hold, otherwise $\mathcal{N}_L$ is no longer closed and the proof breaks down.

All linear functionals in a finite-dimensional Hilbert space are continuous. Actually, there is an isomorphism between any K-dimensional Hilbert space and the Euclidean K-space. See Problems 2.5 and 2.6.

2.1.3 Reproducing Kernel and Non-negative Definite Function

The likelihood part $L(f)$ of the penalized likelihood functional $L(f) + (\lambda/2)J(f)$ usually involves evaluations; thus, for it to be continuous in f, one needs the continuity of the **evaluation functional** $[x]f = f(x)$. Consider a Hilbert space $\mathcal{H}$ of functions on domain $\mathcal{X}$. If the evaluation functional $[x]f = f(x)$ is continuous in $\mathcal{H}$, $\forall x \in \mathcal{X}$, then $\mathcal{H}$ is called a **reproducing kernel Hilbert space.**

By the Riesz representation theorem, there exists $R_x \in \mathcal{H}$, the representer of the evaluation functional $[x](\cdot)$, such that $(R_x, f) = f(x)$, $\forall f \in \mathcal{H}$. The symmetric bivariate function $R(x, y) = R_x(y) = (R_x, R_y)$ has the reproducing property $(R(x, \cdot), f(\cdot)) = f(x)$ and is called the **reproducing kernel** of the space $\mathcal{H}$. The reproducing kernel is unique when it exists (Problem 2.7).

The $\mathcal{L}_2[0, 1]$ space of Example 2.1 is not a reproducing kernel Hilbert space. In fact, since the elements in $\mathcal{L}_2[0, 1]$ are defined by equivalent classes but not individual functions, evaluation is not even well defined. A finite-dimensional Hilbert space is always a reproducing kernel Hilbert space since all linear functionals are continuous.

Example 2.3 (Euclidean space)
Consider again the Euclidean K-space with $(f, g) = f^T g$, with the vectors perceived as functions on $\mathcal{X} = \{1, \ldots, K\}$. The evaluation functional $[x]f = f(x)$ is simply coordinate extraction. Since $f(x) = e_x^T f$, where e_x is the xth unit vector, one has $R_x(y) = I_{[x=y]}$. A bivariate function on $\{1, \ldots, K\}$ can be written as a square matrix, and the reproducing kernel in the Euclidean space is simply the identity matrix. □

A bivariate function $F(x, y)$ on $\mathcal{X}$ is said to be a **non-negative definite function** if $\sum_{i,j} \alpha_i \alpha_j F(x_i, x_j) \geq 0$, $\forall x_i \in \mathcal{X}$, $\forall \alpha_i$ real. For $R(x, y) = R_x(y)$ a reproducing kernel, it is easy to check that

$$\| \textstyle\sum_i \alpha_i R_{x_i} \|^2 = \sum_{i,j} \alpha_i \alpha_j R(x_i, x_j) \geq 0,$$

so $R(x, y)$ is non-negative definite. As a matter of fact, there exists a one-to-one correspondence between reproducing kernel Hilbert spaces and non-negative definite functions, as the following theorem asserts.

Theorem 2.3
For every reproducing kernel Hilbert space $\mathcal{H}$ of functions on $\mathcal{X}$, there corresponds an unique reproducing kernel $R(x, y)$, which is non-negative definite. Conversely, for every non-negative definite function $R(x, y)$ on $\mathcal{X}$, there corresponds a unique reproducing kernel Hilbert space $\mathcal{H}$ that has $R(x, y)$ as its reproducing kernel.

By Theorem 2.3, one may construct a reproducing kernel Hilbert space simply by specifying its reproducing kernel. The following lemma is needed in the proof of the theorem.

Lemma 2.4
Let $R(x, y)$ be any non-negative definite function on $\mathcal{X}$. If

$$\sum_{i=1}^{n} \sum_{j=1}^{n} \alpha_i \alpha_j R(x_i, x_j) = 0,$$

then $\sum_{i=1}^{n} \alpha_i R(x_i, x) = 0$, $\forall x \in \mathcal{X}$.

Proof: Augment the (x_i, α_i) sequence by adding (x_0, α_0), where $x_0 \in \mathcal{X}$ and α_0 real are arbitrary. Since

$$0 \leq \sum_{i=0}^{n} \sum_{j=0}^{n} \alpha_i \alpha_j R(x_i, x_j) = 2\alpha_0 \sum_{i=1}^{n} \alpha_i R(x_i, x_0) + \alpha_0^2 R(x_0, x_0)$$

and $R(x_0, x_0) \geq 0$, it is necessary that $\sum_{i=1}^{n} \alpha_i R(x_i, x_0) = 0$. □

Proof of Theorem 2.3: Only the converse needs a proof. Given $R(x, y)$, write $R_x = R(x, \cdot)$; one starts with the linear space

$$\mathcal{H}^* = \{ f : f = \textstyle\sum_i \alpha_i R_{x_i}, x_i \in \mathcal{X}, \alpha_i \text{ real} \},$$

and defines in $\mathcal{H}^*$ an inner product

$$\Big(\sum_i \alpha_i R_{x_i}, \sum_j \beta_j R_{y_j}\Big) = \sum_{i,j} \alpha_i \beta_j R(x_i, y_j).$$

It is trivial to verify the properties of inner product for such a (f, g), except that $(f, f) = 0$ holds only for $f = 0$, which is proved in Lemma 2.4. It is also easy to verify that $(R_x, f) = f(x)$, $\forall f \in \mathcal{H}^*$.

By the Cauchy-Schwarz inequality,

$$|f(x)| = |(R_x, f)| \leq \sqrt{R(x,x)}\, \|f\|,$$

so convergence in norm implies pointwise convergence. For every Cauchy sequence $\{f_n\}$ in $\mathcal{H}^*$, $\{f_n(x)\}$ is a Cauchy sequence on the real line converging to a limit. Note also that $|\,\|f_n\| - \|f_m\|\,| \leq \|f_n - f_m\|$, so $\{\|f_n\|\}$ has a limit as well. The limit point of $\{f_n\}$ can then be defined by $f(x) = \lim_{n\to\infty} f_n(x)$, $\forall x \in \mathcal{X}$, with $\|f\| = \lim_{n\to\infty} \|f_n\|$. It will be shown shortly that $\|f\|$, thus defined, is unique; that is, for two Cauchy sequences $\{f_n\}$ and $\{g_n\}$ satisfying $\lim_{n\to\infty} f_n(x) = \lim_{n\to\infty} g_n(x)$, $\forall x \in \mathcal{X}$, it is necessary that $\lim_{n\to\infty} \|f_n\| = \lim_{n\to\infty} \|g_n\|$. Adjoining all these limit points of Cauchy sequences to $\mathcal{H}^*$, one obtains a complete linear space $\mathcal{H}$ with the norm $\|f\|$. It is easy to verify that $(f, g) = (\|f+g\|^2 - \|f\|^2 - \|g\|^2)/2$ extends the inner product from $\mathcal{H}^*$ to $\mathcal{H}$ and that $(R_x, f) = f(x)$ holds in $\mathcal{H}$, so $\mathcal{H}$ is a reproducing kernel Hilbert space with $R(x,y)$ as its reproducing kernel.

We now verify the uniqueness of the definition of $\|f\|$ in the completed space, and it suffices to show that for every Cauchy sequence $\{f_n\}$ in $\mathcal{H}^*$ satisfying $\lim_{n\to\infty} f_n(x) = 0$, $\forall x \in \mathcal{X}$, it necessarily holds that $\lim_{n\to\infty} \|f_n\| = 0$. We prove the assertion by negation. Suppose $f_n(x) \to 0$, $\forall x \in \mathcal{X}$, but $\|f_n\|^2 \to 3\delta > 0$. Take $\epsilon \in (0, \delta)$. For n and m sufficiently large, one has $\|f_n\|^2, \|f_m\|^2 > 2\delta$ and $\|f_n - f_m\|^2 < \epsilon$. Fix such an m and write $f_m = \sum_i \alpha_i R_{x_i}$ a finite sum. Since $f_n(x) \to 0$, $\forall x \in \mathcal{X}$, it follows that $\sum_i \alpha_i f_n(x_i) \to 0$. Hence, for n sufficiently large,

$$|(f_n, f_m)| = |(f_n, \textstyle\sum_i \alpha_i R_{x_i})| = |\sum_i \alpha_i f_n(x_i)| < \epsilon.$$

Now,

$$\epsilon > \|f_n - f_m\|^2 = \|f_n\|^2 + \|f_m\|^2 - 2(f_n, f_m) > 4\delta - 2\epsilon > 2\delta,$$

a contradiction.

It remains to be shown that if a space $\tilde{\mathcal{H}}$ has $R(x,y)$ as its reproducing kernel, then $\tilde{\mathcal{H}}$ must be identical to the space $\mathcal{H}$ constructed above. Since $R_x = R(x, \cdot) \in \tilde{\mathcal{H}}$, $\forall x \in \mathcal{X}$, so $\mathcal{H} \subseteq \tilde{\mathcal{H}}$. Now, for any $h \in \tilde{\mathcal{H}} \ominus \mathcal{H}$, by the orthogonality, $h(x) = (R_x, h) = 0$, $\forall x \in \mathcal{X}$, so $\tilde{\mathcal{H}} = \mathcal{H}$. The proof is now complete. □

From the construction in the proof, one can see that the space $\mathcal{H}$ corresponding to R is generated from the "columns" $R_x = R(\cdot, x)$ of R, very much like a vector space generated from the columns of a matrix.

In the sections to follow, we will be constantly decomposing reproducing kernel Hilbert spaces into tensor sums or pasting up larger spaces by taking the tensor sums of smaller ones. The following theorem spells out some of the rules in such operations.

Theorem 2.5
If the reproducing kernel R of a space $\mathcal{H}$ on domain $\mathcal{X}$ can be decomposed into $R = R_0 + R_1$, where R_0 and R_1 are both non-negative definite, $R_0(x,\cdot), R_1(x,\cdot) \in \mathcal{H}$, $\forall x \in \mathcal{X}$, and $(R_0(x,\cdot), R_1(y,\cdot)) = 0$, $\forall x, y \in \mathcal{X}$, then the spaces $\mathcal{H}_0$ and $\mathcal{H}_1$ corresponding respectively to R_0 and R_1 form a tensor sum decomposition of $\mathcal{H}$. Conversely, if R_0 and R_1 are both non-negative definite and $\mathcal{H}_0 \cap \mathcal{H}_1 = \{0\}$, then $\mathcal{H} = \mathcal{H}_0 \oplus \mathcal{H}_1$ has a reproducing kernel $R = R_0 + R_1$.

Proof. By the orthogonality between $R_0(x,\cdot)$ and $R_1(y,\cdot)$,

$$R_0(x,y) = (R_0(x,\cdot), R(y,\cdot)) = (R_0(x,\cdot), R_0(y,\cdot)),$$

so the inner product in $\mathcal{H}_0$ is consistent with that in $\mathcal{H}$; hence, $\mathcal{H}_0$ is a closed linear subspace of $\mathcal{H}$. Now, for every $f \in \mathcal{H}$, let f_0 be the projection of f in $\mathcal{H}_0$ and write $f = f_0 + f_0^c$. Straightforward calculation yields

$$\begin{aligned} f(x) &= (R(x,\cdot), f) \\ &= (R_0(x,\cdot), f_0) + (R_0(x,\cdot), f_0^c) + (R_1(x,\cdot), f_0) + (R_1(x,\cdot), f_0^c) \\ &= f_0(x) + (R_1(x,\cdot), f_0^c), \end{aligned}$$

so $(R_1(x,\cdot), f_0^c) = f(x) - f_0(x) = f_0^c(x)$. This shows that R_1 is the reproducing kernel of $\mathcal{H} \ominus \mathcal{H}_0$; hence, $\mathcal{H} = \mathcal{H}_0 \oplus \mathcal{H}_1$.

For the converse, it is trivial to verify that

$$(R(x,\cdot), f) = (R_0(x,\cdot), f_0)_0 + (R_1(x,\cdot), f_1)_1 = f_0(x) + f_1(x) = f(x),$$

where $f = f_0 + f_1 \in \mathcal{H}$ with $f_0 \in \mathcal{H}_0$ and $f_1 \in \mathcal{H}_1$, and $(\cdot,\cdot)_0$ and $(\cdot,\cdot)_1$ are the inner products in $\mathcal{H}_0$ and $\mathcal{H}_1$, respectively. □

2.2 Smoothing Splines on $\{1, \dots, K\}$

As discussed in Example 2.3, a function on the discrete domain $\mathcal{X} = \{1, \dots, K\}$ is a vector of length K, evaluation is coordinate extraction, and a reproducing kernel can be written as a non-negative definite matrix. A linear functional in a finite-dimensional space is always continuous, so a vector space equipped with an inner product is a reproducing kernel Hilbert space.

Let B be any $K \times K$ non-negative definite matrix. Consider the column space of B, $\mathcal{H}_B = \{f : f = B\mathbf{c} = \sum_j c_j B(\cdot, j)\}$, equipped with the inner product $(f, g) = f^T B g$. The standard eigenvalue decomposition gives

$$B = UDU^T = (U_1, U_2) \begin{pmatrix} D_1 & O \\ O & O \end{pmatrix} \begin{pmatrix} U_1^T \\ U_2^T \end{pmatrix} = U_1 D_1 U_1^T,$$

where the diagonal of D_1 contains the positive eigenvalues of B and the columns of U_1 are the associated eigenvectors. The Moore-Penrose inverse of B has an expression $B^+ = U_1 D_1^{-1} U_1^T$. It is clear that $\mathcal{H}_B = \mathcal{H}_{B^+} = \{f : f = U_1 \mathbf{c}\}$. Now, $B^+ B = U_1 U_1^T$ is the projection matrix onto $\mathcal{H}_B$, so $B^+ B f = f$, $\forall f \in \mathcal{H}_B$. It then follows that

$$[x]f = f(x) = e_x^T f = e_x^T B^+ B f = (B^+ e_x)^T B f,$$

$\forall f \in \mathcal{H}_B$ [i.e., the representer of $[x](\cdot)$ is the xth column of B^+]. Hence, the reproducing kernel is given by $R(x, y) = B^+(x, y)$, where $B^+(x, y)$ is the (x, y)th entry of B^+. The result of Example 2.3 is a trivial special case with $B = I$.

The duality between $(f, g) = f^T B g$ and $R = B^+$ provides a useful insight into the relation between the inner product in a space and the corresponding reproducing kernel: *In a sense, the inner product and the reproducing kernel are inverses of each other.*

Now, consider a decomposition of the reproducing kernel in the Euclidean K-space, $R(x, y) = I_{[x=y]} = 1/K + (I_{[x=y]} - 1/K)$, or in matrix terms, $I = (\mathbf{1}\mathbf{1}^T/K) + (I - \mathbf{1}\mathbf{1}^T/K)$. Since $(\mathbf{1}\mathbf{1}^T/K)(I - \mathbf{1}\mathbf{1}^T/K) = O$, $(R_0(x, \cdot), R_1(y, \cdot)) = 0$, $\forall x, y$. By Theorem 2.5, the decomposition defines a tensor sum decomposition of the space $R^K = \mathcal{H}_0 \oplus \mathcal{H}_1$, where $\mathcal{H}_0 = \{f : f(1) = \cdots = f(K)\}$ and $\mathcal{H}_1 = \{f : \sum_{x=1}^K f(x) = 0\}$. The inner products in $\mathcal{H}_0$ and $\mathcal{H}_1$ have expressions $(f, g)_0 = f^T g = f^T(\mathbf{1}\mathbf{1}^T/K)g$ and $(f, g)_1 = f^T g = f^T(I - \mathbf{1}\mathbf{1}^T/K)g$, respectively, where $\mathbf{1}\mathbf{1}^T/K$ is the Moore-Penrose inverse of $R_0 = \mathbf{1}\mathbf{1}^T/K$ and $I - \mathbf{1}\mathbf{1}^T/K$ is the Moore-Penrose inverse of $R_1 = I - \mathbf{1}\mathbf{1}^T/K$. The decomposition defines a one-way ANOVA decomposition with an averaging operator $Af = \sum_{x=1}^K f(x)/K$. See Problem 2.8 for a construction yielding a one-way ANOVA decomposition with an averaging operator $Af = f(1)$.

Regression on $\mathcal{X} = \{1, \ldots, K\}$ yields the classical one-way ANOVA model. Consider a roughness penalty

$$J(f) = \sum_{x=1}^{K} (f(x) - \bar{f})^2 = f^T \left(I - \frac{\mathbf{1}\mathbf{1}^T}{K} \right) f,$$

where $\bar{f} = \sum_{x=1}^{K} f(x)/K$. The minimizer of

$$\frac{1}{n}\sum_{i=1}^{n}(Y_i - \eta(x_i))^2 + \lambda\sum_{x=1}^{K}(\eta(x) - \bar{\eta})^2 \tag{2.3}$$

defines a shrinkage estimate being shrunk toward a constant. Similarly, if one sets $J(f) = f^T f$, then the minimizer of

$$\frac{1}{n}\sum_{i=1}^{n}(Y_i - \eta(x_i))^2 + \lambda\sum_{x=1}^{K}\eta^2(x) \tag{2.4}$$

defines a shrinkage estimate being shrunk toward zero. Hence, smoothing splines on a discrete domain reduce to shrinkage estimates.

The roughness penalty $\sum_{x=1}^{K}(f(x) - \bar{f})^2$ appears natural for x nominal. For x ordinal, however, one may consider alternatives such as

$$\sum_{x=2}^{K}(f(x) - f(x-1))^2,$$

which have the same null space but use different "scaling" in the penalized contrast space $\mathcal{H}_1 = \{f : \sum_{x=1}^{K} f(x) = 0\}$.

2.3 Polynomial Smoothing Splines on $[0, 1]$

The cubic smoothing spline of §1.1.1 is a special case of the polynomial smoothing splines, the minimizers of

$$\frac{1}{n}\sum_{i=1}^{n}(Y_i - \eta(x_i))^2 + \lambda\int_0^1 (\eta^{(m)})^2 dx, \tag{2.5}$$

in the space $\mathcal{C}^{(m)}[0,1] = \{f : f^{(m)} \in \mathcal{L}_2[0,1]\}$. Equipped with appropriate inner products, the space $\mathcal{C}^{(m)}[0,1]$ can be made a reproducing kernel Hilbert space.

We will present two such constructions and outline an approach to the computation of polynomial smoothing splines. The two constructions yield identical results for univariate smoothing, but provide building blocks satisfying different side conditions for multivariate smoothing with built-in ANOVA decompositions.

2.3.1 A Reproducing Kernel in $\mathcal{C}^{(m)}[0,1]$

For $f \in \mathcal{C}^{(m)}[0,1]$, the standard Taylor expansion gives

$$f(x) = \sum_{\nu=0}^{m-1}\frac{x^\nu}{\nu!}f^{(\nu)}(0) + \int_0^1 \frac{(x-u)_+^{m-1}}{(m-1)!}f^{(m)}(u)du, \tag{2.6}$$

where $(\cdot)_+ = \max(0, \cdot)$. With an inner product

$$(f, g) = \sum_{\nu=0}^{m-1} f^{(\nu)}(0)g^{(\nu)}(0) + \int_0^1 f^{(m)}g^{(m)}dx, \tag{2.7}$$

it can be shown that the representer of evaluation $[x](\cdot)$ is

$$R_x(y) = \sum_{\nu=0}^{m-1} \frac{x^\nu}{\nu!}\frac{y^\nu}{\nu!} + \int_0^1 \frac{(x-u)_+^{m-1}}{(m-1)!}\frac{(y-u)_+^{m-1}}{(m-1)!}du. \tag{2.8}$$

To see this, note that $R_x^{(\nu)}(0) = x^\nu/\nu!$, $\nu = 0, \ldots, m-1$, and that $R_x^{(m)}(y) = (x-y)_+^{m-1}/(m-1)!$. Plugging these into (2.7) with $g = R_x$, one obtains the right-hand side of (2.6), so $(R_x, f) = f(x)$.

The two terms of the reproducing kernel $R(x, y) = R_x(y)$,

$$R_0(x, y) = \sum_{\nu=0}^{m-1} \frac{x^\nu}{\nu!}\frac{y^\nu}{\nu!}, \tag{2.9}$$

and

$$R_1(x, y) = \int_0^1 \frac{(x-u)_+^{m-1}}{(m-1)!}\frac{(y-u)_+^{m-1}}{(m-1)!}du, \tag{2.10}$$

are both non-negative definite themselves, and it is also easy to verify the other conditions of Theorem 2.5. To R_0 there corresponds the space of polynomials $\mathcal{H}_0 = \{f : f^{(m)} = 0\}$ with an inner product $(f, g)_0 = \sum_{\nu=0}^{m-1} f^{(\nu)}(0)g^{(\nu)}(0)$, and to R_1 there corresponds the orthogonal complement of $\mathcal{H}_0$,

$$\mathcal{H}_1 = \big\{f : f^{(\nu)}(0) = 0, \nu = 0, \ldots, m-1, \textstyle\int_0^1 (f^{(m)})^2 dx < \infty\big\}, \tag{2.11}$$

with an inner product $(f, g)_1 = \int_0^1 f^{(m)}g^{(m)}dx$. The space $\mathcal{H}_0$ can be further decomposed into the tensor sum of m subspaces of monomials $\{f : f \propto (\cdot)^\nu\}$ with inner products $f^{(\nu)}(0)g^{(\nu)}(0)$ and reproducing kernels $(x^\nu/\nu!)(y^\nu/\nu!)$, $\nu = 0, \ldots, m-1$.

Setting $m = 1$, one has $R_0(x, y) = 1$ and

$$R_1(x, y) = \int_0^1 I_{[u<x]}I_{[u<y]}du = x \wedge y, \tag{2.12}$$

where $x \wedge y = \min(x, y)$. This setting is useful for the computation of a linear smoothing spline, the minimizer of

$$\frac{1}{n}\sum_{i=1}^n (Y_i - \eta(x_i))^2 + \lambda \int_0^1 \dot{\eta}^2 dx. \tag{2.13}$$

Setting $m = 2$, one has $R_0(x,y) = 1 + xy$ and

$$R_1(x,y) = \int_0^1 (x-u)_+(y-u)_+ du$$
$$= (x \wedge y)^2(3(x \vee y) - (x \wedge y))/6, \tag{2.14}$$

where $x \vee y = \max(x,y)$. The latter formula can be used in the computation of a cubic smoothing spline.

For $m = 1$, the tensor sum decomposition characterized by $R = R_0 + R_1 = [1] + [x \wedge y]$ naturally defines a one-way ANOVA decomposition with an averaging operator $Af = f(0)$, where the corresponding $\mathcal{H}_0$ spans the "mean" space and $\mathcal{H}_1$ spans the "contrast" space; see §1.3.1 for discussions on ANOVA decomposition and averaging operator.

For $m = 2$, the same ANOVA decomposition is characterized by the kernel decomposition

$$R = R_{00} + [R_{01} + R_1] = [1] + [xy + \{(x \wedge y)^2(3(x \vee y) - (x \wedge y))/6\}],$$

where $R_0 = 1 + xy$ is further decomposed into the sum of $R_{00} = 1$ and $R_{01} = xy$. The kernel R_{00} generates the "mean" space and the kernels R_{01} and R_1 together generate the "contrast" space, with R_{01} contributing to the "parametric contrast" and R_1 to the "nonparametric contrast."

2.3.2 Computation of Polynomial Smoothing Splines

Given the sampling points x_i, $i = 1, \dots, n$, in (2.5) and noting that the space $\{f : f = \sum_{i=1}^n \alpha_i R_1(x_i, \cdot)\}$ is a closed linear subspace of $\mathcal{H}_1$ given in (2.11), one may write $\eta \in \mathcal{C}^{(m)}[0,1]$ as

$$\eta(x) = \sum_{\nu=0}^{m-1} d_\nu \frac{x^\nu}{\nu!} + \sum_{i=1}^n c_i R_1(x_i, x) + \rho(x), \tag{2.15}$$

where c_i and d_ν are real coefficients, R_1 is given in (2.10), and

$$\rho \in \mathcal{H}_1 \ominus \{f : f = \textstyle\sum_{i=1}^n c_i R_1(x_i, \cdot)\}.$$

By orthogonality, $\rho(x_i) = (R_1(x_i, \cdot), \rho) = 0$, $i = 1, \dots, n$. Denoting by S the $n \times m$ matrix with the (i,ν)th entry $x_i^\nu/\nu!$ and by Q the $n \times n$ matrix with the (i,j)th entry $R_1(x_i, x_j)$, (2.5) can be written as

$$(\mathbf{Y} - S\mathbf{d} - Q\mathbf{c})^T(\mathbf{Y} - S\mathbf{d} - Q\mathbf{c}) + n\lambda \mathbf{c}^T Q \mathbf{c} + n\lambda(\rho, \rho), \tag{2.16}$$

where the fact that $\int_0^1 R_1^{(m)}(x_i, x) R_1^{(m)}(x_j, x) dx = R_1(x_i, x_j)$ is used. Note that ρ only appears in the third term in (2.16), which is minimized at $\rho = 0$. Hence, a polynomial smoothing spline resides in a space

$$\mathcal{H}_0 \oplus \{f : f = \textstyle\sum_{i=1}^n c_i R_1(x_i, \cdot)\},$$

of finite dimension, and so can be computed via the minimization of the first two terms of (2.16) with respect to $\mathbf{c}$ and $\mathbf{d}$.

In this approach to the computation of polynomial smoothing splines, one needs the reproducing kernel R_1 that corresponds to a space $\mathcal{H}_1$ in which the roughness penalty $\int_0^1 (f^{(m)})^2 dx$ is a full square norm, plus a basis that spans the null space of the penalty.

2.3.3 Another Reproducing Kernel in $\mathcal{C}^{(m)}[0, 1]$

The bilinear form $\int_0^1 f^{(m)} g^{(m)} dx$ is a semi-inner-product in $\mathcal{C}^{(m)}[0, 1]$, which can be augmented to a full inner product by the addition of an inner product in its null space, the space $\{f : f^{(m)} = 0\}$ of polynomials up to order $m - 1$. In §2.3.1, we used $\sum_{\nu=0}^{m-1} f^{(\nu)}(0) g^{(\nu)}(0)$ as the inner product in $\{f : f^{(m)} = 0\}$. In this section, we will use a different inner product, $\sum_{\nu=0}^{m-1} (\int_0^1 f^{(\nu)} dx)(\int_0^1 g^{(\nu)} dx)$, in $\{f : f^{(m)} = 0\}$, and derive the reproducing kernel associated with

$$(f, g) = \sum_{\nu=0}^{m-1} \left(\int_0^1 f^{(\nu)} dx \right) \left(\int_0^1 g^{(\nu)} dx \right) + \int_0^1 f^{(m)} g^{(m)} dx, \qquad (2.17)$$

which defines an inner product different from that in (2.7).

The sought-after reproducing kernel can most conveniently be expressed in terms of the functions

$$k_r(x) = -\left(\sum_{\mu=-\infty}^{-1} + \sum_{\mu=1}^{\infty} \right) \frac{\exp(2\pi \mathbf{i} \mu x)}{(2\pi \mathbf{i} \mu)^r}, \qquad r = 1, 2, \ldots, \qquad (2.18)$$

where $\mathbf{i} = \sqrt{-1}$. It is easy to check that for $r > 1$, k_r is well defined and continuous on the real line, and for $r = 1$, it is well defined and continuous at noninteger points; see Problem 2.9(a). It is also easy to verify that $k_r(x)$ is real-valued and is periodic with period 1; see Problem 2.9(b). It can be seen that $k_r^{(p)} = k_{r-p}$, $p = 1, \ldots, r - 2$ and that $k_r^{(r-1)}(x) = k_1(x)$ for x not an integer. It is known that $k_1(x) = x - 0.5$ on $(0, 1)$ (Problem 2.9(c)), and we define $k_0 = 1$. The k_r functions are actually scaled Bernoulli polynomials, $k_r(x) = B_r(x)/r!$; see Abramowitz and Stegun (1964, Chapter 26) for a comprehensive list of results concerning the Bernoulli polynomials $B_r(x)$.

From the properties listed above, it is easy to verify that $\int_0^1 k_\mu^{(\nu)} dx = \delta_{\mu,\nu}$, $\mu, \nu = 0, \ldots, m - 1$, where $\delta_{\mu,\nu}$ is the Kronecker delta. It then follows that k_ν, $\nu = 0, \ldots, m - 1$, form an orthonormal basis of $\mathcal{H}_0 = \{f : f^{(m)} = 0\}$ under the inner product $(f, g)_0 = \sum_{\nu=0}^{m-1} (\int_0^1 f^{(\nu)} dx)(\int_0^1 g^{(\nu)} dx)$ and that

$$R_0(x, y) = \sum_{\nu=0}^{m-1} k_\nu(x) k_\nu(y) \qquad (2.19)$$

is the reproducing kernel in $\mathcal{H}_0$; see Problem 2.5(c) for the definition of orthonormal basis. In fact, $\mathcal{H}_0$ can be further decomposed into the tensor sum of m subspaces $\{f : f \propto k_\nu\}$ with inner products $(\int_0^1 f^{(\nu)}dx)(\int_0^1 g^{(\nu)}dx)$ and reproducing kernels $k_\nu(x)k_\nu(y)$, $\nu = 0, \ldots, m-1$, respectively.

We now show that in the space

$$\mathcal{H}_1 = \{f : \textstyle\int_0^1 f^{(\nu)}dx = 0, \nu = 0, \ldots, m-1, f^{(m)} \in \mathcal{L}_2[0,1]\} \tag{2.20}$$

with a square norm $(f,g)_1 = \int_0^1 f^{(m)}g^{(m)}dx$, the function

$$R_x(y) = k_m(x)k_m(y) + (-1)^{m-1}k_{2m}(x-y) \tag{2.21}$$

is the representer of evaluation $[x](\cdot)$. From the properties of k_r, it is easy to check that $\int_0^1 R_x^{(\nu)}(y)dy = 0$, $\nu = 0, \ldots, m-1$, and that $R_x^{(m)}(y) = k_m(x) - k_m(x-y) \in \mathcal{L}_2[0,1]$, so $R_x \in \mathcal{H}_1$ for $\mathcal{H}_1$ given in (2.20). Integrating by parts, and using the periodicity of k_r, $r > 1$, and the fact that $\int_0^1 f^{(\nu)}dx = 0$, $\nu = 0, \ldots, m-1$, one can show that

$$\begin{aligned}(R_x, f)_1 &= \int_0^1 (k_m(x) - k_m(x-y))f^{(m)}(y)dy \\ &= -\int_0^1 k_{m-1}(x-y)f^{(m-1)}(y)dy \\ &= \cdots = -\int_0^1 k_1(x-y)\dot{f}(y)dy;\end{aligned} \tag{2.22}$$

see Problem 2.10. Now, since

$$k_1(x-y) = \begin{cases} x - y - 0.5 = k_1(x) - y, & y \in (0,x), \\ (1 + x - y) - 0.5 = k_1(x) - y + 1, & y \in (x,1), \end{cases}$$

straightforward calculation yields

$$\begin{aligned}&-\int_0^1 k_1(x-y)\dot{f}(y)dy \\ &\qquad = -\int_0^1 k_1(x)\dot{f}(y)dy + \int_0^1 y\dot{f}(y)dy - \int_x^1 \dot{f}(y)dy \\ &\qquad = 0 + f(1) - (f(1) - f(x)) = f(x).\end{aligned}$$

This proves that

$$R_1(x,y) = k_m(x)k_m(y) + (-1)^{m-1}k_{2m}(x-y) \tag{2.23}$$

is the reproducing kernel of $\mathcal{H}_1$ given in (2.20).

Obviously, $\mathcal{H}_0 \cap \mathcal{H}_1 = \{0\}$, so by the converse of Theorem 2.5, $\mathcal{C}^{(m)}[0,1] = \mathcal{H}_0 \oplus \mathcal{H}_1$ has the reproducing kernel $R = R_0 + R_1$. The identity

$$f(x) = \sum_{\nu=0}^{m-1} k_\nu(x) \int_0^1 f^{(\nu)}(y)dy + \int_0^1 (k_m(x) - k_m(x-y)) f^{(m)}(y)dy, \quad (2.24)$$

$\forall f \in \mathcal{C}^{(m)}[0,1]$, may be called a generalized Taylor expansion, where the scaled Bernoulli polynomials $k_\nu(x)$ play the role of the scaled monomials $x^\nu/\nu!$ in the standard Taylor expansion (2.6). The standard Taylor expansion is asymmetric with respect to the domain $[0,1]$, in the sense that a swapping of the two ends 0 and 1 would change its composition entirely, whereas the generalized Taylor expansion of (2.24) is symmetric with respect to the domain.

The computation of polynomial smoothing splines as outlined in §2.3.2 can also be performed by using the R_1 of (2.23) instead of that of (2.10). Also, one may use any basis $\{\phi_\nu\}_{\nu=0}^{m-1}$ of the subspace $\mathcal{H}_0$ in the place of $\{x^\nu/\nu!\}_{\nu=0}^{m-1}$ in the expression of η given in (2.15). The coefficients c_i and d_ν will be different when different ϕ_ν and R_1 are used, but the function estimate

$$\eta(x) = \sum_{\nu=0}^{m-1} d_\nu \phi_\nu(x) + \sum_{i=1}^{n} c_i R_1(x_i, x)$$

will remain the same regardless of the choices of ϕ_ν and R_1.

When $m = 1$, $R_0(x,y) = 1$ and

$$R_1(x,y) = k_1(x)k_1(y) + k_2(x-y). \quad (2.25)$$

When $m = 2$, $R_0(x,y) = 1 + k_1(x)k_1(y)$ and

$$R_1(x,y) = k_2(x)k_2(y) - k_4(x-y). \quad (2.26)$$

The R_1 in (2.25) and (2.26) can be used in the computation of linear and cubic smoothing splines in lieu of those in (2.12) and (2.14). To calculate R_1 in (2.25) and (2.26), one has, on $x \in [0,1]$,

$$\begin{aligned} k_2(x) &= \frac{1}{2}\left(k_1^2(x) - \frac{1}{12}\right), \\ k_4(x) &= \frac{1}{24}\left(k_1^4(x) - \frac{k_1^2(x)}{2} + \frac{7}{240}\right), \end{aligned} \quad (2.27)$$

where $k_1(x) = x - 0.5$; see Problem 2.11. Note that k_2 and k_4 are symmetric with respect to 0.5 on $[0,1]$, so $k_2(x-y) = k_2(|x-y|)$ and $k_4(x-y) = k_4(|x-y|)$, for $x, y \in [0,1]$.

For $m = 1$, the tensor sum decomposition characterized by $R = R_0 + R_1 = [1] + [k_1(x)k_1(y) + k_2(x-y)]$ defines a one-way ANOVA decomposition with an averaging operator $Af = \int_0^1 f dx$, where the corresponding $\mathcal{H}_0$ spans the "mean" space and $\mathcal{H}_1$ spans the "contrast" space.

For $m = 2$, the same ANOVA decomposition is characterized by the kernel decomposition

$$R = R_{00} + [R_{01} + R_1] = [1] + [k_1(x)k_1(y) + \{k_2(x)k_2(y) - k_4(x-y)\}],$$

where $R_0 = 1 + k_1(x)k_1(y)$ is further decomposed into the sum of $R_{00} = 1$ and $R_{01} = k_1(x)k_1(y)$. The kernel R_{00} generates the "mean" space and the kernels R_{01} and R_1 together generate the "contrast" space, with R_{01} contributing to the "parametric contrast" and R_1 to the "nonparametric contrast."

2.4 Smoothing Splines on Product Domains

To incorporate the ANOVA decomposition introduced in §1.3.2 for the estimation of a multivariate function, one may construct a tensor product reproducing kernel Hilbert space. Based on Theorem 2.3, the construction of the space is done through the construction of the reproducing kernel, which uses reproducing kernels on the marginal domains. One-way ANOVA decompositions on the marginal domains naturally induce an ANOVA decomposition on the product domain.

We begin with some general discussion of tensor product reproducing kernel Hilbert spaces, where it is shown that the products of reproducing kernels on the marginal domains form reproducing kernels on the product domain. The construction is then illustrated with marginal domains $\{1, \dots, K\}$ and $[0, 1]$, using the (marginal) reproducing kernels introduced in §2.2 and §2.3. Besides serving the illustrative purposes, the sample constructions will also be employed in later chapters in simulations and case studies.

2.4.1 Tensor Product Reproducing Kernel Hilbert Spaces

A convenient approach to the construction of reproducing kernel Hilbert spaces on a product domain $\prod_{\gamma=1}^{\Gamma} \mathcal{X}_\gamma$ is by taking the tensor product of spaces constructed on the marginal domains $\mathcal{X}_\gamma$. The construction builds on the following theorem.

Theorem 2.6
For $R_{\langle 1 \rangle}(x_{\langle 1 \rangle}, y_{\langle 1 \rangle})$ non-negative definite on $\mathcal{X}_1$ and $R_{\langle 2 \rangle}(x_{\langle 2 \rangle}, y_{\langle 2 \rangle})$ non-negative definite on $\mathcal{X}_2$, $R(x, y) = R_{\langle 1 \rangle}(x_{\langle 1 \rangle}, y_{\langle 1 \rangle})R_{\langle 2 \rangle}(x_{\langle 2 \rangle}, y_{\langle 2 \rangle})$ is non-negative definite on $\mathcal{X} = \mathcal{X}_1 \times \mathcal{X}_2$.

Proof. It suffices to show that, for two non-negative definite matrices A and B of the same size, their entrywise product, $A \circ B$, is necessarily non-negative definite. By elementary matrix theory, A and B are non-negative definite if and only if there exist vectors a_i and b_j such that $A = \sum_i a_i a_i^T$ and $B = \sum_j b_j b_j^T$. Now,

$$A \circ B = \Big(\sum_i a_i a_i^T\Big) \circ \Big(\sum_j b_j b_j^T\Big) = \sum_{i,j} (a_i a_i^T) \circ (b_j b_j^T) = \sum_{i,j} (a_i \circ b_j)(a_i \circ b_j)^T,$$

so $A \circ B$ is non-negative definite. □

By Theorem 2.3, every non-negative definite function R on domain $\mathcal{X}$ corresponds to a reproducing kernel Hilbert space with R as its reproducing kernel. Given $\mathcal{H}_{\langle 1 \rangle}$ on $\mathcal{X}_1$ with reproducing kernel $R_{\langle 1 \rangle}$ and $\mathcal{H}_{\langle 2 \rangle}$ on $\mathcal{X}_2$ with reproducing kernel $R_{\langle 2 \rangle}$, $R = R_{\langle 1 \rangle} R_{\langle 2 \rangle}$ is non-negative definite on $\mathcal{X}_1 \times \mathcal{X}_2$ by Theorem 2.6. The reproducing kernel Hilbert space corresponding to such an R is called the **tensor product space** of $\mathcal{H}_{\langle 1 \rangle}$ and $\mathcal{H}_{\langle 2 \rangle}$, and is denoted by $\mathcal{H}_{\langle 1 \rangle} \otimes \mathcal{H}_{\langle 2 \rangle}$. The operation extends to multiple-term products recursively.

Suppose one has reproducing kernel Hilbert spaces $\mathcal{H}_{\langle \gamma \rangle}$ on domains $\mathcal{X}_\gamma$, $\gamma = 1, \dots, \Gamma$, respectively. Further, assume that the spaces have one-way ANOVA decompositions built in via the tensor sum decompositions $\mathcal{H}_{\langle \gamma \rangle} = \mathcal{H}_{0\langle \gamma \rangle} \oplus \mathcal{H}_{1\langle \gamma \rangle}$, where $\mathcal{H}_{0\langle \gamma \rangle} = \{f : f \propto 1\}$ has a reproducing kernel $R_{0\langle \gamma \rangle} \propto 1$ and $\mathcal{H}_{1\langle \gamma \rangle}$ has a reproducing kernel $R_{1\langle \gamma \rangle}$ satisfying side conditions $A_\gamma R_{1\langle \gamma \rangle}(x_{\langle \gamma \rangle}, \cdot) = 0$, $\forall x_{\langle \gamma \rangle} \in \mathcal{X}_\gamma$, where A_γ are the averaging operators defining the one-way ANOVA decompositions on $\mathcal{X}_\gamma$. The tensor product space $\mathcal{H} = \otimes_{\gamma=1}^{\Gamma} \mathcal{H}_{\langle \gamma \rangle}$ has a tensor sum decomposition

$$\mathcal{H} = \bigotimes_{\gamma=1}^{\Gamma} (\mathcal{H}_{0\langle \gamma \rangle} \oplus \mathcal{H}_{1\langle \gamma \rangle}) = \bigoplus_{\mathcal{S}} \Big\{ \Big(\bigotimes_{\gamma \in \mathcal{S}} \mathcal{H}_{1\langle \gamma \rangle} \Big) \otimes \Big(\bigotimes_{\gamma \notin \mathcal{S}} \mathcal{H}_{0\langle \gamma \rangle} \Big) \Big\} = \bigoplus_{\mathcal{S}} \mathcal{H}_{\mathcal{S}}, \quad (2.28)$$

which parallels (1.7) on page 7, where the summation is over all subsets $\mathcal{S} \subseteq \{1, \dots, \Gamma\}$. The term $\mathcal{H}_{\mathcal{S}}$ has a reproducing kernel $R_{\mathcal{S}} \propto \prod_{\gamma \in \mathcal{S}} R_{1\langle \gamma \rangle}$, and the projection of $f \in \mathcal{H}$ in $\mathcal{H}_{\mathcal{S}}$ is the $f_{\mathcal{S}}$ appearing in (1.7). The minimizer of $L(f) + (\lambda/2)J(f)$ in a tensor product reproducing kernel Hilbert space is called a **tensor product smoothing spline**. Examples of the construction follow.

2.4.2 Reproducing Kernel Hilbert Spaces on $\{1, \dots, K\}^2$

Set $A_\gamma f = \sum_{x_{\langle \gamma \rangle}=1}^{K_\gamma} f(x)/K_\gamma$ on discrete domains $\mathcal{X}_\gamma = \{1, \dots, K_\gamma\}$, $\gamma = 1, 2$. The marginal reproducing kernels that define the one-way ANOVA decomposition on $\mathcal{X}_\gamma$ can be taken as $R_{0\langle \gamma \rangle}(x_{\langle \gamma \rangle}, y_{\langle \gamma \rangle}) = 1/K_\gamma$ and

$$R_{1\langle \gamma \rangle}(x_{\langle \gamma \rangle}, y_{\langle \gamma \rangle}) = I_{[x_{\langle \gamma \rangle} = y_{\langle \gamma \rangle}]} - 1/K_\gamma,$$

$\gamma = 1, 2$, as given in §2.2.

Subspace	Reproducing Kernel
$\mathcal{H}_{0\langle 1\rangle} \otimes \mathcal{H}_{0\langle 2\rangle}$	$(\mathbf{1}_{K_1}\mathbf{1}_{K_1}^T/K_1) \otimes (\mathbf{1}_{K_2}\mathbf{1}_{K_2}^T/K_2)$
$\mathcal{H}_{0\langle 1\rangle} \otimes \mathcal{H}_{1\langle 2\rangle}$	$(\mathbf{1}_{K_1}\mathbf{1}_{K_1}^T/K_1) \otimes (I_{K_2} - \mathbf{1}_{K_2}\mathbf{1}_{K_2}^T/K_2)$
$\mathcal{H}_{1\langle 1\rangle} \otimes \mathcal{H}_{0\langle 2\rangle}$	$(I_{K_1} - \mathbf{1}_{K_1}\mathbf{1}_{K_1}^T/K_1) \otimes (\mathbf{1}_{K_2}\mathbf{1}_{K_2}^T/K_2)$
$\mathcal{H}_{1\langle 1\rangle} \otimes \mathcal{H}_{1\langle 2\rangle}$	$(I_{K_1} - \mathbf{1}_{K_1}\mathbf{1}_{K_1}^T/K_1) \otimes (I_{K_2} - \mathbf{1}_{K_2}\mathbf{1}_{K_2}^T/K_2)$

TABLE 2.1. Product Reproducing Kernels on $\{1,\dots,K_1\} \times \{1,\dots,K_2\}$

A function on $\{1,\dots,K_1\} \times \{1,\dots,K_2\}$ can be written as a vector of length K_1K_2,

$$f = (f(1,1),\dots,f(1,K_2),\dots,f(K_1,1),\dots,f(K_1,K_2))^T,$$

and a reproducing kernel as a $(K_1K_2) \times (K_1K_2)$ matrix. Using matrix notation, the products of the marginal reproducing kernels $R_{0\langle\gamma\rangle}$ and $R_{1\langle\gamma\rangle}$ given above and the subspaces they correspond to are listed in Table 2.1, where $\mathbf{1}_K$ is of length K, I_K is of size $K \times K$, and, as a matrix operator, $\otimes$ denotes the Kronecker product of matrices. The corresponding inner products are defined by the Moore-Penrose inverses of these matrices, which are themselves because they are idempotent. The decomposition of (2.28) is seen to be

$$\begin{aligned}\mathcal{H} &= (\mathcal{H}_{0\langle 1\rangle} \oplus \mathcal{H}_{1\langle 1\rangle}) \otimes (\mathcal{H}_{0\langle 2\rangle} \oplus \mathcal{H}_{1\langle 2\rangle})\\ &= (\mathcal{H}_{0\langle 1\rangle} \otimes \mathcal{H}_{0\langle 2\rangle}) \oplus (\mathcal{H}_{1\langle 1\rangle} \otimes \mathcal{H}_{0\langle 2\rangle})\\ &\quad \oplus (\mathcal{H}_{0\langle 1\rangle} \otimes \mathcal{H}_{1\langle 2\rangle}) \oplus (\mathcal{H}_{1\langle 1\rangle} \otimes \mathcal{H}_{1\langle 2\rangle})\\ &= \mathcal{H}_{\{\}} \oplus \mathcal{H}_{\{1\}} \oplus \mathcal{H}_{\{2\}} \oplus \mathcal{H}_{\{1,2\}},\end{aligned} \tag{2.29}$$

where $\mathcal{H}_{\{\}}$ spans the constant, $\mathcal{H}_{\{1\}}$ spans the $x_{\langle 1\rangle}$ main effect, $\mathcal{H}_{\{2\}}$ spans the $x_{\langle 2\rangle}$ main effect, and $\mathcal{H}_{\{1,2\}}$ spans the interaction.

If one would like to use the averaging operator $Af = f(1)$ on a marginal domain $\{1,\dots,K\}$, the K-dimensional vector space may be decomposed alternatively as

$$\mathcal{H}_0 \oplus \mathcal{H}_1 = \{f : f(1) = \cdots = f(K)\} \oplus \{f : f(1) = 0\},$$

with the reproducing kernels given by $R_0 = 1$ and $R_1(x,y) = I_{[x=y\neq 1]}$; see Problem 2.8.

2.4.3 Reproducing Kernel Hilbert Spaces on $[0,1]^2$

Set $Af = \int_0^1 f dx$ on $[0,1]$. The tensor product reproducing kernel Hilbert spaces on $[0,1]^2$ can be constructed using the reproducing kernels (2.19) and (2.23) derived in §2.3.3.

Example 2.4 (Tensor product linear spline)
Setting $m = 1$ in §2.3.3, one has

$$\begin{aligned}\{f : \dot{f} \in \mathcal{L}_2[0,1]\} &= \{f : f \propto 1\} \oplus \{f : \textstyle\int_0^1 f dx = 0, \dot{f} \in \mathcal{L}_2[0,1]\} \\ &= \mathcal{H}_0 \oplus \mathcal{H}_1,\end{aligned}$$

with reproducing kernels $R_0(x,y) = 1$ and $R_1(x,y) = k_1(x)k_1(y)+k_2(x-y)$. This marginal space can be used on both axes to construct a tensor product reproducing kernel Hilbert space with the structure of (2.28), with averaging operators $A_\gamma f = \int_0^1 f dx_{\langle\gamma\rangle}$, $\gamma = 1, 2$. The reproducing kernels and the corresponding inner products in the subspaces are listed in Table 2.2. □

Example 2.5 (Tensor product cubic spline)
Setting $m = 2$ in §2.3.3, one has

$$\begin{aligned}\{f : \ddot{f} \in \mathcal{L}_2[0,1]\} &= \{f : f \propto 1\} \oplus \{f : f \propto k_1\} \\ &\quad \oplus \{f : \textstyle\int_0^1 f dx = \int_0^1 \dot{f} dx = 0, \ddot{f} \in \mathcal{L}_2[0,1]\} \\ &= \mathcal{H}_{00} \oplus \mathcal{H}_{01} \oplus \mathcal{H}_1,\end{aligned}$$

where $\mathcal{H}_{01} \oplus \mathcal{H}_1$ forms the contrast in a one-way ANOVA decomposition with an averaging operator $Af = \int_0^1 f dx$. The corresponding reproducing kernels are $R_{00}(x,y) = 1$, $R_{01}(x,y) = k_1(x)k_1(y)$, and $R_1(x,y) = k_2(x)k_2(y) - k_4(x-y)$. Note that $\int_0^1 R_{01}(x,y)dy = \int_0^1 R_1(x,y)dy = 0$, $\forall x \in [0,1]$. Using this space on both marginal domains, one can construct a tensor product space with nine tensor sum terms. The subspace $\mathcal{H}_{00\langle 1\rangle} \otimes \mathcal{H}_{00\langle 2\rangle}$ spans the constant term in (1.7) on page 7, the subspaces $\mathcal{H}_{00\langle 1\rangle} \otimes (\mathcal{H}_{01\langle 2\rangle} \oplus \mathcal{H}_{1\langle 2\rangle})$ and $(\mathcal{H}_{01\langle 1\rangle} \oplus \mathcal{H}_{1\langle 1\rangle}) \otimes \mathcal{H}_{00\langle 2\rangle}$ span the main effects, and the subspace $(\mathcal{H}_{01\langle 1\rangle} \oplus \mathcal{H}_{1\langle 1\rangle}) \otimes (\mathcal{H}_{01\langle 2\rangle} \oplus \mathcal{H}_{1\langle 2\rangle})$ spans the interaction. The reproducing kernels and the corresponding inner products in some of the subspaces are listed in Table 2.3. The separation of $\mathcal{H}_{01}$ and $\mathcal{H}_1$ is intended to facilitate adequate numerical treatment of the different components; it is not needed for the characterization of the ANOVA decomposition in (2.28). □

For the averaging operator $Af = f(0)$, similar tensor product reproducing kernel Hilbert spaces can be constructed using the marginal spaces described in §2.3.1; details are to be worked out in Problem 2.13. Note that it is not necessary to use the same marginal space on both axes. Actually, the choice of the order m and that of the averaging operator Af on different axes are unrelated to each other. Although the reproducing kernels of §2.3.1 and §2.3.3 lead to identical polynomial smoothing splines for univariate smoothing on $[0,1]$, they do yield different tensor product smoothing splines on $[0,1]^2$, as the corresponding roughness penalties are different.

Subspace	Reproducing Kernel	Inner Product
$\mathcal{H}_{0\langle 1\rangle} \otimes \mathcal{H}_{0\langle 2\rangle}$	1	$(\int_0^1 \int_0^1 f)(\int_0^1 \int_0^1 g)$
$\mathcal{H}_{0\langle 1\rangle} \otimes \mathcal{H}_{1\langle 2\rangle}$	$k_1(x_{\langle 2\rangle})k_1(y_{\langle 2\rangle}) + k_2(x_{\langle 2\rangle} - y_{\langle 2\rangle})$	$\int_0^1 (\int_0^1 \dot{f}_{\langle 2\rangle} dx_{\langle 1\rangle})(\int_0^1 \dot{g}_{\langle 2\rangle} dx_{\langle 1\rangle}) dx_{\langle 2\rangle}$
$\mathcal{H}_{1\langle 1\rangle} \otimes \mathcal{H}_{0\langle 2\rangle}$	$k_1(x_{\langle 1\rangle})k_1(y_{\langle 1\rangle}) + k_2(x_{\langle 1\rangle} - y_{\langle 1\rangle})$	$\int_0^1 (\int_0^1 \dot{f}_{\langle 1\rangle} dx_{\langle 2\rangle})(\int_0^1 \dot{g}_{\langle 1\rangle} dx_{\langle 2\rangle}) dx_{\langle 1\rangle}$
$\mathcal{H}_{1\langle 1\rangle} \otimes \mathcal{H}_{1\langle 2\rangle}$	$[k_1(x_{\langle 1\rangle})k_1(y_{\langle 1\rangle}) + k_2(x_{\langle 1\rangle} - y_{\langle 1\rangle})][k_1(x_{\langle 2\rangle})k_1(y_{\langle 2\rangle}) + k_2(x_{\langle 2\rangle} - y_{\langle 2\rangle})]$	$\int_0^1 \int_0^1 \ddot{f}_{\langle 12\rangle} \ddot{g}_{\langle 12\rangle}$

TABLE 2.2. Reproducing Kernels and Inner Products in Example 2.4

Subspace	Reproducing Kernel	Inner Product
$\mathcal{H}_{00\langle 1\rangle} \otimes \mathcal{H}_{00\langle 2\rangle}$	1	$(\int_0^1 \int_0^1 f)(\int_0^1 \int_0^1 g)$
$\mathcal{H}_{01\langle 1\rangle} \otimes \mathcal{H}_{00\langle 2\rangle}$	$k_1(x_{\langle 1\rangle})k_1(y_{\langle 1\rangle})$	$(\int_0^1 \int_0^1 \dot{f}_{\langle 1\rangle})(\int_0^1 \int_0^1 \dot{g}_{\langle 1\rangle})$
$\mathcal{H}_{01\langle 1\rangle} \otimes \mathcal{H}_{01\langle 2\rangle}$	$k_1(x_{\langle 1\rangle})k_1(y_{\langle 1\rangle})k_1(x_{\langle 2\rangle})k_1(y_{\langle 2\rangle})$	$(\int_0^1 \int_0^1 \ddot{f}_{\langle 12\rangle})(\int_0^1 \int_0^1 \ddot{g}_{\langle 12\rangle})$
$\mathcal{H}_{1\langle 1\rangle} \otimes \mathcal{H}_{00\langle 2\rangle}$	$k_2(x_{\langle 1\rangle})k_2(y_{\langle 1\rangle}) - k_4(x_{\langle 1\rangle} - y_{\langle 1\rangle})$	$\int_0^1 (\int_0^1 \ddot{f}_{\langle 11\rangle} dx_{\langle 2\rangle})(\int_0^1 \ddot{g}_{\langle 11\rangle} dx_{\langle 2\rangle}) dx_{\langle 1\rangle}$
$\mathcal{H}_{1\langle 1\rangle} \otimes \mathcal{H}_{01\langle 2\rangle}$	$[k_2(x_{\langle 1\rangle})k_2(y_{\langle 1\rangle}) - k_4(x_{\langle 1\rangle} - y_{\langle 1\rangle})]k_1(x_{\langle 2\rangle})k_1(y_{\langle 2\rangle})$	$\int_0^1 (\int_0^1 f^{(3)}_{\langle 112\rangle} dx_{\langle 2\rangle})(\int_0^1 g^{(3)}_{\langle 112\rangle} dx_{\langle 2\rangle}) dx_{\langle 1\rangle}$
$\mathcal{H}_{1\langle 1\rangle} \otimes \mathcal{H}_{1\langle 2\rangle}$	$[k_2(x_{\langle 1\rangle})k_2(y_{\langle 1\rangle}) - k_4(x_{\langle 1\rangle} - y_{\langle 1\rangle})][k_2(x_{\langle 2\rangle})k_2(y_{\langle 2\rangle}) - k_4(x_{\langle 2\rangle} - y_{\langle 2\rangle})]$	$\int_0^1 \int_0^1 f^{(4)}_{\langle 1122\rangle} g^{(4)}_{\langle 1122\rangle}$

TABLE 2.3. Reproducing Kernels and Inner Products in Example 2.5

2.4.4 Reproducing Kernel Hilbert Spaces on $\{1, \ldots, K\} \times [0, 1]$

Setting $A_1 f = \sum_{x_{\langle 1 \rangle}=1}^{K} f(x)/K$ on $\mathcal{X}_1 = \{1, \ldots, K\}$ and $A_2 f = \int_0^1 f dx_{\langle 2 \rangle}$ on $\mathcal{X}_2 = [0, 1]$, tensor product spaces with the structure of (2.28) built in can be constructed using the marginal spaces used in §2.4.2 and §2.4.3.

Example 2.6
One construction of a tensor product space is by using $R_{0\langle 1 \rangle}(x_{\langle 1 \rangle}, y_{\langle 1 \rangle}) = 1/K$ and $R_{1\langle 1 \rangle}(x_{\langle 1 \rangle}, y_{\langle 1 \rangle}) = I_{[x_{\langle 1 \rangle}=y_{\langle 1 \rangle}]} - 1/K$ on $\mathcal{X}_1$ and $R_{0\langle 2 \rangle}(x_{\langle 2 \rangle}, y_{\langle 2 \rangle}) = 1$ and $R_{1\langle 2 \rangle}(x_{\langle 2 \rangle}, y_{\langle 2 \rangle}) = k_1(x_{\langle 2 \rangle})k_1(y_{\langle 2 \rangle}) + k_2(x_{\langle 2 \rangle} - y_{\langle 2 \rangle})$ on $\mathcal{X}_2$. The reproducing kernels and the corresponding inner products in the subspaces are listed in Table 2.4. □

Example 2.7
Using $R_{0\langle 1 \rangle} = 1/K$ and $R_{1\langle 1 \rangle} = I_{[x_{\langle 1 \rangle}=y_{\langle 1 \rangle}]} - 1/K$ on $\mathcal{X}_1$ and $R_{00\langle 2 \rangle} = 1$, $R_{01\langle 2 \rangle} = k_1(x_{\langle 2 \rangle})k_1(y_{\langle 2 \rangle})$, and $R_{1\langle 2 \rangle} = k_2(x_{\langle 2 \rangle})k_2(y_{\langle 2 \rangle}) - k_4(x_{\langle 2 \rangle} - y_{\langle 2 \rangle})$ on $\mathcal{X}_2$, one can construct a tensor product space with six tensor sum terms. The subspace $\mathcal{H}_{0\langle 1 \rangle} \otimes \mathcal{H}_{00\langle 2 \rangle}$ spans the constant, $\mathcal{H}_{0\langle 1 \rangle} \otimes (\mathcal{H}_{01\langle 2 \rangle} \oplus \mathcal{H}_{1\langle 2 \rangle})$ and $\mathcal{H}_{1\langle 1 \rangle} \otimes \mathcal{H}_{00\langle 2 \rangle}$ span the main effects, and $\mathcal{H}_{1\langle 1 \rangle} \otimes (\mathcal{H}_{01\langle 2 \rangle} \oplus \mathcal{H}_{1\langle 2 \rangle})$ spans the interaction. The reproducing kernels and the corresponding inner products in the subspaces are listed in Table 2.5. □

2.4.5 Multiple-Term Reproducing Kernel Hilbert Spaces: General Form

The examples of tensor product reproducing kernel Hilbert spaces on product domains presented above all contain multiple tensor sum terms. In general, a multiple-term reproducing kernel Hilbert space can be written as $\mathcal{H} = \oplus_\beta \mathcal{H}_\beta$, where β is a generic index, with subspaces $\mathcal{H}_\beta$ having inner products $(f_\beta, g_\beta)_\beta$ and reproducing kernels R_β, where f_β is the projection of f in $\mathcal{H}_\beta$. It is often convenient to write $(f, g)_\beta$ for $(f_\beta, g_\beta)_\beta$, which can be formally defined as a semi-inner-product in $\mathcal{H}$ satisfying $(f - f_\beta, f - f_\beta)_\beta = 0$.

The subspaces $\mathcal{H}_\beta$ are independent modules, and the within-module metrics implied by the inner products $(f_\beta, g_\beta)_\beta$ are not necessarily comparable between the modules. Allowing for intermodule rescaling of the metrics, an inner product in $\mathcal{H}$ can be specified via

$$J(f, g) = \sum_\beta \theta_\beta^{-1} (f_\beta, g_\beta)_\beta, \tag{2.30}$$

Subspace	Reproducing Kernel	Inner Product
$\mathcal{H}_{0\langle 1\rangle} \otimes \mathcal{H}_{0\langle 2\rangle}$	1/K	$(\sum_{x_{\langle 1\rangle}=1}^{K} \int_0^1 f)(\sum_{x_{\langle 1\rangle}=1}^{K} \int_0^1 g)/K$
$\mathcal{H}_{0\langle 1\rangle} \otimes \mathcal{H}_{1\langle 2\rangle}$	$[k_1(x_{\langle 2\rangle})k_1(y_{\langle 2\rangle}) + k_2(x_{\langle 2\rangle} - y_{\langle 2\rangle})]/K$	$\int_0^1 (\sum_{x_{\langle 1\rangle}=1}^{K} \dot{f}_{\langle 2\rangle})(\sum_{x_{\langle 1\rangle}=1}^{K} \dot{g}_{\langle 2\rangle})/K$
$\mathcal{H}_{1\langle 1\rangle} \otimes \mathcal{H}_{0\langle 2\rangle}$	$I_{[x_{\langle 1\rangle}=y_{\langle 1\rangle}]} - 1/K$	$\sum_{x_{\langle 1\rangle}=1}^{K} (\int_0^1 (I - A_1)f)(\int_0^1 (I - A_1)g)$
$\mathcal{H}_{1\langle 1\rangle} \otimes \mathcal{H}_{1\langle 2\rangle}$	$(I_{[x_{\langle 1\rangle}=y_{\langle 1\rangle}]} - 1/K)[k_1(x_{\langle 2\rangle})k_1(y_{\langle 2\rangle}) + k_2(x_{\langle 2\rangle} - y_{\langle 2\rangle})]$	$\int_0^1 \sum_{x_{\langle 1\rangle}=1}^{K} (I - A_1)\dot{f}_{\langle 2\rangle}(I - A_1)\dot{g}_{\langle 2\rangle}$

TABLE 2.4. Reproducing Kernels and Inner Products in Example 2.6

Subspace	Reproducing Kernel	Inner Product
$\mathcal{H}_{0\langle 1\rangle} \otimes \mathcal{H}_{00\langle 2\rangle}$	1/K	$(\sum_{x_{\langle 1\rangle}=1}^{K} \int_0^1 f)(\sum_{x_{\langle 1\rangle}=1}^{K} \int_0^1 g)/K$
$\mathcal{H}_{0\langle 1\rangle} \otimes \mathcal{H}_{01\langle 2\rangle}$	$k_1(x_{\langle 2\rangle})k_1(y_{\langle 2\rangle})/K$	$(\sum_{x_{\langle 1\rangle}=1}^{K} \int_0^1 \dot{f}_{\langle 2\rangle})(\sum_{x_{\langle 1\rangle}=1}^{K} \int_0^1 \dot{g}_{\langle 2\rangle})/K$
$\mathcal{H}_{0\langle 1\rangle} \otimes \mathcal{H}_{1\langle 2\rangle}$	$[k_2(x_{\langle 2\rangle})k_2(y_{\langle 2\rangle}) - k_4(x_{\langle 2\rangle} - y_{\langle 2\rangle})]/K$	$\int_0^1 (\sum_{x_{\langle 1\rangle}=1}^{K} \ddot{f}_{\langle 22\rangle})(\sum_{x_{\langle 1\rangle}=1}^{K} \ddot{g}_{\langle 22\rangle})/K$
$\mathcal{H}_{1\langle 1\rangle} \otimes \mathcal{H}_{00\langle 2\rangle}$	$I_{[x_{\langle 1\rangle}=y_{\langle 1\rangle}]} - 1/K$	$\sum_{x_{\langle 1\rangle}=1}^{K} (\int_0^1 (I - A_1)f)(\int_0^1 (I - A_1)g)$
$\mathcal{H}_{1\langle 1\rangle} \otimes \mathcal{H}_{01\langle 2\rangle}$	$(I_{[x_{\langle 1\rangle}=y_{\langle 1\rangle}]} - 1/K)k_1(x_{\langle 2\rangle})k_1(y_{\langle 2\rangle})$	$\sum_{x_{\langle 1\rangle}=1}^{K} (\int_0^1 (I - A_1)\dot{f}_{\langle 2\rangle})(\int_0^1 (I - A_1)\dot{g}_{\langle 2\rangle})$
$\mathcal{H}_{1\langle 1\rangle} \otimes \mathcal{H}_{1\langle 2\rangle}$	$(I_{[x_{\langle 1\rangle}=y_{\langle 1\rangle}]} - 1/K)[k_2(x_{\langle 2\rangle})k_2(y_{\langle 2\rangle}) + k_4(x_{\langle 2\rangle} - y_{\langle 2\rangle})]$	$\int_0^1 \sum_{x_{\langle 1\rangle}=1}^{K} (I - A_1)\ddot{f}_{\langle 22\rangle}(I - A_1)\ddot{g}_{\langle 22\rangle}$

TABLE 2.5. Reproducing Kernels and Inner Products in Example 2.7

where $\theta_\beta \in (0, \infty)$ are tunable parameters. The reproducing kernel associated with (2.30) is $R_J = \sum_\beta \theta_\beta R_\beta$, as

$$J(R_J(x,\cdot), f) = \sum_\beta \theta_\beta^{-1}(\theta_\beta R_\beta(x,\cdot), f_\beta)_\beta = \sum_\beta f_\beta(x) = f(x).$$

When some of the θ_β are set to ∞ in (2.30), $J(f,g)$ defines a semi-inner-product in $\mathcal{H} = \oplus_\beta \mathcal{H}_\beta$. Such a semi-inner-product may be used to specify $J(f) = J(f,f)$ for use in $L(f) + (\lambda/2)J(f)$. Subspaces not contributing to $J(f)$ form the null space of $J(f)$, $\mathcal{N}_J = \{f : J(f) = 0\}$. Subspaces contributing to $J(f)$ form the space $\mathcal{H}_J = \mathcal{H} \ominus \mathcal{N}_J$, in which $J(f,g)$ is a full inner product.

Observing $Y_i = \eta(x_i) + \epsilon_i$, where $x_i \in \mathcal{X}$ is a product domain and $\epsilon_i \sim N(0, \sigma^2)$, one may estimate η via minimizing

$$\frac{1}{n}\sum_{i=1}^{n}(Y_i - \eta(x_i))^2 + \lambda J(\eta), \tag{2.31}$$

where $J(f) = J(f,f)$ is as given above. The minimizer of (2.31) defines a smoothing spline on $\mathcal{X}$. The computation strategy outlined in §2.3.2 readily applies here, with the subspaces $\mathcal{H}_0$ and $\mathcal{H}_1$ in §2.3.2 replaced by $\mathcal{N}_J$ and $\mathcal{H}_J$, respectively.

When some of the θ_β are set to 0 in $J(f) = J(f,f)$, the corresponding f_β are not allowed in the estimate. One simply eliminates the corresponding $\mathcal{H}_\beta$ from the tensor sum.

Note that for the computation of a smoothing spline, all that one needs are a basis of $\mathcal{N}_J$ and the reproducing kernel R_J associated with $J(f)$ in $\mathcal{H}_J = \mathcal{H} \ominus \mathcal{N}_J$. In particular, the explicit form of $J(f)$ is *not* needed.

Example 2.8

Consider the construction of Example 2.5 on $\mathcal{X} = [0,1]^2$. Denote in this example $\mathcal{H}_{\nu,\mu} = \mathcal{H}_{\nu\langle 1\rangle} \otimes \mathcal{H}_{\mu\langle 2\rangle}$, $\nu, \mu = 00, 01, 1$, with inner products $(f,g)_{\nu,\mu}$ and reproducing kernels $R_{\nu,\mu} = R_{\nu\langle 1\rangle} R_{\mu\langle 2\rangle}$. One may set

$$\begin{aligned} J(f,g) &= \theta_{1,00}^{-1}(f,g)_{1,00} + \theta_{1,01}^{-1}(f,g)_{1,01} \\ &\quad + \theta_{00,1}^{-1}(f,g)_{00,1} + \theta_{01,1}^{-1}(f,g)_{01,1} + \theta_{1,1}^{-1}(f,g)_{1,1} \end{aligned}$$

and minimize (2.31) in $\mathcal{H} = \oplus_{\nu,\mu}\mathcal{H}_{\nu,\mu}$. The null space of $J(f) = J(f,f)$ is

$$\begin{aligned} \mathcal{N}_J &= \mathcal{H}_{00,00} \oplus \mathcal{H}_{01,00} \oplus \mathcal{H}_{00,01} \oplus \mathcal{H}_{01,01} \\ &= \text{span}\{\phi_{00,00}, \phi_{01,00}, \phi_{00,01}, \phi_{01,01}\} \\ &= \text{span}\{1, k_1(x_{\langle 1\rangle}), k_1(x_{\langle 2\rangle}), k_1(x_{\langle 1\rangle})k_1(x_{\langle 2\rangle})\}, \end{aligned}$$

where the basis functions $\phi_{\nu,\mu}$ are explicitly specified. The minimizer of (2.31) in $\mathcal{H} = \oplus_{\nu,\mu}\mathcal{H}_{\nu,\mu}$ has an expression

$$\eta(x) = \sum_{\nu,\mu=00,01} d_{\nu,\mu}\phi_{\nu,\mu}(x) + \sum_{i=1}^{n} c_i R_J(x_i, x),$$

where

$$R_J = \theta_{1,00}R_{1,00} + \theta_{1,01}R_{1,01} + \theta_{00,1}R_{00,1} + \theta_{01,1}R_{01,1} + \theta_{1,1}R_{1,1}.$$

The projections of η in $\mathcal{H}_{\nu,\mu}$ are readily available from the expression. For example, $\eta_{01,00} = d_{01,00}\phi_{01,00}(x)$ and $\eta_{01,1} = \sum_{i=1}^n c_i\theta_{01,1}R_{01,1}(x_i, x)$.

To fit an additive model, one may set

$$J(f,g) = \theta_{1,00}^{-1}(f,g)_{1,00} + \theta_{00,1}^{-1}(f,g)_{00,1}$$

and minimize (2.31) in $\mathcal{H}_a = \mathcal{H}_{00,00} \oplus \mathcal{H}_{01,00} \oplus \mathcal{H}_{1,00} \oplus \mathcal{H}_{00,01} \oplus \mathcal{H}_{00,1}$. The null space is now

$$\mathcal{N}_J = \mathcal{H}_{00,00} \oplus \mathcal{H}_{01,00} \oplus \mathcal{H}_{00,01} = \mathrm{span}\{\phi_{00,00}, \phi_{01,00}, \phi_{00,01}\},$$

and $\mathcal{H}_J = \mathcal{H}_{1,00} \oplus \mathcal{H}_{00,1}$ with a reproducing kernel

$$R_J = \theta_{1,00}R_{1,00} + \theta_{00,1}R_{00,1}.$$

The spaces $\mathcal{H}_{01,01}$, $\mathcal{H}_{1,01}$, $\mathcal{H}_{01,1}$, and $\mathcal{H}_{1,1}$ are eliminated from $\mathcal{H}_a$. □

2.5 Bayes Model

Penalized likelihood estimation in a reproducing kernel Hilbert space $\mathcal{H}$ with the penalty $J(f)$ a square (semi) norm is equivalent to a certain empirical Bayes model with a Gaussian prior. The prior has a diffuse component in the null space $\mathcal{N}_J$ of $J(f)$ and a proper component in $\mathcal{H}_J = \mathcal{H} \ominus \mathcal{N}_J$ with mean zero and a covariance function proportional to the reproducing kernel R_J in $\mathcal{H}_J$. The Bayes model may also be perceived as a mixed effect model, with the fixed effects residing in $\mathcal{N}_J$ and the random effects residing in $\mathcal{H}_J$.

We start the discussion with the familiar shrinkage estimates on discrete domains, followed by the polynomial smoothing splines on $[0, 1]$. The calculus is seen to depend only on the null space $\mathcal{N}_J$ of $J(f)$ and the reproducing kernel R_J in its orthogonal complement $\mathcal{H}_J = \mathcal{H} \ominus \mathcal{N}_J$, hence applies to smoothing splines in general. The general results are noted concerning the general multiple-term smoothing splines of §2.4.5.

2.5.1 Shrinkage Estimates as Bayes Estimates

Consider the classical one-way ANOVA model with independent observations $Y_i \sim N(\eta(x_i), \sigma^2)$, $i = 1, \ldots, n$, where $x_i \in \{1, \ldots, K\}$. With a prior $\eta \sim N(0, bI)$, it is easy to see that the posterior mean of η is given by the minimizer of

$$\frac{1}{\sigma^2}\sum_{i=1}^n (Y_i - \eta(x_i))^2 + \frac{1}{b}\sum_{x=1}^K \eta^2(x). \tag{2.32}$$

Setting $b = \sigma^2/n\lambda$, (2.32) is equivalent to (2.4) of §2.2.

Now, consider $\eta = \alpha\mathbf{1} + \eta_1$, with independent priors $\alpha \sim N(0, \tau^2)$ for the mean and $\eta_1 \sim N(0, b(I - \mathbf{1}\mathbf{1}^T/K))$ for the contrast. Note that $\eta_1^T\mathbf{1} = 0$ almost surely and that $\bar{\eta} = \sum_{x=1}^K \eta(x)/K = \alpha$. The posterior mean of η is given by the minimizer of

$$\frac{1}{\sigma^2}\sum_{i=1}^n (Y_i - \eta(x_i))^2 + \frac{1}{\tau^2}\bar{\eta}^2 + \frac{1}{b}\sum_{x=1}^K (\eta(x) - \bar{\eta})^2. \tag{2.33}$$

Letting $\tau^2 \to \infty$ and setting $b = \sigma^2/n\lambda$, (2.33) reduces to (2.3) of §2.2. In the limit, α is said to have a diffuse prior. This setting may also be considered as a mixed effect model, with $\alpha\mathbf{1}$ being the fixed effect and η_1 being the random effect.

Next let us look at a two-way ANOVA model on $\{1, \dots, K_1\} \times \{1, \dots, K_2\}$ using the notation of §2.4.2. Assume that $\eta = \eta_\emptyset + \eta_1 + \eta_2 + \eta_{1,2}$ has four independent components, with priors $\eta_\emptyset \sim N(0, b\theta_\emptyset R_\emptyset)$, $\eta_1 \sim N(0, b\theta_1 R_1)$, $\eta_2 \sim N(0, b\theta_2 R_2)$, and $\eta_{1,2} \sim N(0, b\theta_{1,2} R_{1,2})$, where $R_\emptyset = R_{0\langle 1\rangle}R_{0\langle 2\rangle}$, $R_1 = R_{1\langle 1\rangle}R_{0\langle 2\rangle}$, $R_2 = R_{0\langle 1\rangle}R_{1\langle 2\rangle}$, and $R_{1,2} = R_{1\langle 1\rangle}R_{1\langle 2\rangle}$, as given in Table 2.1. Note that R_β's are orthogonal to each other and that a η_β resides in the column space of R_β almost surely. The posterior mean of η is given by the minimizer of

$$\frac{1}{\sigma^2}\sum_{i=1}^n (Y_i - \eta(x_i))^2 + \frac{1}{b}\sum_\beta \theta_\beta^{-1}\eta^T R_\beta^+ \eta. \tag{2.34}$$

Setting $b = \sigma^2/n\lambda$ and $J(f) = \sum_\beta \theta_\beta^{-1} f^T R_\beta^+ f$, (2.34) reduces to (2.31) of §2.4.5, which defines a bivariate smoothing spline on a discrete product domain. A $\theta_\beta = \infty$ in $J(f)$ puts η_β in $\mathcal{N}_J$, which is equivalent to a diffuse prior, or a fixed effect in a mixed effect model. To obtain the additive model, one simply eliminates $\eta_{1,2}$ by setting $\theta_{1,2} = 0$.

2.5.2 Polynomial Smoothing Splines as Bayes Estimates

Consider $\eta = \eta_0 + \eta_1$ on $[0, 1]$, with η_0 and η_1 having independent Gaussian priors with mean zero and covariance functions,

$$E[\eta_0(x)\eta_0(y)] = \tau^2 R_0(x, y) = \tau^2 \sum_{\nu=0}^{m-1} \frac{x^\nu}{\nu!}\frac{y^\nu}{\nu!},$$
$$E[\eta_1(x)\eta_1(y)] = bR_1(x, y) = b\int_0^1 \frac{(x-u)_+^{m-1}}{(m-1)!}\frac{(y-u)_+^{m-1}}{(m-1)!}du,$$

where R_0 and R_1 are taken from (2.9) and (2.10) of §2.3.1. Observing $Y_i \sim N(\eta(x_i), \sigma^2)$, the joint distribution of $\mathbf{Y}$ and $\eta(x)$ is normal with mean zero and a covariance matrix

$$\begin{pmatrix} bQ + \tau^2 SS^T + \sigma^2 I & b\boldsymbol{\xi} + \tau^2 S\boldsymbol{\phi} \\ b\boldsymbol{\xi}^T + \tau^2 \boldsymbol{\phi}^T S^T & bR_1(x,x) + \tau^2 \boldsymbol{\phi}^T \boldsymbol{\phi} \end{pmatrix}, \tag{2.35}$$

where Q is $n \times n$ with the (i,j)th entry $R_1(x_i, x_j)$, S is $n \times m$ with the (i,ν)th entry $x_i^{\nu-1}/(\nu-1)!$, $\boldsymbol{\xi}$ is $n \times 1$ with the ith entry $R_1(x_i, x)$, and $\boldsymbol{\phi}$ is $m \times 1$ with the νth entry $x^{\nu-1}/(\nu-1)!$. Using a standard result on multivariate normal distribution (see, e.g., Johnson and Wichern (1992, Result 4.6)), the posterior mean of $\eta(x)$ is seen to be

$$\begin{aligned} E[\eta(x)|\mathbf{Y}] &= (b\boldsymbol{\xi}^T + \tau^2 \boldsymbol{\phi}^T S^T)(bQ + \tau^2 SS^T + \sigma^2 I)^{-1}\mathbf{Y} \\ &= \boldsymbol{\xi}^T (Q + \rho SS^T + n\lambda I)^{-1}\mathbf{Y} \\ &\quad + \boldsymbol{\phi}^T \rho S^T (Q + \rho SS^T + n\lambda I)^{-1}\mathbf{Y}, \end{aligned} \tag{2.36}$$

where $\rho = \tau^2/b$ and $n\lambda = \sigma^2/b$.

Lemma 2.7
Suppose M is symmetric and nonsingular and S is of full column rank.

$$\lim_{\rho\to\infty} (\rho SS^T + M)^{-1} = M^{-1} - M^{-1}S(S^T M^{-1} S)^{-1} S^T M^{-1}, \tag{2.37}$$

$$\lim_{\rho\to\infty} \rho S^T (\rho SS^T + M)^{-1} = (S^T M^{-1} S)^{-1} S^T M^{-1}. \tag{2.38}$$

Proof: It can be verified that (Problem 2.17)

$$\begin{aligned} &(\rho SS^T + M)^{-1} = \\ &\quad M^{-1} - M^{-1}S(S^T M^{-1}S)^{-1}(I + \rho^{-1}(S^T M^{-1} S)^{-1})^{-1} S^T M^{-1}. \end{aligned} \tag{2.39}$$

Equation (2.37) follows trivially from (2.39). Substituting (2.39) into the left-hand side of (2.38), some algebra leads to

$$\begin{aligned} \rho S^T (\rho SS^T + M)^{-1} &= \rho(I - (I + \rho^{-1}(S^T M^{-1} S)^{-1})^{-1}) S^T M^{-1} \\ &= (S^T M^{-1} S)^{-1}(I + \rho^{-1}(S^T M^{-1}S)^{-1})^{-1} S^T M^{-1}. \end{aligned}$$

Letting $\rho \to \infty$ yields (2.38). □

Setting $\rho \to \infty$ in (2.36) and applying Lemma 2.7, the posterior mean $E[\eta(x)|\mathbf{Y}]$ is of the form $\boldsymbol{\xi}^T \mathbf{c} + \boldsymbol{\phi}^T \mathbf{d}$, with the coefficients given by

$$\begin{aligned} \mathbf{c} &= (M^{-1} - M^{-1}S(S^T M^{-1} S)^{-1} S^T M^{-1})\mathbf{Y}, \\ \mathbf{d} &= (S^T M^{-1} S)^{-1} S^T M^{-1}\mathbf{Y}, \end{aligned} \tag{2.40}$$

where $M = Q + n\lambda I$.

Theorem 2.8
The polynomial smoothing spline of (2.5) is the posterior mean of $\eta = \eta_0 + \eta_1$, *where* η_0 *diffuses in* $\text{span}\{x^{\nu-1}, \nu = 1, \ldots, m\}$ *and* η_1 *has a Gaussian process prior with mean zero and a covariance function*

$$bR_1(x, y) = b \int_0^1 \frac{(x-u)_+^{m-1}}{(m-1)!} \frac{(y-u)_+^{m-1}}{(m-1)!} du,$$

where $b = \sigma^2/n\lambda$.

Proof: The only thing that remains to be verified is that **c** and **d** in (2.40) minimize (2.16) on page 34. Differentiating (2.16) with respect to **c** and **d** and setting the derivatives to 0, one gets

$$\begin{aligned} Q\{(Q + n\lambda I)\mathbf{c} + S\mathbf{d} - \mathbf{Y}\} &= 0, \\ S^T\{Q\mathbf{c} + S\mathbf{d} - \mathbf{Y}\} &= 0. \end{aligned} \tag{2.41}$$

It is easy to verify that **c** and **d** given in (2.40) satisfy (2.41). □

2.5.3 Smoothing Splines as Bayes Estimates: General Form

Besides the choices of covariance functions R_0 and R_1, nothing is specific to polynomial smoothing splines in the derivation of §2.5.2. In general, consider a reproducing kernel Hilbert space $\mathcal{H} = \oplus_{\beta=0}^p \mathcal{H}_\beta$ on a domain $\mathcal{X}$ with an inner product

$$(f, g) = \sum_{\beta=0}^p \theta_\beta^{-1}(f, g)_\beta = \sum_{\beta=0}^p \theta_\beta^{-1}(f_\beta, g_\beta)_\beta$$

and a reproducing kernel

$$R(x, y) = \sum_{\beta=0}^p \theta_\beta R_\beta(x, y),$$

where $(f, g)_\beta$ is an inner product in $\mathcal{H}_\beta$ with a reproducing kernel R_β, f_β is the projection of f in $\mathcal{H}_\beta$, and $\mathcal{H}_0$ is finite dimensional. Observing $Y_i \sim N(\eta(x_i), \sigma^2)$, a smoothing spline on $\mathcal{X}$ can be defined as the minimizer of the functional

$$\frac{1}{n}\sum_{i=1}^n (Y_i - \eta(x_i))^2 + \lambda \sum_{\beta=1}^p \theta_\beta^{-1}(\eta, \eta)_\beta \tag{2.42}$$

in $\mathcal{H}$; see also (2.31) of §2.4.5. A smoothing spline thus defined is a Bayes estimate of $\eta = \sum_{\beta=0}^p \eta_\beta$, where η_0 has a diffuse prior in $\mathcal{H}_0$ and η_β, $\beta = 1, \ldots, p$, have mean zero Gaussian process priors on $\mathcal{X}$ with covariance functions $E[\eta_\beta(x)\eta_\beta(y)] = b\theta_\beta R_\beta(x, y)$, independent of each other, where $b = \sigma^2/n\lambda$. Treated as a mixed effect model, η_0 contains the fixed effects and η_β, $\beta = 1, \ldots, p$, are the random effects.

2.6 Minimization of Penalized Functional

As an optimization object, analytical properties of the penalized likelihood functional $L(f) + (\lambda/2)J(f)$ can be studied under general functional analytical conditions such as the continuity, convexity, and differentiability of $L(f)$ and $J(f)$. Among such properties are the existence of the minimizer and the equivalence of penalized optimization and constrained optimization.

We first show that the penalized likelihood estimate exists as long as the maximum likelihood estimate uniquely exists in the null space $\mathcal{N}_J$ of $J(f)$. We then prove that the minimization of $L(f) + (\lambda/2)J(f)$ is equivalent to the minimization of $L(f)$ subject to a constraint of the form $J(f) \leq \rho$ for some $\rho \geq 0$, and quantify the relation between ρ and λ.

2.6.1 Existence of Minimizer

A functional $A(f)$ in a linear space $\mathcal{L}$ is said to be **convex** if for $f, g \in \mathcal{L}$, $A(\alpha f + (1-\alpha)g) \leq \alpha A(f) + (1-\alpha)A(g)$, $\forall \alpha \in (0,1)$; the convexity is strict if the equality holds only for $f = g$.

Theorem 2.9 (Existence)
Suppose $L(f)$ is a continuous and convex functional in a Hilbert space $\mathcal{H}$ and $J(f)$ is a square (semi) norm in $\mathcal{H}$ with a null space $\mathcal{N}_J$, of finite dimension. If $L(f)$ has a unique minimizer in $\mathcal{N}_J$, then $L(f) + (\lambda/2)J(f)$ has a minimizer in $\mathcal{H}$.

The minus log likelihood $L(f|\text{data})$ in (1.3) is usually convex in f, as will be verified on a case-by-case basis in later chapters. The quadratic functional $J(f)$ is convex; see Problem 2.18. A minimizer of $L(f)$ is unique in $\mathcal{N}_J$ if the convexity is strict in it, which is often the case.

Without loss of generality, one may set $\lambda = 2$ in the theorem. The proof of the theorem builds on the following two lemmas, with $L(f)$ and $J(f)$ in the lemmas being the same as those in Theorem 2.9.

Lemma 2.10
If a continuous and convex functional $A(f)$ has a unique minimizer in $\mathcal{N}_J$, then it has a minimizer in the cylinder area $C_\rho = \{f : f \in \mathcal{H}, J(f) \leq \rho\}$, $\forall \rho \in (0, \infty)$.

Lemma 2.11
If $L(f)+J(f)$ has a minimizer in $C_\rho = \{f : f \in \mathcal{H}, J(f) \leq \rho\}$, $\forall \rho \in (0, \infty)$, then it has a minimizer in $\mathcal{H}$.

The rest of the section are the proofs.

Proof of Lemma 2.10: Let $\|\cdot\|_0$ be the norm in $\mathcal{N}_J$, and f_0 be the unique minimizer of $A(f)$ in $\mathcal{N}_J$. By Theorem 4 of Tapia and Thompson (1978, page 162), $A(f)$ has a minimizer in a "rectangle"

$$R_{\rho,\gamma} = \{f : f \in \mathcal{H}, J(f) \leq \rho, \|f - f_0\|_0 \leq \gamma\}.$$

Now, if the lemma is not true [i.e., that $A(f)$ has no minimizer in C_ρ for some ρ], then a minimizer f_γ of $A(f)$ in $R_{\rho,\gamma}$ must satisfy $\|f_\gamma - f_0\|_0 = \gamma$. By the convexity of $A(f)$ and the fact that $A(f_\gamma) \leq A(f_0)$,

$$A(\alpha f_\gamma + (1-\alpha)f_0) \leq \alpha A(f_\gamma) + (1-\alpha)A(f_0) \leq A(f_0), \tag{2.43}$$

for $\alpha \in (0,1)$. Now, take a sequence $\gamma_i \to \infty$ and set $\alpha_i = \gamma_i^{-1}$, and write $\alpha_i f_{\gamma_i} + (1-\alpha_i)f_0 = f_i^o + f_i^*$, where $f_i^o \in \mathcal{N}_J$ and $f_i^* \in \mathcal{H} \ominus \mathcal{N}_J$. It is easy to check that $\|f_i^o - f_0\|_0 = 1$ and that $J(f_i^*) \leq \alpha_i^2 \rho$. Since $\mathcal{N}_J$ is finite dimensional, $\{f_i^o\}$ has a convergent subsequence converging to, say, $f_1 \in \mathcal{N}_J$, and $\|f_1 - f_0\|_0 = 1$. It is apparent that $f_i^* \to 0$. By the continuity of $A(f)$ and (2.43), $A(f_1) \leq A(f_0)$, which contradicts the fact that f_0 uniquely minimizes $A(f)$ in $\mathcal{N}_J$. Hence, $\|f_\gamma - f_0\|_0 = \gamma$ cannot hold for all $\gamma \in (0,\infty)$. This completes the proof. □

Proof of Lemma 2.11: Without loss of generality we assume $L(0) = 0$. If the lemma is not true, then a minimizer f_ρ of $L(f) + J(f)$ in C_ρ must fall on the boundary of C_ρ for every ρ [i.e., $J(f_\rho) = \rho$, $\forall \rho \in (0,\infty)$]. By the convexity of $L(f)$,

$$L(\alpha f_\rho) \leq \alpha L(f_\rho), \tag{2.44}$$

for $\alpha \in (0,1)$. By the definition of f_ρ,

$$L(f_\rho) + J(f_\rho) \leq L(\alpha f_\rho) + J(\alpha f_\rho). \tag{2.45}$$

Combining (2.44) and (2.45) and substituting $J(f_\rho) = \rho$, one obtains

$$L(\alpha f_\rho)/\alpha + \rho \leq L(\alpha f_\rho) + \alpha^2 \rho,$$

which, after some algebra, yields

$$L(\alpha f_\rho) \leq -\alpha(1+\alpha)\rho. \tag{2.46}$$

Now, choose $\alpha = \rho^{-1/2}$. Since $J(\alpha f_\rho) = 1$, (2.46) leads to

$$L(f_1) \leq -(\rho^{1/2} + 1),$$

which is impossible for large enough ρ. This proves the lemma. □

Proof of Theorem 2.9: Applying Lemma 2.10 on $A(f) = L(f) + J(f)$ leads to the condition of Lemma 2.11, and the lemma, in turn, yields the theorem. □

2.6.2 Penalized and Constrained Optimization

For a functional $A(f)$ in a linear space $\mathcal{L}$, define $A_{f,g}(\alpha) = A(f + \alpha g)$ as functions of α real indexed by $f, g \in \mathcal{L}$. If $\dot{A}_{f,g}(0)$ exists and is linear in g, $\forall f, g \in \mathcal{L}$, $A(f)$ is said to be **Fréchet differentiable** in $\mathcal{L}$, and $\dot{A}_{f,g}(0)$ is the **Fréchet derivative** of A at f in the direction of g.

Theorem 2.12
Suppose $L(f)$ is continuous, convex, and Fréchet differentiable in a Hilbert space $\mathcal{H}$, and $J(f)$ is a square (semi) norm in $\mathcal{H}$. If f^* minimizes $L(f)$ in $C_{\rho} = \{f : f \in \mathcal{H}, J(f) \leq \rho\}$, then f^* minimizes $L(f) + (\lambda/2)J(f)$ in $\mathcal{H}$, where the Lagrange multiplier relates to ρ via $\lambda = -\rho^{-1}\dot{L}_{f^*,f_1^*}(0) \geq 0$, with f_1^* being the projection of f^* in $\mathcal{H}_J = \mathcal{H} \ominus \mathcal{N}_J$. Conversely, if f^o minimizes $L(f) + (\lambda/2)J(f)$ in $\mathcal{H}$, where $\lambda > 0$, then f^o minimizes $L(f)$ in $\{f : f \in \mathcal{H}, J(f) \leq J(f^o)\}$.

The minus log likelihood $L(f|\text{data})$ in (1.3) is usually Fréchet differentiable, as will be verified on a case-by-case basis in later chapters.

Proof of Theorem 2.12: If $J(f^*) < \rho$, then by the convexity of $L(f)$, f^* is a global minimizer of $L(f)$, so the result holds with $\lambda = \dot{L}_{f^*,f_1^*}(0) = 0$.

In general, $J(f^*) = \rho$; thus, f^* minimizes $L(f)$ on the boundary contour $C_{\rho}^o = \{f : f \in \mathcal{H}, J(f) = \rho\}$. It is easy to verify that $\dot{J}_{f,g}(0) = 2J(f,g)$, where $J(f,g)$ is the (semi) inner product associated with $J(f)$. The space tangent to the contour C_{ρ}^o at f^* is thus $\mathcal{G} = \{g : J(f^*, g) = J(f_1^*, g) = 0\}$.

Pick an arbitrary $g \in \mathcal{G}$. When $J(g) = 0$, $f^* + \alpha g \in C_{\rho}^o$. Since

$$0 \leq L(f^* + \alpha g) - L(f^*) = \alpha \dot{L}_{f^*,g}(0) + o(\alpha),$$

one has $\dot{L}_{f^*,g}(0) = 0$. When $J(g) \neq 0$, without loss of generality one may scale g so that $J(g) = \rho$; then, $\sqrt{1-\alpha^2}f^* + \alpha g \in C_{\rho}^o$. Now, write $\gamma = (\sqrt{1-\alpha^2} - 1)/\alpha$. By the linearity of $\dot{L}_{f,g}(0)$ in g, one has

$$\begin{aligned}
0 &\leq L(\sqrt{1-\alpha^2}f^* + \alpha g) - L(f^*) \\
&= L(f^* + \alpha(\gamma f^* + g)) - L(f^*) \\
&= \alpha\gamma \dot{L}_{f^*,f^*}(0) + \alpha \dot{L}_{f^*,g}(0) + o(\alpha) \\
&= \alpha \dot{L}_{f^*,g}(0) + o(\alpha),
\end{aligned}$$

where $\alpha\gamma = \sqrt{1-\alpha^2} - 1 = O(\alpha^2) = o(\alpha)$; so, again, $\dot{L}_{f^*,g}(0) = 0$.

It is easy to see that $J(f_1^*) = \rho$ and that $\mathcal{G}^c = \text{span}\{f_1^*\}$. Now, every $f \in \mathcal{H}$ has an unique decomposition $f = \beta f_1^* + g$, with β real and $g \in \mathcal{G}$; hence,

$$\begin{aligned}
\dot{L}_{f^*,f}(0) + \frac{\lambda}{2}\dot{J}_{f^*,f}(0) &= \dot{L}_{f^*,\beta f_1^*}(0) + \dot{L}_{f^*,g}(0) + \lambda J(f^*, \beta f_1^* + g) \\
&= \beta \dot{L}_{f^*,f_1^*}(0) + \beta\lambda\rho. \qquad (2.47)
\end{aligned}$$

With $\lambda = -\rho^{-1}\dot{L}_{f^*,f_1^*}(0)$, (2.47) is annihilated for all $f \in \mathcal{H}$; thus, f^* minimizes $L(f) + (\lambda/2)J(f)$. Finally, note that $L(f^* - \alpha f_1^*) \geq L(f^*)$ for $\alpha \in (0,1)$, so $\dot{L}_{f^*,f_1^*}(0) \leq 0$. The converse is straightforward and is left as an exercise (Problem 2.21). □

2.7 Bibliographic Notes

Section 2.1

The theory of Hilbert space is at the core of many advanced analysis courses. The elementary materials presented in §2.1.1 provide a minimal exposition for our need. An excellent treatment of vector spaces can be found in Rao (1973, Chapter 1). Proofs of the Riesz representation theorem can be found in many references, of different levels of abstraction; the one given in §2.1.2 was taken from Akhiezer and Glazman (1961). The theory of reproducing kernel Hilbert space was developed by Aronszajn (1950), which remains the primary reference on the subject. The exposition in §2.1.3 is minimally sufficient to serve our need.

Section 2.2

Shrinkage estimates are among basic techniques in classical decision theory and Bayesian statistics; see, e.g., Lehmann and Casella (1998, §5.5). The interpretation of shrinkage estimates as smoothing splines on discrete domains has not previously appeared in the literature. Vector spaces are much more familiar to statisticians than reproducing kernel Hilbert spaces, and this section is intended to help the reader to gain further insights into entities in a reproducing kernel Hilbert space.

Section 2.3

The space $\mathcal{C}^{(m)}[0,1]$ with the inner product (2.7) and the representer of evaluation (2.8) derived from the standard Taylor expansion are standard results found in numerical analysis literature; see, e.g., Schumaker (1981, Chapter 8). The reproducing kernel (2.21) of $\mathcal{C}^{(m)}[0,1]$ associated with the inner product (2.17) was derived by Craven and Wahba (1979) and was used more often than (2.8) as marginal kernels in tensor product smoothing splines on $[0,1]^\Gamma$. Results concerning Bernoulli polynomials can be found in Abramowitz and Stegun (1964, Chapter 26).

The computational strategy outlined in §2.3.2 was derived by Kimeldorf and Wahba (1971) in the setting of Chebyshev splines, of which the polynomial smoothing splines of (2.5) are special cases; see §4.3.2 for Chebyshev splines. For many years, however, the device was not used much in actual numerical computation. The reasons were multifold. First, algorithms based on (2.16) are of order $O(n^3)$, whereas $O(n)$ algorithms exist for polynomial smoothing splines; see §3.4 and §3.8. Second, portable numerical linear algebra software and powerful desktop computing were not available until much later. Since the late 1980s, generic algorithms and software have been developed based on (2.16) for the computation of smoothing splines, univariate and multivariate alike; see §3.4 for details.

Section 2.4

A comprehensive treatment of tensor product reproducing kernel Hilbert spaces can be found in Aronszajn (1950), where Theorem 2.6 was referenced as a classical result of I. Schur. The proof given here was suggested by Liqing Yan.

The idea of tensor product smoothing splines was conceived by Barry (1986) and Wahba (1986). Dozens of references appeared in the literature since then, among which Chen (1991), Gu and Wahba (1991b, 1993a, 1993b), Wahba, Wang, Gu, Klein, and Klein (1995), and Gu (1995a, 1996, 1998b) registered notable innovations in the theory and practice of the tensor product spline technique. The materials of §§2.4.3—2.4.5 are scattered in these references. The materials of §2.4.2, however, have not appeared in the smoothing literature.

Section 2.5

The Bayes model of polynomial smoothing splines was first observed by Kimeldorf and Wahba (1970a, 1970b). The materials of §2.5.2 and §2.5.3 are mainly taken from Wahba (1978, 1983). The elementary materials of §2.5.1 in the familiar discrete setting provide insights into the general results. In Bayesian statistics, such models are more specifically referred to as empirical Bayes models; see, e.g., Berger (1985, §4.5).

Section 2.6

The existence of penalized likelihood estimates has been discussed by many authors in various settings; see, e.g., Tapia and Thompson (1978, Chapter 4) and Silverman (1982). The general result of Theorem 2.9 and the elementary proof are taken from Gu and Qiu (1993).

The relation between penalized optimization and constrained optimization in the context of natural polynomial splines was noted by Schoenberg (1964), where $L(f)$ was a least squares functional. The general result of Theorem 2.12 was adapted from the discussion of Gill, Murray, and Wright (1981, §3.4) on constrained nonlinear optimization.

2.8 Problems

Section 2.1

2.1 Prove the Cauchy-Schwarz inequality of (2.1).

2.2 Prove the triangle inequality of (2.2).

2.3 Let $\mathcal{H}$ be a Hilbert space and $\mathcal{G} \subset \mathcal{H}$ a closed linear subspace. For every $f \in \mathcal{H}$, prove that the projection of f in $\mathcal{G}$, $f_{\mathcal{G}} \in \mathcal{G}$, that satisfies

$$\|f - f_{\mathcal{G}}\| = \inf_{g \in \mathcal{G}} \|f - g\|$$

uniquely exists.

(a) Show that there exists a sequence $\{g_n\} \subset \mathcal{G}$ such that

$$\lim_{n \to \infty} \|f - g_n\| = \delta = \inf_{g \in \mathcal{G}} \|f - g\|.$$

(b) Show that

$$\|g_m - g_n\|^2 = 2\|f - g_m\|^2 + 2\|f - g_n\|^2 - 4\|f - \frac{g_m + g_n}{2}\|^2.$$

Since $\lim_{m,n \to \infty} \|f - \frac{g_m + g_n}{2}\| = \delta$, $\{g_n\}$ is a Cauchy sequence.

(c) Show the uniqueness of $f_{\mathcal{G}}$ using the triangle inequality.

2.4 Given Hilbert spaces $\mathcal{H}_0$ and $\mathcal{H}_1$ satisfying $\mathcal{H}_0 \cap \mathcal{H}_1 = \{0\}$, prove that the space $\mathcal{H} = \{f : f = f_0 + f_1, f_0 \in \mathcal{H}_0, f_1 \in \mathcal{H}_1\}$ with an inner product $(f, g) = (f_0, g_0)_0 + (f_1, g_1)_1$ is a Hilbert space, where $f = f_0 + f_1$, $g = g_0 + g_1$, $f_0, g_0 \in \mathcal{H}_0$, $f_1, g_1 \in \mathcal{H}_1$, and $(\cdot,\cdot)_0$ and $(\cdot,\cdot)_1$ are the inner products in $\mathcal{H}_0$ and $\mathcal{H}_1$, respectively. Prove that $\mathcal{H}_0$ and $\mathcal{H}_1$ are the orthogonal complements of each other as closed linear subspaces of $\mathcal{H}$.

2.5 The isomorphism between a K-dimensional Hilbert space $\mathcal{H}$ and the Euclidean K-space is outlined in the following steps:

(a) Take any $\phi \in \mathcal{H}^0 = \mathcal{H}$ nonzero, denote $\phi_1 = \phi/\|\phi\|$, and obtain

$$\mathcal{H}^1 = \mathcal{H}^0 \ominus \{f : f = \alpha\phi_1, \alpha \text{ real}\}.$$

Prove that $\mathcal{H}^1$ contains nonzero elements if $K > 1$.

(b) Repeat step (a) for $\mathcal{H}^{i-1}$, $i = 2, \ldots, K$, to obtain ϕ_i and

$$\mathcal{H}^i = \mathcal{H}^{i-1} \ominus \{f : f = \alpha\phi_i, \alpha \text{ real}\}.$$

Prove that $\mathcal{H}^{K-1} = \{f : f = \alpha\phi_K, \alpha \text{ real}\}$, so $\mathcal{H}^K = \{0\}$.

(c) Verify that $(\phi_i, \phi_j) = \delta_{i,j}$, where $\delta_{i,j}$ is the Kronecker delta. The elements ϕ_i, $i = 1, \dots, K$, are said to form an orthonormal basis of $\mathcal{H}$. For every $f \in \mathcal{H}$, there is a unique representation $f = \sum_{i=1}^{K} \alpha_i \phi_i$, where α_i are real coefficients.

(d) Prove that the mapping $f \leftrightarrow \boldsymbol{\alpha}$, where $\boldsymbol{\alpha}$ are the coefficients of f, defines an isomorphism between $\mathcal{H}$ and the Euclidean space.

2.6 Prove that in a Euclidean space, every linear functional is continuous.

2.7 Prove that the reproducing kernel of a Hilbert space, when it exists, is unique.

Section 2.2

2.8 On $\mathcal{X} = \{1, \dots, K\}$, the constructions of reproducing kernel Hilbert spaces outlined below yield a one-way ANOVA decomposition with an averaging operator $Af = f(1)$.

(a) Verify that the reproducing kernel $R_0 = 1 = \mathbf{1}\mathbf{1}^T$ generates the space $\mathcal{H}_0 = \{f : f(1) = \cdots = f(K)\}$ with an inner product $(f, g)_0 = f^T(\mathbf{1}\mathbf{1}^T/K^2)g$.

(b) Verify that the reproducing kernel $R_1 = I_{[x=y\neq 1]} = (I - e_1e_1^T)$ generates the space $\mathcal{H}_1 = \{f : f(1) = 0\}$ with an inner product $(f, g)_1 = f^T(I - e_1e_1^T)g$, where e_1 is the first unit vector.

(c) Note that $\mathcal{H}_0 \cap \mathcal{H}_1 = \{0\}$, so $\mathcal{H}_0 \oplus \mathcal{H}_1$ is well defined and has the reproducing kernel $R_0 + R_1$. With the expressions given in (a) and (b), however, one in general has $(f_1, f_1)_0 \neq 0$ for $f_1 \in \mathcal{H}_1$ and $(f_0, f_0)_1 \neq 0$ for $f_0 \in \mathcal{H}_0$. Nevertheless, $f = \mathbf{1}e_1^T f$ for $f \in \mathcal{H}_0$, so one may write $(f, g)_0 = f^T(e_1e_1^T)g$. Similarly, as $f = (I - \mathbf{1}e_1^T)f$ for $f \in \mathcal{H}_1$, one may write $(f, g)_1 = f^T(I - e_1\mathbf{1}^T)(I - \mathbf{1}e_1^T)g$. Verify the new expressions of $(f, g)_0$ and $(f, g)_1$. Check that with the new expressions, $(f_1, f_1)_0 = 0$, $\forall f_1 \in \mathcal{H}_1$, and that $(f_0, f_0)_1 = 0$, $\forall f_0 \in \mathcal{H}_0$, so the inner product in $\mathcal{H}_0 \oplus \mathcal{H}_1$ can be written as $(f, g) = (f, g)_0 + (f, g)_1$ with the new expressions.

(d) Verify that $(\mathbf{1}\mathbf{1}^T + I - e_1e_1^T)^{-1} = e_1e_1^T + (I - e_1\mathbf{1}^T)(I - \mathbf{1}e_1^T)$ [i.e., the reproducing kernel $R_0 + R_1$ and the inner product $(f, g)_0 + (f, g)_1$ are inverses of each other].

Section 2.3

2.9 Consider the function $k_r(x)$ of (2.18).

(a) Prove that the infinite series converges for $r > 1$ on the real line and for $r = 1$ at noninteger points.

(b) Prove that $k_r(x)$ is real-valued.

(c) Prove that $k_1(x) = x - 0.5$ on $x \in (0, 1)$.

2.10 Prove (2.22) through integration by parts. Note that k_r, $r > 1$, are periodic with period 1 and that $\int_0^1 f^{(\nu)} dx = 0$, $\nu = 0, \ldots, m-1$.

2.11 Derive the expressions of $k_2(x)$ and $k_4(x)$ on $[0, 1]$ as given in (2.27) by successive integration from $k_1(x) = x - .5$. Note that for $r > 1$, $dk_r/dx = k_{r-1}$ and $k_r(0) = k_r(1)$.

Section 2.4

2.12 On $\mathcal{X} = \{1, \ldots, K_1\} \times \{1, \ldots, K_2\}$, construct tensor product reproducing kernel Hilbert spaces with the structure of (2.28)

(a) with $A_1 f = f(1, x_{\langle 2 \rangle})$ and $A_2 f = f(x_{\langle 1 \rangle}, 1)$

(b) with $A_1 f = f(1, x_{\langle 2 \rangle})$ and $A_2 f = \sum_{x_{\langle 2 \rangle}=1}^{K_2} f(x)/K_2$.

2.13 On $\mathcal{X} = [0, 1]^2$, construct tensor product reproducing kernel Hilbert spaces with the structure of (2.28)

(a) with $A_1 f = f(0, x_{\langle 2 \rangle})$ and $A_2 f = f(x_{\langle 1 \rangle}, 0)$, using (2.9) and (2.10) with $m = 1, 2$

(b) with $A_1 f = f(0, x_{\langle 2 \rangle})$ and $A_2 f = \int_0^1 f dx_{\langle 2 \rangle}$, using (2.9), (2.10), (2.19), and (2.23), with $m = 1, 2$.

2.14 On $\mathcal{X} = \{1, \ldots, K\} \times [0, 1]$, construct tensor product reproducing kernel Hilbert spaces with the structure of (2.28)

(a) with $A_1 f = f(1, x_{\langle 2 \rangle})$ and $A_2 f = f(x_{\langle 1 \rangle}, 0)$

(b) with $A_1 f = f(1, x_{\langle 2 \rangle})$ and $A_2 f = \int_0^1 f dx_{\langle 2 \rangle}$

(c) with $A_1 f = \sum_{x_{\langle 1 \rangle}=1}^{K} f(x)/K$ and $A_2 f = f(x_{\langle 1 \rangle}, 0)$.

2.15 To compute the tensor product smoothing splines of Example 2.8, one may use the strategy outlined in §2.3.2.

(a) Specify the matrices S and Q in (2.16), for both the full model and the additive model.

(b) Decompose the expression of $\eta(x)$ into those of the constant, the main effects, and the interaction.

2.16 In parallel to Example 2.8 and Problem 2.15, work out the corresponding details for computing the tensor product smoothing splines on $\{1,\ldots,K\} \times [0,1]$, using the construction of Example 2.7.

Section 2.5

2.17 Verify (2.39).

Section 2.6

2.18 Prove that a quadratic functional $J(f)$ is convex.

2.19 Let $A(f)$ be a strictly convex functional in a Hilbert space $\mathcal{H}$. Prove that if the minimizer of $A(f)$ exists in $\mathcal{H}$, then it is also unique.

2.20 Consider a strictly convex continuous function $f(x)$ on $(-\infty,\infty)^2$. Prove that if $f_1(x_{\langle 1\rangle}) = f(x_{\langle 1\rangle},0)$ has a minimizer, then $f(x)+x_{\langle 2\rangle}^2$ has a unique minimizer.

2.21 Prove that if f^o minimizes $L(f)+\lambda J(f)$, where $\lambda>0$, then f^o minimizes $L(f)$ subject to $J(f)\leq J(f^o)$.

3

Regression with Gaussian-Type Responses

For regression with Gaussian responses, $L(f) + (\lambda/2)J(f)$ reduces to the familiar penalized least squares functional. Among topics of primary interest are the selection of smoothing parameters, the computation of the estimates, the asymptotic convergence of the estimates, and various data analytical tools.

The main focus of this chapter is on the development of generic computational and data analytical tools for the general multiple-term smoothing splines as formulated in §2.4.5. After a brief review of elementary facts in §3.1, we discuss (§3.2) three popular scores for smoothing parameter selection in detail, namely an unbiased estimate of relative loss, the generalized cross-validation, and the restricted maximum likelihood under the Bayes model of §2.5. In §3.3, we derive the Bayesian confidence intervals of Wahba (1983) and briefly discuss their across-the-function coverage property. Generic algorithms implementing these tools are described in §3.4, followed by illustrations of the accompanying software in §3.5. Heuristic diagnostics are introduced in §3.6 for the identifiability and practical significance of terms in multiple-term models. Real-data examples are presented in §3.7. Also presented (§3.8) are selected fast algorithms for problems admitting structures through alternative formulations, such as the $O(n)$ algorithm for the univariate polynomial splines.

The asymptotic convergence of penalized least squares estimates will be discussed in Chapter 8, along with that of penalized likelihood estimates in other settings.

3.1 Preliminaries

Observing $Y_i = \eta(x_i) + \epsilon_i$, $i = 1, \ldots, n$, with $\epsilon_i \sim N(0, \sigma^2)$, the minus log likelihood functional $L(f)$ in $L(f) + (\lambda/2)J(f)$ of (1.3) reduces to the least squares functional proportional to $\sum_{i=1}^n (Y_i - f(x_i))^2$. As discussed in §2.4.5 and §2.5.3, the general form of penalized least squares functional in a reproducing kernel Hilbert space $\mathcal{H} = \oplus_{\beta=0}^p \mathcal{H}_\beta$ can be written as

$$\frac{1}{n}\sum_{i=1}^n (Y_i - \eta(x_i))^2 + \lambda J(\eta), \tag{3.1}$$

where $J(f) = J(f,f) = \sum_{\beta=1}^p \theta_\beta^{-1}(f,f)_\beta$ and $(f,g)_\beta$ are inner products in $\mathcal{H}_\beta$ with reproducing kernels $R_\beta(x,y)$. The penalty is seen to be

$$\lambda J(f) = \lambda \sum_{\beta=1}^p \theta_\beta^{-1}(f,f)_\beta,$$

with λ and θ_β as smoothing parameters. This is an overparameterization, as what really matter are the ratios λ/θ_β. It is conceivable to fix one of the θ_β, but we opt to preserve the symmetry and we do want to keep a λ up front. The bilinear form $J(f,g) = \sum_{\beta=1}^p \theta_\beta^{-1}(f,g)_\beta$ is an inner product in $\oplus_{\beta=1}^p \mathcal{H}_\beta$, with a reproducing kernel $R_J(x,y) = \sum_{\beta=1}^p \theta_\beta R_\beta(x,y)$ and a null space $\mathcal{N}_J = \mathcal{H}_0$ of finite dimension, say m. By the arguments of §2.3.2, the minimizer η_λ of (3.1) has the expression

$$\eta(x) = \sum_{\nu=1}^m d_\nu \phi_\nu(x) + \sum_{i=1}^n c_i R_J(x_i, x) = \boldsymbol{\phi}^T \mathbf{d} + \boldsymbol{\xi}^T \mathbf{c}, \tag{3.2}$$

where $\{\phi_\nu\}_{\nu=1}^m$ is a basis of $\mathcal{N}_J = \mathcal{H}_0$, $\boldsymbol{\xi}$ and $\boldsymbol{\phi}$ are vectors of functions, and $\mathbf{c}$ and $\mathbf{d}$ are vectors of real coefficients. The estimation then reduces to the minimization of

$$(\mathbf{Y} - S\mathbf{d} - Q\mathbf{c})^T(\mathbf{Y} - S\mathbf{d} - Q\mathbf{c}) + n\lambda \mathbf{c}^T Q \mathbf{c} \tag{3.3}$$

with respect to $\mathbf{c}$ and $\mathbf{d}$, where S is $n \times m$ with the (i,ν)th entry $\phi_\nu(x_i)$ and Q is $n \times n$ with the (i,j)th entry $R_J(x_i, x_j)$. See also (2.16) on page 34.

The least squares functional $\sum_{i=1}^n (Y_i - f(x_i))^2$ is continuous and convex in $\mathcal{H}$, and when S is of full column rank, the convexity is strict in $\mathcal{N}_J$. Also, (3.1) is strictly convex in $\mathcal{H}$ when S is of full column rank. See Problem 3.1. By Theorem 2.9, the minimizer η_λ of (3.1) uniquely exists as long as it uniquely exists in $\mathcal{N}_J$, which requires S to be of full column rank. When Q is singular, (3.3) may have multiple solutions for $\mathbf{c}$ and $\mathbf{d}$, all that satisfy (2.41) on page 49. All the solutions, however, yield the same function estimate η_λ through (3.2). For definiteness in the numerical calculation, we shall compute a particular solution of (3.3) by solving the linear system

$$\begin{aligned}(Q + n\lambda I)\mathbf{c} + S\mathbf{d} &= \mathbf{Y}, \\ S^T\mathbf{c} &= 0.\end{aligned} \tag{3.4}$$

It is easy to check that (3.4) has a unique solution that satisfies (2.41) (Problem 3.2).

Suppose S is of full column rank. Let

$$S = FR^* = (F_1, F_2)\begin{pmatrix} R \\ O \end{pmatrix} = F_1 R \tag{3.5}$$

be the QR-decomposition of S with F orthogonal and R upper-triangular; see, e.g., Golub and Van Loan (1989, §5.2) for QR-decomposition. From $S^T\mathbf{c} = 0$, one has $F_1^T\mathbf{c} = 0$, so $\mathbf{c} = F_2F_2^T\mathbf{c}$. Premultiplying the first equation of (3.4) by F_2^T and F_1^T, simple algebra leads to

$$\begin{aligned}\mathbf{c} &= F_2(F_2^TQF_2 + n\lambda I)^{-1}F_2^T\mathbf{Y}, \\ \mathbf{d} &= R^{-1}(F_1^T\mathbf{Y} - F_1^TQ\mathbf{c}).\end{aligned} \tag{3.6}$$

Denote the fitted values by $\hat{\mathbf{Y}} = (\eta_\lambda(x_1), \ldots, \eta_\lambda(x_n))^T$ and the residuals by $\mathbf{e} = \mathbf{Y} - \hat{\mathbf{Y}}$. Some algebra yields

$$\begin{aligned}\hat{\mathbf{Y}} &= Q\mathbf{c} + S\mathbf{d} \\ &= (F_1F_1^T + F_2F_2^TQF_2(F_2^TQF_2 + n\lambda I)^{-1}F_2^T)\mathbf{Y} \\ &= (I - F_2(I - F_2^TQF_2(F_2^TQF_2 + n\lambda I)^{-1})F_2^T)\mathbf{Y} \\ &= (I - n\lambda F_2(F_2^TQF_2 + n\lambda I)^{-1}F_2^T)\mathbf{Y}.\end{aligned}$$

The symmetric matrix

$$A(\lambda) = I - n\lambda F_2(F_2^TQF_2 + n\lambda I)^{-1}F_2^T \tag{3.7}$$

is known as the smoothing matrix associated with (3.1), which has all its eigenvalues in the range $[0, 1]$ (Problem 3.3). It is easy to see from (3.4) that $\mathbf{e} = (I - A(\lambda))\mathbf{Y} = n\lambda\mathbf{c}$. Using formula (2.40) on page 48 for $\mathbf{c}$, the smoothing matrix can alternatively be written as

$$A(\lambda) = I - n\lambda(M^{-1} - M^{-1}S(S^TM^{-1}S)^{-1}S^TM^{-1}), \tag{3.8}$$

where $M = Q + n\lambda I$.

When $\epsilon_i \sim N(0, \sigma^2/w_i)$ with w_i known, $L(f) + (\lambda/2)J(f)$ of (1.3) reduces to a penalized weighted least squares functional

$$\frac{1}{n}\sum_{i=1}^{n} w_i(Y_i - \eta(x_i))^2 + \lambda J(\eta). \tag{3.9}$$

The counter part of (3.4) is

$$\begin{aligned} (Q_w + n\lambda I)\mathbf{c}_w + S_w \mathbf{d} &= \mathbf{Y}_w, \\ S_w^T \mathbf{c}_w &= 0, \end{aligned} \tag{3.10}$$

where $Q_w = W^{1/2} Q W^{1/2}$, $\mathbf{c}_w = W^{-1/2}\mathbf{c}$, $S_w = W^{1/2} S$, and $\mathbf{Y}_w = W^{1/2}\mathbf{Y}$, for $W = \text{diag}(w_i)$; see Problem 3.4. Write $\hat{\mathbf{Y}}_w = W^{1/2}\hat{\mathbf{Y}} = A_w(\lambda)\mathbf{Y}_w$ and $\mathbf{e}_w = \mathbf{Y}_w - \hat{\mathbf{Y}}_w$; it is easy to see that $\mathbf{e}_w = n\lambda \mathbf{c}_w$ and that

$$A_w(\lambda) = I - n\lambda F_2 (F_2^T Q_w F_2 + n\lambda I)^{-1} F_2^T, \tag{3.11}$$

where $F_2^T F_2 = I$ and $F_2 S_w = 0$. Parallel to (3.8), one also has

$$A_w(\lambda) = I - n\lambda (M_w^{-1} - M_w^{-1} S_w (S_w^T M_w^{-1} S_w)^{-1} S_w^T M_w^{-1}), \tag{3.12}$$

where $M_w = Q_w + n\lambda I$.

Other than the claim that the least squares functional is proportional to the log likelihood, the normality of ϵ_i has not been used so far. Indeed, many of the results to be presented only require moment conditions of ϵ_i. This is reflected in the title of the chapter, where we advertise Gaussian-*type* responses instead of strict Gaussian responses.

3.2 Smoothing Parameter Selection

With varying smoothing parameters λ and θ_β, the minimizer η_λ of (3.1) defines a family of possible estimates. In practice, one has to choose some specific estimate from the family, which calls for effective methods for smoothing parameter selection.

We introduce three scores that are in popular use for smoothing parameter selection in the context. The first score, which assumes a known variance σ^2, is an unbiased estimate of a relative loss. The second score, the generalized cross-validation of Craven and Wahba (1979), targets the same loss without assuming a known σ^2. These scores are presented along with their asymptotic justifications. The third score is derived from the Bayes model of §2.5 through restricted maximum likelihood, which is of appeal to some but is not designed to minimize any particular loss. Parallel scores for weighted and replicated data are also presented. The empirical performance of the three methods is illustrated through simple simulations.

To keep the notation simple, we only make the dependence of various entities on the smoothing parameter λ explicit and suppress their dependence on θ_β. The derivations and proofs apply without change to the general case, with both λ and θ_β tunable.

3.2.1 Unbiased Estimate of Relative Loss

As an estimate of η based on data collected from the sampling points x_i, $i = 1, \ldots, n$, the performance of η_λ can be assessed via the loss function

$$L(\lambda) = \frac{1}{n}\sum_{i=1}^{n}(\eta_\lambda(x_i) - \eta(x_i))^2. \tag{3.13}$$

This is not to be confused with the log likelihood functional $L(f)$, which will not appear again in this chapter except in Problem 3.1. The λ that minimizes $L(\lambda)$ represents the ideal choice one would like to make given the data and will be referred to as the optimal smoothing parameter.

Write $\mathbf{Y} = \boldsymbol{\eta} + \boldsymbol{\epsilon}$, where $\boldsymbol{\eta} = (\eta(x_1), \ldots, \eta(x_n))^T$. It is easy to verify that

$$\begin{aligned} L(\lambda) &= \frac{1}{n}(A(\lambda)\mathbf{Y} - \boldsymbol{\eta})^T(A(\lambda)\mathbf{Y} - \boldsymbol{\eta}) \\ &= \frac{1}{n}\boldsymbol{\eta}^T(I - A(\lambda))^2\boldsymbol{\eta} - \frac{2}{n}\boldsymbol{\eta}^T(I - A(\lambda))A(\lambda)\boldsymbol{\epsilon} + \frac{1}{n}\boldsymbol{\epsilon}^T A^2(\lambda)\boldsymbol{\epsilon}. \end{aligned}$$

Define

$$U(\lambda) = \frac{1}{n}\mathbf{Y}^T(I - A(\lambda))^2\mathbf{Y} + 2\frac{\sigma^2}{n}\mathrm{tr}A(\lambda). \tag{3.14}$$

Simple algebra yields

$$\begin{aligned} U(\lambda) = &\frac{1}{n}(A(\lambda)\mathbf{Y} - \boldsymbol{\eta})^T(A(\lambda)\mathbf{Y} - \boldsymbol{\eta}) + \frac{1}{n}\boldsymbol{\epsilon}^T\boldsymbol{\epsilon} \\ &+ \frac{2}{n}\boldsymbol{\eta}^T(I - A(\lambda))\boldsymbol{\epsilon} - \frac{2}{n}(\boldsymbol{\epsilon}^T A(\lambda)\boldsymbol{\epsilon} - \sigma^2\mathrm{tr}A(\lambda)). \end{aligned}$$

It follows that

$$\begin{aligned} &U(\lambda) - L(\lambda) - n^{-1}\boldsymbol{\epsilon}^T\boldsymbol{\epsilon} \\ &\qquad = \frac{2}{n}\boldsymbol{\eta}^T(I - A(\lambda))\boldsymbol{\epsilon} - \frac{2}{n}(\boldsymbol{\epsilon}^T A(\lambda)\boldsymbol{\epsilon} - \sigma^2\mathrm{tr}A(\lambda)). \end{aligned} \tag{3.15}$$

It is easy to see that $U(\lambda)$ is an unbiased estimate of the relative loss $L(\lambda) + n^{-1}\boldsymbol{\epsilon}^T\boldsymbol{\epsilon}$.

Denote the risk function by

$$R(\lambda) = E[L(\lambda)] = \frac{1}{n}\boldsymbol{\eta}^T(I - A(\lambda))^2\boldsymbol{\eta} + \frac{\sigma^2}{n}\mathrm{tr}A^2(\lambda), \tag{3.16}$$

where the first term represents the "bias" in the estimation and the second term represents the "variance." Under a condition

Condition 3.2.1 $nR(\lambda) \to \infty$ as $n \to \infty$,

one can establish the consistency of $U(\lambda)$. Condition 3.2.1 is a very mild one, as one would not expect nonparametric estimation to deliver a parametric convergence rate of $O(n^{-1})$. See §4.2.3 and Chapter 8.

Theorem 3.1
Assume independent noise ϵ_i with mean zero, a common variance σ^2, and uniformly bounded fourth moments. If Condition 3.2.1 holds, then

$$U(\lambda) - L(\lambda) - n^{-1}\boldsymbol{\epsilon}^T\boldsymbol{\epsilon} = o_p(L(\lambda)).$$

Note that $n^{-1}\boldsymbol{\epsilon}^T\boldsymbol{\epsilon}$ does not depend on λ, so the minimizer of $U(\lambda)$ approximately minimizes $L(\lambda)$ when Condition 3.2.1 holds uniformly in a neighborhood of the optimal λ.

Proof of Theorem 3.1: From (3.15), it suffices to show that

$$L(\lambda) - R(\lambda) = o_p(R(\lambda)), \tag{3.17}$$

$$\frac{1}{n}\boldsymbol{\eta}^T(I - A(\lambda))\boldsymbol{\epsilon} = o_p(R(\lambda)), \tag{3.18}$$

$$\frac{1}{n}(\boldsymbol{\epsilon}^T A(\lambda)\boldsymbol{\epsilon} - \sigma^2 \text{tr}A(\lambda)) = o_p(R(\lambda)). \tag{3.19}$$

We will show (3.17), (3.18), and (3.19) only for the case with ϵ_i normal here, leaving the more tedious general case to Problem 3.5. Let $A(\lambda) = PDP^T$ be the eigenvalue decomposition of $A(\lambda)$, where P is orthogonal and D is diagonal with diagonal entries d_i, $i = 1, \dots, n$. It is seen that the eigenvalues d_i are in the range $[0, 1]$; see Problem 3.3. Write $\tilde{\boldsymbol{\eta}} = P^T\boldsymbol{\eta}$ and $\tilde{\boldsymbol{\epsilon}} = P^T\boldsymbol{\epsilon}$. It follows that

$$L(\lambda) = \frac{1}{n}\sum_{i=1}^{n}\{(1 - d_i)^2\tilde{\eta}_i^2 - 2d_i(1 - d_i)\tilde{\eta}_i\tilde{\epsilon}_i + d_i^2\tilde{\epsilon}_i^2\},$$

$$R(\lambda) = \frac{1}{n}\sum_{i=1}^{n}\{(1 - d_i)^2\tilde{\eta}_i^2 + d_i^2\sigma^2\}.$$

To see (3.17), note that

$$\text{Var}[L(\lambda)] = \frac{1}{n^2}\sum_{i=1}^{n}\{4d_i^2(1 - d_i)^2\tilde{\eta}_i^2\sigma^2 + 2d_i^4\sigma^4\} \le \frac{4\sigma^2}{n}R(\lambda) = o(R^2(\lambda)).$$

Similarly, (3.18) follows from

$$\text{Var}\left[\frac{1}{n}\boldsymbol{\eta}^T(I - A(\lambda))\boldsymbol{\epsilon}\right] = \frac{1}{n^2}\sum_{i=1}^{n}(1 - d_i)^2\tilde{\eta}_i^2\sigma^2 = o(R^2(\lambda)),$$

and (3.19) follows from $E[\boldsymbol{\epsilon}^T A(\lambda)\boldsymbol{\epsilon}] = \sigma^2\text{tr}A(\lambda)$ and

$$\text{Var}\left[\frac{1}{n}\boldsymbol{\epsilon}^T A(\lambda)\boldsymbol{\epsilon}\right] = \frac{2}{n^2}\sum_{i=1}^{n}d_i^2\sigma^4 = o(R^2(\lambda)).$$

The proof is thus complete for the case with $\epsilon_i \sim N(0, \sigma^2)$. $\square$

3.2.2 Cross-Validation and Generalized Cross-Validation

To use $U(\lambda)$ as defined in (3.14), one needs to know the sampling variance σ^2, which is impractical in many applications. The problem can be circumvented, however, by using the method of cross-validation.

The method of cross-validation aims at the prediction error at the sampling points. If an independent validation data set were available with $Y_i^* = \eta(x_i) + \epsilon_i^*$, then an intuitive strategy for the selection of λ would be to minimize $n^{-1} \sum_{i=1}^n (\eta_\lambda(x_i) - Y_i^*)^2$. Lacking an independent validation data set, an alternative strategy is to cross-validate, that is, to minimize

$$V_0(\lambda) = \frac{1}{n} \sum_{i=1}^n (\eta_\lambda^{[i]}(x_i) - Y_i)^2, \tag{3.20}$$

where $\eta_\lambda^{[k]}$ is the minimizer of the "delete-one" functional

$$\frac{1}{n} \sum_{i \neq k} (Y_i - \eta(x_i))^2 + \lambda J(\eta). \tag{3.21}$$

Instead of solving (3.21) n times, one can perform the delete-one operation analytically with the assistance of the following lemma.

Lemma 3.2
The minimizer $\eta_\lambda^{[k]}$ of the "delete-one" functional (3.21) minimizes the full data functional (3.1) with $\tilde{Y}_k = \eta_\lambda^{[k]}(x_k)$ replacing Y_k.

Proof: For all $\eta \neq \eta_\lambda^{[k]}$,

$$\begin{aligned}
\frac{1}{n}\Big((\tilde{Y}_k - \eta_\lambda^{[k]}(x_k))^2 &+ \sum_{i \neq k} (Y_i - \eta_\lambda^{[k]}(x_i))^2 \Big) + \lambda J(\eta_\lambda^{[k]}) \\
&= \frac{1}{n} \sum_{i \neq k} (Y_i - \eta_\lambda^{[k]}(x_i))^2 + \lambda J(\eta_\lambda^{[k]}) \\
&< \frac{1}{n} \sum_{i \neq k} (Y_i - \eta(x_i))^2 + \lambda J(\eta) \\
&\leq \frac{1}{n} \Big((\tilde{Y}_k - \eta(x_k))^2 + \sum_{i \neq k} (Y_i - \eta(x_i))^2 \Big) + \lambda J(\eta).
\end{aligned}$$

The lemma follows. □

The fitted values $\hat{\mathbf{Y}} = A(\lambda)\mathbf{Y}$ are linear in $\mathbf{Y}$. By Lemma 3.2, it is easy to see that

$$\eta_\lambda(x_i) - \eta_\lambda^{[i]}(x_i) = a_{i,i}(Y_i - \eta_\lambda^{[i]}(x_i)),$$

where $a_{i,i}$ is the (i,i)th entry of $A(\lambda)$. Solving for $\eta_\lambda^{[i]}(x_i)$, one has

$$\eta_\lambda^{[i]}(x_i) = \frac{\eta_\lambda(x_i) - a_{i,i}Y_i}{1 - a_{i,i}}.$$

It then follows that

$$\eta_\lambda^{[i]}(x_i) - Y_i = \frac{\eta_\lambda(x_i) - Y_i}{1 - a_{i,i}}.$$

Hence,

$$V_0(\lambda) = \frac{1}{n}\sum_{i=1}^{n} \frac{(Y_i - \eta_\lambda(x_i))^2}{(1 - a_{i,i})^2}. \tag{3.22}$$

It is rarely the case that all sampling points contribute equally to the estimation of $\eta(x)$. To adjust for such an imbalance, it might pay to consider alternative scores with unequal weights,

$$\tilde{V}(\lambda) = \frac{1}{n}\sum_{i=1}^{n} w_i \frac{(Y_i - \eta_\lambda(x_i))^2}{(1 - a_{i,i})^2}.$$

With the choice of $w_i = (1-a_{i,i})^2/[n^{-1}\mathrm{tr}(I - A(\lambda))]^2$ [i.e., substituting $a_{i,i}$ in (3.22) by its average $n^{-1}\sum_{i=1}^{n} a_{i,i}$], one obtains the generalized cross-validation (GCV) score of Craven and Wahba (1979),

$$V(\lambda) = \frac{n^{-1}\mathbf{Y}^T(I - A(\lambda))^2\mathbf{Y}}{[n^{-1}\mathrm{tr}(I - A(\lambda))]^2}. \tag{3.23}$$

A desirable property of the GCV score $V(\lambda)$ is its invariance to an orthogonal transform of $\mathbf{Y}$. Under an extra condition

Condition 3.2.2 $\{n^{-1}\mathrm{tr}A(\lambda)\}^2/n^{-1}\mathrm{tr}A^2(\lambda) \to 0$ as $n \to \infty$,

$V(\lambda)$ can be shown to be a consistent estimate of the relative loss. Condition 3.2.2 generally holds in most settings of interest; see Craven and Wahba (1979) and Li (1986) for details. See also §4.2.3.

Theorem 3.3
Assume independent noise ϵ_i with mean zero, a common variance σ^2, and uniformly bounded fourth moments. If Conditions 3.2.1 and 3.2.2 hold, then

$$V(\lambda) - L(\lambda) - n^{-1}\boldsymbol{\epsilon}^T\boldsymbol{\epsilon} = o_p(L(\lambda)).$$

Proof: Write $\mu = n^{-1}\mathrm{tr}A(\lambda)$ and $\tilde{\sigma}^2 = n^{-1}\boldsymbol{\epsilon}^T\boldsymbol{\epsilon}$. Note that $n^{-1}\mathrm{tr}A^2(\lambda) < 1$, so Condition 3.2.2 implies that $\mu \to 0$. Straightforward algebra yields

$$\begin{aligned} V(\lambda) - L(\lambda) - \tilde{\sigma}^2 &= \frac{1}{(1-\mu)^2}[U(\lambda) - 2\sigma^2\mu - (L(\lambda) + \tilde{\sigma}^2)(1-\mu)^2] \\ &= \frac{U(\lambda) - L(\lambda) - \tilde{\sigma}^2}{(1-\mu)^2} + \frac{(2-\mu)\mu L(\lambda)}{(1-\mu)^2} \\ &\quad - \frac{\mu^2\tilde{\sigma}^2}{(1-\mu)^2} + \frac{2\mu(\tilde{\sigma}^2 - \sigma^2)}{(1-\mu)^2}. \end{aligned}$$

The first term is $o_p(L(\lambda))$ by Theorem 3.1. The second term is $o_p(L(\lambda))$ since $\mu \to 0$. By Condition 3.2.2, $\mu^2 = o_p(L(\lambda))$, so the third term is $o_p(L(\lambda))$. Combining this with $\tilde{\sigma}^2 - \sigma^2 = O_p(n^{-1/2}) = o_p(L^{1/2}(\lambda))$, one obtains $o_p(L(\lambda))$ for the fourth term. □

When the conditions of Theorem 3.3 hold uniformly in a neighborhood of the optimal λ, the minimizers λ_u of $U(\lambda)$ and λ_v of $V(\lambda)$ should be close to each other. Differentiating $U(\lambda)$ and setting the derivative to zero, one gets

$$\frac{d}{d\lambda}\mathbf{Y}^T(I - A(\lambda))^2\mathbf{Y} = -2\sigma^2\frac{d}{d\lambda}\mathrm{tr}A(\lambda). \tag{3.24}$$

Differentiating $V(\lambda)$ and setting the derivative to zero, one similarly has

$$\frac{d}{d\lambda}\mathbf{Y}^T(I - A(\lambda))^2\mathbf{Y} = -2\frac{\mathbf{Y}^T(I - A(\lambda))^2\mathbf{Y}}{\mathrm{tr}(I - A(\lambda))}\frac{d}{d\lambda}\mathrm{tr}A(\lambda). \tag{3.25}$$

Setting $\lambda_u = \lambda_v$ by equating (3.24) and (3.25) and solving for σ^2, one obtains a variance estimate

$$\hat{\sigma}_v^2 = \frac{\mathbf{Y}^T(I - A(\lambda_v))^2\mathbf{Y}}{\mathrm{tr}(I - A(\lambda_v))}. \tag{3.26}$$

The consistency of the variance estimate $\hat{\sigma}_v^2$ is established below.

Theorem 3.4
If Conditions 3.2.1 and 3.2.2 hold uniformly in a neighborhood of the optimal λ, then the variance estimate $\hat{\sigma}_v^2$ of (3.26) is consistent.

Proof: By Theorems 3.1 and 3.3 and (3.17),

$$o_p(R(\lambda_v)) = V(\lambda_v) - U(\lambda_v) = \hat{\sigma}_v^2/(1-\mu) - \hat{\sigma}_v^2(1-\mu) - 2\sigma^2\mu,$$

where $\mu = n^{-1}\mathrm{tr}A(\lambda_v)$, as in the proof of Theorem 3.3. Solving for σ^2, one has

$$\sigma^2 = \hat{\sigma}_v^2\frac{1-\mu/2}{1-\mu} + o_p(\mu^{-1}R(\lambda_v)) = \hat{\sigma}_v^2(1+o(1)) + o_p(\mu^{-1}R(\lambda_v)).$$

It remains to show that $\mu^{-1}R(\lambda_v) = O(1)$. In the neighborhood of the optimal λ, the "bias" term and the "variance" term of $R(\lambda)$ should be of the same order, so $R(\lambda) = O(n^{-1}\mathrm{tr}A^2(\lambda))$. Since the eigenvalues of $A(\lambda)$ are in the range of $[0,1]$, $\mathrm{tr}A^2(\lambda)/\mathrm{tr}A(\lambda) \leq 1$. Now,

$$\mu^{-1}R(\lambda) = \mu^{-1}n^{-1}\mathrm{tr}A^2(\lambda)\{R(\lambda)/n^{-1}\mathrm{tr}A^2(\lambda)\} = O(1).$$

This completes the proof. □

It is easy to see that any estimate of the form $\hat{\sigma}_v^2(1 + o_p(1))$ is also consistent. The consistency of $\hat{\sigma}_v^2(1 + o_p(1))$ may also be obtained directly from Theorem 3.3 and the fact that $L(\lambda) = o_p(1)$.

3.2.3 Restricted Maximum Likelihood Under Bayes Model

Another popular approach to smoothing parameter selection in the context is the restricted maximum likelihood (REML) estimate under the Bayes model of §2.5, which can be perceived as a mixed effect model.

Under the Bayes model, one observes $Y_i = \eta(x_i) + \epsilon_i$ with $\epsilon_i \sim N(0, \sigma^2)$ and $\eta(x) = \sum_{\nu=1}^{m} d_\nu \phi_\nu(x) + \eta_1(x)$, where $\eta_1(x)$ is a mean zero Gaussian process with a covariance function $E[\eta_1(x)\eta_1(y)] = bR_J(x, y)$. To eliminate the nuisance parameters d_ν, a common practice is to consider the likelihood of the contrasts $\mathbf{Z} = F_2^T \mathbf{Y}$, where F_2 is as in (3.5) on page 61. The minus log (restricted) likelihood of σ^2 and b based on the restricted data $\mathbf{Z}$ is seen to be

$$\begin{aligned} &\frac{1}{2}\mathbf{Z}^T(bQ^* + \sigma^2 I)^{-1}\mathbf{Z} + \frac{1}{2}\log|bQ^* + \sigma^2 I| \\ &\qquad = \frac{1}{2b}\mathbf{Z}^T(Q^* + n\lambda I)^{-1}\mathbf{Z} + \frac{1}{2}\log|Q^* + n\lambda I| + \frac{n-m}{2}\log b, \quad (3.27) \end{aligned}$$

where $Q^* = F_2^T Q F_2$ and $n\lambda = \sigma^2/b$; see Problem 3.6. Minimizing (3.27) with respect to b, one gets

$$\hat{b} = \frac{\mathbf{Z}^T(Q^* + n\lambda I)^{-1}\mathbf{Z}}{n-m},$$

with λ to be estimated by the minimizer of the profile minus log likelihood,

$$\frac{1}{2}\log|Q^* + n\lambda I| + \frac{n-m}{2}\log(\hat{b}). \quad (3.28)$$

From (3.7), one has

$$\mathbf{Z}^T(Q^* + n\lambda I)^{-1}\mathbf{Z} = (n\lambda)^{-1}\mathbf{Y}^T(I - A(\lambda))\mathbf{Y}$$

and

$$|Q^* + n\lambda I| = (n\lambda)^{n-m}|I - A(\lambda)|_+^{-1},$$

where $|B|_+$ denotes the product of positive eigenvalues of B. With some algebra, a monotone transform of (3.28) gives

$$M(\lambda) = \frac{n^{-1}\mathbf{Y}^T(I - A(\lambda))\mathbf{Y}}{|I - A(\lambda)|_+^{1/(n-m)}}, \quad (3.29)$$

whose minimizer λ_m is called the generalized maximum likelihood (GML) estimate of λ by Wahba (1985). The corresponding variance estimate is then

$$\hat{\sigma}_m^2 = \frac{\mathbf{Y}^T(I - A(\lambda_m))\mathbf{Y}}{n-m}. \quad (3.30)$$

As $n \to \infty$, it was shown by Wahba (1985) that $\lambda_m = o_p(\lambda_v)$ for η "super-smooth" (in the sense that η satisfies smoothness conditions more stringent than $J(\eta) < \infty$) and that $\lambda_m = O_p(\lambda_v)$ otherwise; see §4.2.3. Hence, asymptotically, GML tends to deliver rougher estimates than GCV.

3.2.4 Weighted and Replicated Data

For weighted data with $E[\epsilon_i^2] = \sigma^2/w_i$, it is appropriate to replace the loss function $L(\lambda)$ of (3.13) by its weighted version

$$L_w(\lambda) = \frac{1}{n}\sum_{i=1}^{n} w_i(\eta_\lambda(x_i) - \eta(x_i))^2. \tag{3.31}$$

The unbiased estimate of relative loss is now

$$U_w(\lambda) = \frac{1}{n}\mathbf{Y}_w^T(I - A_w(\lambda))^2\mathbf{Y}_w + 2\frac{\sigma^2}{n}\text{tr}A_w(\lambda), \tag{3.32}$$

where $\mathbf{Y}_w = W^{1/2}\mathbf{Y}$ for $W = \text{diag}(w_i)$ and $A_w(\lambda)$ is as given in (3.12). The corresponding GCV score is

$$V_w(\lambda) = \frac{n^{-1}\mathbf{Y}_w^T(I - A_w(\lambda))^2\mathbf{Y}_w}{[n^{-1}\text{tr}(I - A_w(\lambda))]^2}. \tag{3.33}$$

The following theorem establishes the consistency of $U_w(\lambda)$ and $V_w(\lambda)$ as estimates of the relative loss $L_w(\lambda)+n^{-1}\boldsymbol{\epsilon}^T W\boldsymbol{\epsilon}$, with the proof easily adapted from the proofs of Theorems 3.1 and 3.3; see Problem 3.7.

Theorem 3.5
Suppose the scaled noise $\sqrt{w_i}\epsilon_i$ are independent with mean zero, a common variance σ^2, and uniformly bounded fourth moments. Denote $R_w(\lambda) = E[L_w(\lambda)]$. If $nR_w(\lambda) \to \infty$ and $\{n^{-1}trA_w(\lambda)\}^2/n^{-1}trA_w^2(\lambda) \to 0$ as $n \to \infty$, then

$$U_w(\lambda) - L_w(\lambda) - n^{-1}\boldsymbol{\epsilon}^T W\boldsymbol{\epsilon} = o_p(L_w(\lambda)),$$
$$V_w(\lambda) - L_w(\lambda) - n^{-1}\boldsymbol{\epsilon}^T W\boldsymbol{\epsilon} = o_p(L_w(\lambda)).$$

For the restricted maximum likelihood under the Bayes model, one can start with the contrasts of $\mathbf{Y}_w$ and derive the corresponding GML score

$$M_w(\lambda) = \frac{n^{-1}\mathbf{Y}_w^T(I - A_w(\lambda))\mathbf{Y}_w}{|I - A_w(\lambda)|_+^{1/(n-m)}}. \tag{3.34}$$

Now, suppose one observes replicated data $Y_{i,j} = \eta(x_i) + \epsilon_{i,j}$, where $j = 1, \dots, w_i$, $i = 1, \dots, n$, and $\epsilon_{i,j} \sim N(0, \sigma^2)$. The penalized unweighted least squares functional

$$\frac{1}{n}\sum_{i=1}^{n}\sum_{j=1}^{w_i}(Y_{i,j} - \eta(x_i))^2 + \lambda J(\eta) \tag{3.35}$$

is equivalent to the penalized weighted least squares functional

$$\frac{1}{n}\sum_{i=1}^{n} w_i(\bar{Y}_i - \eta(x_i))^2 + \lambda J(\eta), \tag{3.36}$$

where $\bar{Y}_i = \sum_{j=1}^{w_i} Y_{i,j}/w_i$; see Problem 3.8(a). Let $\tilde{\mathbf{Y}}$ be the response vector in (3.35) of length $N = \sum_{i=1}^{n} w_i$ and $\tilde{A}(\lambda)$ be the corresponding smoothing matrix, and let $\mathbf{Y}_w$ be the weighted response vector in (3.36) of length n with the ith entry $\sqrt{w_i}\bar{Y}_i$ and $A_w(\lambda)$ be the corresponding smoothing matrix as given in (3.11). It can be shown that $\mathbf{Y}_w = W^{-1/2}P^T\tilde{\mathbf{Y}}$ and

$$I - \tilde{A}(\lambda) = PW^{-1/2}(I - A_w(\lambda))W^{-1/2}P^T + F_3F_3^T,$$

where $P = \text{diag}(\mathbf{1}_{w_i})$ is of size $N \times n$ and F_3 is orthogonal of size $N \times (N-n)$ satisfying $F_3^T P = O$; see Problem 3.8. It follows that

$$\begin{aligned} \tilde{\mathbf{Y}}^T(I_N - \tilde{A}(\lambda))^p\tilde{\mathbf{Y}} &= \mathbf{Y}_w^T(I_n - A_w(\lambda))^p\mathbf{Y}_w + (N-n)\tilde{\sigma}^2, \qquad p = 1,2, \\ \text{tr}(I_N - \tilde{A}(\lambda)) &= \text{tr}(I_n - A_w(\lambda)) + (N-n), \end{aligned}$$

where the sizes of the identity matrices are marked by the subscripts N and n and $\tilde{\sigma}^2 = \sum_{i=1}^{n}\sum_{j=1}^{w_i}(Y_{i,j} - \bar{Y}_i)^2/(N-n)$. It is easy to see that $\text{tr}\tilde{A}(\lambda) = \text{tr}A_w(\lambda)$ and $|I_N - \tilde{A}(\lambda)|_+ = |I_n - A_w(\lambda)|_+$. Hence, the $U(\lambda)$, $V(\lambda)$, and $M(\lambda)$ scores associated with (3.35) can be expressed in terms of $\mathbf{Y}_w$ and $A_w(\lambda)$ as

$$U(\lambda) = \frac{1}{N}\mathbf{Y}_w^T(I_n - A_w(\lambda))^2\mathbf{Y}_w + 2\frac{\sigma^2}{N}\text{tr}A_w(\lambda) + \frac{N-n}{N}\tilde{\sigma}^2, \tag{3.37}$$

$$V(\lambda) = \frac{N^{-1}[\mathbf{Y}_w^T(I_n - A_w(\lambda))^2\mathbf{Y}_w + (N-n)\tilde{\sigma}^2]}{[1 - N^{-1}\text{tr}A_w(\lambda)]^2}, \tag{3.38}$$

$$M(\lambda) = \frac{N^{-1}[\mathbf{Y}_w^T(I_n - A_w(\lambda))\mathbf{Y}_w + (N-n)\tilde{\sigma}^2]}{|I_n - A_w(\lambda)|_+^{1/(N-m)}}. \tag{3.39}$$

It is clear that $U(\lambda)$ of (3.37) is equivalent to $U_w(\lambda)$ of (3.32), but $V(\lambda)$ of (3.38) and $V_w(\lambda)$ of (3.33) are different, so are $M(\lambda)$ of (3.39) and $M_w(\lambda)$ of (3.34). Note that the information concerning σ^2 contained in $\tilde{\sigma}^2$ is ignored in $V_w(\lambda)$ and $M_w(\lambda)$.

The numerical treatment through (3.4) on page 61 is immune to possible singularity of Q, so one usually can ignore the presence of replicated data. When n is substantially smaller than N, however, the computation via (3.36) can result in substantial savings; see §3.4 for the cost of computation. Also, a fast algorithm for the computation of L-splines of §4.3 assumes distinctive x_i's; see §4.3.5.

3.2.5 Empirical Performance

We now illustrate the practical performance of the methods discussed above through some simple simulation. One hundred replicates of samples of size $n = 100$ were generated from $Y_i = \eta(x_i) + \epsilon_i$, $x_i = (i - 0.5)/n$, $i = 1, \ldots, n$, where

$$\eta(x) = 1 + 3\sin(2\pi x - \pi)$$

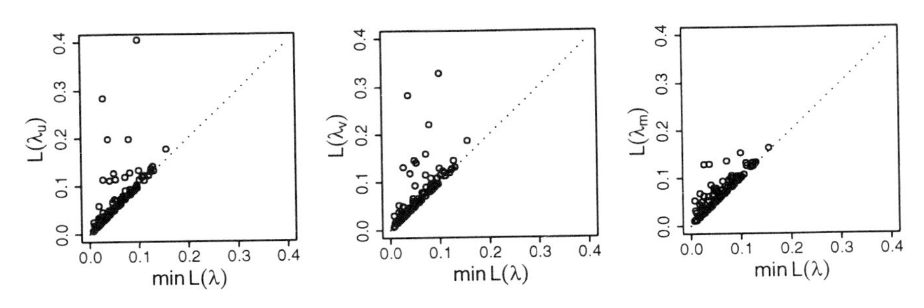

FIGURE 3.1. Performance of $U(\lambda)$, $V(\lambda)$, and $M(\lambda)$ in Simulation: $n = 100$. Left: Loss achieved by $U(\lambda)$ of (3.14). Center: Loss achieved by $V(\lambda)$ of (3.23). Right: Loss achieved by $M(\lambda)$ of (3.29).

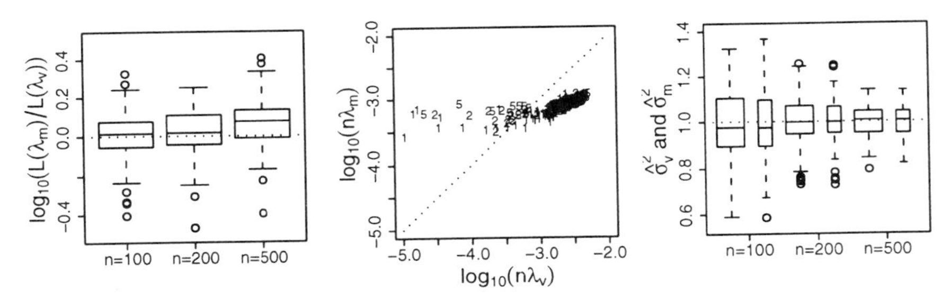

FIGURE 3.2. Comparison of $V(\lambda)$ Versus $M(\lambda)$ in Simulation. Center: Symbols "1," "2," and "5" indicate replicates with $n = 100$, $n = 200$, and $n = 500$, respectively. Right: $\hat{\sigma}_v^2$ are in wider boxes, $\hat{\sigma}_m^2$ are in thinner boxes, $\sigma^2 = 1$.

and $\epsilon_i \sim N(0,1)$. Cubic smoothing splines were calculated with λ minimizing $U(\lambda)$, $V(\lambda)$, and $M(\lambda)$, and with λ on the grid $\log_{10} n\lambda = (-6)(0.1)(0)$. The mean square error $L(\lambda) = n^{-1}\sum_{i=1}^{n}(\eta_\lambda(x_i) - \eta(x_i))^2$ was calculated for all the estimates, from which the optimal λ was located. The losses $L(\lambda_u)$, $L(\lambda_v)$, and $L(\lambda_m)$ are plotted against the minimum $L(\lambda)$ on the grid for all the replicates in Figure 3.1, where a point on the dotted line indicates a perfect selection by the empirical method. All of the methods appeared to perform well most of the time, with occasional wild failures found in $L(\lambda_u)$ and $L(\lambda_v)$ but not in $L(\lambda_m)$.

To empirically investigate the asymptotic behavior of $V(\lambda)$ versus that of $M(\lambda)$, part of the simulation was repeated for sample sizes $n = 200$ and $n = 500$, each with 100 replicates. Plotted in Figure 3.2 are the relative efficacy $\log_{10} L(\lambda_m)/L(\lambda_v)$ of λ_v over λ_m, the comparison of the magnitudes of λ_v versus λ_m, and the performance of the variance estimates $\hat{\sigma}_v^2$ and $\hat{\sigma}_m^2$. It appeared that $L(\lambda_v)$ came ahead of $L(\lambda_m)$ more often than the other way around, and the frequency increased as n increased. The magnitude of λ_m indeed came below that of λ_v in general, as predicted by the asymptotic analysis of Wahba (1985), but λ_v was severely undersmoothing in a few cases, which actually were responsible for its occasional wild failures seen in Figure 3.1. The performances of the variance estimates were reasonably good and did improve as n increased. The variance estimates $\hat{\sigma}_v^2$ and $\hat{\sigma}_m^2$ were actually within 1.5% of each other in all but eight $n = 100$ replicates, three $n = 200$ replicates, and two $n = 500$ replicates.

3.3 Bayesian Confidence Intervals

Point estimate alone is often insufficient in practical applications, as it lacks an assessment of the estimation precision. Lacking the parametric sampling distribution, however, an adequately justified interval estimate is a rarity in nonparametric function estimation. An exception to this is the Bayesian confidence intervals of Wahba (1983), which are derived from the Bayes model of §2.5.

We derive the posterior mean and the posterior variance of $\eta(x)$ and those of its components under the Bayes model, which form the basis for the construction of the interval estimates. The posterior variance permits a somewhat simpler expression on the sampling points, which we will also explore. Despite their derivation from the Bayes model, the interval estimates demonstrate a certain across-the-function coverage property for η fixed and smooth, which makes them comparable to the standard parametric confidence intervals. The practical performance of the interval estimates is illustrated through simple simulation. Parallel results for weighted data are also briefly noted.

3.3.1 Posterior Distribution

Consider $\eta = \eta_0 + \eta_1$, where η_0 and η_1 have independent mean zero Gaussian process priors with covariances $E[\eta_0(x)\eta_0(y)] = \tau^2 \sum_{\nu=1}^{m} \phi_\nu(x)\phi_\nu(y)$ and $E[\eta_1(x)\eta_1(y)] = bR_J(x,y)$, respectively. From (2.35) on page 48 and a standard result on multivariate normal distribution (see, e.g., Johnson and Wichern (1992, Result 4.6)), the conditional variance of $\eta(x)$ given $Y_i = \eta(x_i) + \epsilon_i$ is seen to be

$$\begin{aligned} &bR_J(x,x) + \tau^2 \boldsymbol{\phi}^T \boldsymbol{\phi} - (b\boldsymbol{\xi}^T + \tau^2 \boldsymbol{\phi}^T S^T) \\ &\qquad\qquad \times (bQ + \tau^2 SS^T + \sigma^2 I)^{-1} (b\boldsymbol{\xi} + \tau^2 S\boldsymbol{\phi}) \\ &\quad = b\{R_J(x,x) + \rho\boldsymbol{\phi}^T\boldsymbol{\phi} \\ &\qquad - (\boldsymbol{\xi}^T + \rho\boldsymbol{\phi}^T S^T)(Q + \rho SS^T + n\lambda I)^{-1}(\boldsymbol{\xi} + \rho S\boldsymbol{\phi})\} \\ &\quad = b\{R_J(x,x) + \boldsymbol{\phi}^T(\rho I - \rho^2 S^T(\rho SS^T + M)^{-1}S)\boldsymbol{\phi} \\ &\qquad - 2\boldsymbol{\phi}^T(\rho S^T(\rho SS^T + M)^{-1})\boldsymbol{\xi} - \boldsymbol{\xi}^T(\rho SS^T + M)^{-1}\boldsymbol{\xi}\}, \end{aligned} \tag{3.40}$$

where $\boldsymbol{\xi}$ is $n \times 1$ with the ith entry $R_J(x_i, x)$, Q is $n \times n$ with the (i,j)th entry $R_J(x_i, x_j)$, $\boldsymbol{\phi}$ is $m \times 1$ with the νth entry $\phi_\nu(x)$, S is $n \times m$ with the (i,ν)th entry $\phi_\nu(x_i)$, $\rho = \tau^2/b$, $n\lambda = \sigma^2/b$, and $M = Q + n\lambda I$. Setting $\rho \to \infty$ in (3.40), one obtains the following theorem.

Theorem 3.6
Let $\eta = \eta_0 + \eta_1$, where η_0 has a diffuse prior in $\mathrm{span}\{\phi_\nu, \nu = 1, \dots, \mathrm{m}\}$ *and η_1 has a mean zero Gaussian process prior with covariance function $E[\eta_1(x)\eta_1(y)] = bR_J(x,y)$. Observing $Y_i = \eta(x_i) + \epsilon_i$, $i = 1, \dots, n$, where $\epsilon_i \sim N(0, \sigma^2)$, the posterior variance of $\eta(x)$ satisfies*

$$b^{-1}\mathrm{Var}[\eta(x)|\mathbf{Y}] = R_J(x,x) + \boldsymbol{\phi}^T(S^TM^{-1}S)^{-1}\boldsymbol{\phi} - 2\boldsymbol{\phi}^T\mathbf{d}_\xi - \boldsymbol{\xi}^T\mathbf{c}_\xi, \quad (3.41)$$

where

$$\begin{aligned} \mathbf{c}_\xi &= (M^{-1} - M^{-1}S(S^TM^{-1}S)^{-1}S^TM^{-1})\boldsymbol{\xi}, \\ \mathbf{d}_\xi &= (S^TM^{-1}S)^{-1}S^TM^{-1}\boldsymbol{\xi}. \end{aligned} \quad (3.42)$$

The proof of Theorem 3.6 follows readily from Lemma 2.7 of §2.5.2 and the following lemma.

Lemma 3.7
Suppose M is symmetric and nonsingular and S is of full column rank.

$$\lim_{\rho\to\infty} \rho I - \rho^2 S^T(\rho SS^T + M)^{-1}S = (S^TM^{-1}S)^{-1}. \quad (3.43)$$

Proof: From (2.39) on page 48, one has

$$\begin{aligned} S^T(\rho SS^T + M)^{-1}S &= (I - (I + \rho^{-1}(S^TM^{-1}S)^{-1})^{-1})S^TM^{-1}S \\ &= \rho^{-1}(I + \rho^{-1}(S^TM^{-1}S)^{-1})^{-1}, \end{aligned}$$

so

$$\begin{aligned} \rho I - \rho^2 S^T(\rho SS^T + M)^{-1}S &= \rho(I - (I + \rho^{-1}(S^TM^{-1}S)^{-1})^{-1}) \\ &= (I + \rho^{-1}(S^TM^{-1}S)^{-1})^{-1}(S^TM^{-1}S)^{-1}. \end{aligned}$$

The lemma follows. □

Now, consider the multiple-term model of §2.5.3 in $\mathcal{H} = \oplus_{\beta=0}^p \mathcal{H}_\beta$,

$$\eta(x) = \sum_{\nu=1}^m \psi_\nu(x) + \sum_{\beta=1}^p \eta_\beta(x),$$

where ψ_ν have diffuse priors in $\mathrm{span}\{\phi_\nu\}$ with $\{\phi_\nu\}_{\nu=1}^m$ a basis of $\mathcal{H}_0$ and $\eta_\beta(x)$ have independent Gaussian process priors with mean zero and covariance functions $b\theta_\beta R_\beta(x,y)$. Remember that the model may also be perceived as a mixed effect model, with ψ_ν, $\nu = 1, \dots, m$, being the fixed effects and η_β, $\beta = 1, \dots, p$, being the random effects.

Theorem 3.8
Under the multiple-term model specified above, observing $Y_i = \eta(x_i) + \epsilon_i$, $\epsilon_i \sim N(0, \sigma^2)$, $i = 1, \dots, n$, the posterior means and covariances of the fixed effects ψ_ν and the random effects η_β are as follows:

$$E[\psi_\nu(x)|\mathbf{Y}] = \phi_\nu(x)\mathbf{e}_\nu^T \mathbf{d}, \tag{3.44}$$
$$E[\eta_\beta(x)|\mathbf{Y}] = \boldsymbol{\xi}_\beta^T \mathbf{c}, \tag{3.45}$$
$$b^{-1}\text{Cov}[\psi_\nu(x), \psi_\mu(x)|\mathbf{Y}] = \phi_\nu(x)\phi_\mu(x)\mathbf{e}_\nu^T (S^T M^{-1} S)^{-1} \mathbf{e}_\mu, \tag{3.46}$$
$$b^{-1}\text{Cov}[\phi_\nu(x), \eta_\beta(x)|\mathbf{Y}] = -\phi_\nu \mathbf{e}_\nu^T \mathbf{d}_\beta, \tag{3.47}$$
$$b^{-1}\text{Cov}[\eta_\beta(x), \eta_\gamma(x)|\mathbf{Y}] = \theta_\beta R_\beta(x, x)\delta_{\beta,\gamma} - \mathbf{c}_\beta^T \boldsymbol{\xi}_\gamma, \tag{3.48}$$

where $\mathbf{c}$ *and* $\mathbf{d}$ *are as given in (2.40),* $\mathbf{e}_\nu$ *is the* ν*th unit vector of size* $m \times 1$*,* $\boldsymbol{\xi}_\beta$ *is* $n \times 1$ *with the* i*th entry* $\theta_\beta R_\beta(x_i, x)$*, and*

$$\begin{aligned} \mathbf{c}_\beta &= (M^{-1} - M^{-1}S(S^T M^{-1} S)^{-1} S^T M^{-1})\boldsymbol{\xi}_\beta, \\ \mathbf{d}_\beta &= (S^T M^{-1} S)^{-1} S^T M^{-1} \boldsymbol{\xi}_\beta. \end{aligned} \tag{3.49}$$

The proof of the theorem is straightforward but tedious following the lines of the proofs of Theorems 2.8 and 3.6; see Problem 3.9.

For weighted data with weights w_i, one simply replaces, in the formulas appearing in Theorems 3.6 and 3.8, S by $W^{1/2}S$, $M = Q + n\lambda I$ by $M_w = W^{1/2}QW^{1/2} + n\lambda I$, $\boldsymbol{\xi}_\beta$ by $W^{1/2}\boldsymbol{\xi}_\beta$, and $\mathbf{c}$, $\mathbf{c}_\xi$, and $\mathbf{c}_\beta$ by $W^{-1/2}\mathbf{c}$, $W^{-1/2}\mathbf{c}_\xi$, and $W^{-1/2}\mathbf{c}_\beta$, respectively, where $W = \text{diag}(w_i)$; see Problem 3.10.

The results of Theorems 2.8, 3.6, and 3.8 can be used to construct interval estimates of $\eta(x)$, of its components $\psi_\nu(x)$ and $\eta_\beta(x)$, and of their linear combinations.

3.3.2 Confidence Intervals on Sampling Points

At a sampling point x_i, $\boldsymbol{\phi}^T$ is the ith row of S and $\boldsymbol{\xi}$ is the ith column of Q. Write $B = S(S^T M^{-1} S)^{-1} S^T$. It is easy to check that $b^{-1}\text{Var}[\eta(x_i)|\mathbf{Y}]$ as given in Theorem 3.6 is the (i,i)th entry of the matrix

$$Q + B - BM^{-1}Q - QM^{-1}B - Q(M^{-1} - M^{-1}BM^{-1})Q. \tag{3.50}$$

Note that $QM^{-1} = M^{-1}Q = I - n\lambda M^{-1}$. Following straightforward but tedious algebra, (3.50) simplifies to

$$n\lambda(I - n\lambda(M^{-1} - M^{-1}BM^{-1})) = n\lambda A(\lambda),$$

where the last equation is from (3.8); see Problem 3.11. With b and $\sigma^2 = (n\lambda)b$ known, the $100(1-\alpha)\%$ confidence interval of $\eta(x_i)$ based on the posterior distribution is thus

$$\eta_\lambda(x_i) \pm z_{\alpha/2}\sigma\sqrt{a_{i,i}}, \tag{3.51}$$

where η_λ is the minimizer of (3.1) and $a_{i,i}$ is the (i,i)th entry of the smoothing matrix $A(\lambda)$ given in (3.8).

For weighted data with weights w_i, it can be shown that $b^{-1}\text{Var}[\eta(x_i)|\mathbf{Y}]$ is the (i,i)th entry of $n\lambda W^{-1/2}A_w(\lambda)W^{-1/2}$, where $A_w(\lambda)$ is given in (3.12); see Problem 3.12.

3.3.3 Across-the-Function Coverage

Despite its derivation from the Bayes model, the interval estimates of (3.51), when used with the GCV smoothing parameter λ_v and the corresponding variance estimate $\hat{\sigma}_v^2$, demonstrate a certain across-the-function coverage property for η fixed and smooth, as was illustrated by Wahba (1983).

Over the sampling points, define the average coverage proportion

$$\text{ACP}(\alpha) = \frac{1}{n}\#\{i : |\eta_{\lambda_v}(x_i) - \eta(x_i)| < z_{\alpha/2}\hat{\sigma}_v\sqrt{a_{i,i}}\}.$$

Simulation results in Wahba (1983) suggest that for n large,

$$E[\text{ACP}(\alpha)] \approx 1 - \alpha, \tag{3.52}$$

where the expectation is with respect to ϵ_i in $Y_i = \eta(x_i) + \epsilon_i$ with $\eta(x)$ fixed and smooth. Note that the construction of the intervals is pointwise but the coverage property is across-the-function. Heuristic arguments in support of (3.52) can be found in Wahba (1983). A more rigorous treatment for smoothing splines on $[0,1]$ was given by Nychka (1988). It remains unclear whether a general treatment is possible, however.

For the components $\psi_\nu(x)$ and $\eta_\beta(x)$ and their linear combinations, one may likewise define the corresponding average coverage proportion. The counterpart of (3.52) for componentwise intervals appears less plausible, however, as the simulations of Gu and Wahba (1993b) suggest.

To put (3.52) in perspective, consider some parametric model $\eta(x) = f(x,\boldsymbol{\beta})$ with $f(x,\boldsymbol{\beta})$ known up to the parameters $\boldsymbol{\beta}$. The standard large sample confidence interval for $\eta(x)$, $f(x,\hat{\boldsymbol{\beta}}) \pm z_{\alpha/2}\hat{\sigma}_{f(x,\hat{\boldsymbol{\beta}})}$, has the pointwise coverage property

$$P(|f(x,\hat{\boldsymbol{\beta}}) - \eta(x)| < z_{\alpha/2}\hat{\sigma}_{f(x,\hat{\boldsymbol{\beta}})}) \approx 1 - \alpha. \tag{3.53}$$

The property (3.52) is weaker than (3.53), but (3.53) does imply (3.52). Hence, the intervals satisfying (3.52) can be compared with the standard confidence intervals in parametric models on the basis of the across-the-function coverage property.

For the replicates in the simulation of §3.2.5, ACP(α) was also calculated for $\alpha = 0.05, 0.10$. The results are summarized in Table 3.1.

α	$n = 100$	$n = 200$	$n = 500$
0.05	0.943	0.958	0.962
0.10	0.897	0.915	0.911

TABLE 3.1. Empirical ACP in Simulation.

3.4 Computation: Generic Algorithms

For the estimation tools developed in §3.2 and §3.3 to be practical, one needs efficient algorithms for the minimization of $U(\lambda)$, $V(\lambda)$, or $M(\lambda)$ with respect to the smoothing parameters. Generic algorithms based on the linear system (3.4) are the topic of this section. From discussions in §§3.1—3.3 concerning weighted data, it is clear that the same algorithms are applicable to the penalized weighted least squares problem (3.9) through the linear system (3.10). Special algorithms for problems with certain structures are to be found in §3.8.

Fixing the smoothing parameters, one needs $n^3/3 + O(n^2)$ floating-point operations, or flops, to calculate η_λ. This serves as a benchmark to measure the relative efficiency of the practical algorithms to follow. With only λ tunable, one needs about four times as many flops to execute the algorithm of §3.4.2 to minimize $U(\lambda)$, $V(\lambda)$, or $M(\lambda)$. With λ and θ_β, $\beta = 1, \ldots, p$, all tunable, the iterative algorithm of §3.4.3 takes $4pn^3/3 + O(n^2)$ flops per iteration and needs about 5 to 10 iterations to converge on most problems. The algorithms are largely based on standard numerical linear algebra procedures, of which details, including the flop counts, can be found in Golub and Van Loan (1989).

As in previous sections, we suppress from the notation the dependence of entities on θ_β, except in §3.4.3.

3.4.1 Algorithm for Fixed Smoothing Parameters

Fixing the smoothing parameters λ and θ_β hidden in Q, the calculation of $\mathbf{c}$ and $\mathbf{d}$ in (3.6) is straightforward using standard numerical linear algebra procedures.

For $\mathbf{c}$, one calculates the Cholesky decomposition $(F_2^T Q F_2 + n\lambda I) = G^T G$, where G is upper-triangular, solves for $\mathbf{u}$ from $G\mathbf{u} = F_2^T \mathbf{Y}$ by back substitution and for $\mathbf{v}$ from $G^T \mathbf{v} = \mathbf{u}$ by forward substitution, then $\mathbf{c} = F_2 \mathbf{v}$; for $\mathbf{d}$, one simply solves $R\mathbf{d} = (F_1^T \mathbf{Y} - F_1^T Q F_2 \mathbf{v})$ by back substitution. See, e.g., Golub and Van Loan (1989, §4.2, §3.1) for Cholesky decomposition and forward and back substitutions.

The calculation of the Cholesky decomposition takes $n^3/3 + O(n^2)$ flops, and the rest of the computation, including the QR-decomposition $S = FR^* = (F_1, F_2)\binom{R}{O}$ and the formation of $F^T Q F$, takes $O(n^2)$ flops. This algorithm is rarely used in practice, since it is inadequate to fix the smoothing parameters, but its flop count serves as a benchmark to measure the relative efficiency of the practical algorithms to follow.

3.4.2 Algorithm for Single Smoothing Parameter

We now present an algorithm for the minimization of $U(\lambda)$, $V(\lambda)$, or $M(\lambda)$ as functions of a single smoothing parameter λ. The algorithm employs a one-time $O(n^3)$ matrix decomposition to introduce a certain banded structure, with which the evaluations of $U(\lambda)$, $V(\lambda)$, or $M(\lambda)$ become negligible $O(n)$ operations. The algorithm also serves as a building block in the algorithm for multiple smoothing parameters, to be discussed in §3.4.3.

Algorithm 3.1
Given S, Q, $\mathbf{Y}$, and possibly σ^2 as inputs, perform the following steps to minimize $U(\lambda)$, $V(\lambda)$, or $M(\lambda)$, and return the associated coefficients $\mathbf{c}$, $\mathbf{d}$:

1. Initialization:
   (a) Compute the QR-decomposition $S = FR^* = (F_1, F_2)\left(\begin{smallmatrix} R \\ O \end{smallmatrix}\right)$.
   (b) Compute $F^T\mathbf{Y}$, F^TQF, from which $\mathbf{z} = F_2^T\mathbf{Y}$, $Q^* = F_2^TQF_2$, $F_1^T\mathbf{Y}$, and $F_1^TQF_2$ can be extracted.

2. Tridiagonalization and minimization:
   (a) Compute $Q^* = UTU^T$, where U is orthogonal and T is tridiagonal.
   (b) Compute $\mathbf{x} = U^T\mathbf{z}$.
   (c) Minimize one of the following scores:

$$U^*(\lambda) = \frac{1}{n}\mathbf{x}^T(T + n\lambda I)^{-2}\mathbf{x} - \frac{2\sigma^2}{n}(n\lambda)\mathrm{tr}(T + n\lambda I)^{-1}, \quad (3.54)$$

$$V(\lambda) = \frac{n^{-1}\mathbf{x}^T(T + n\lambda I)^{-2}\mathbf{x}}{[n^{-1}\mathrm{tr}(T + n\lambda I)^{-1}]^2}, \quad (3.55)$$

$$M(\lambda) = \frac{n^{-1}\mathbf{x}^T(T + n\lambda I)^{-1}\mathbf{x}}{|T + n\lambda I|^{-1/(n-M)}}, \quad (3.56)$$

   with respect to λ.

3. Compute return values:
   (a) Compute $\mathbf{v} = U(T + n\lambda I)^{-1}\mathbf{x}$ at the selected λ.
   (b) Return $\mathbf{c} = F_2\mathbf{v}$ and $\mathbf{d} = R^{-1}(F_1^T\mathbf{Y} - F_1^TQF_2\mathbf{v})$.

Note that $U^*(\lambda) = U(\lambda) - 2\sigma^2$ and that

$$I - A(\lambda) = (n\lambda)F_2(F_2^TQF_2 + n\lambda I)^{-1}F_2^T = (n\lambda)F_2U(T + n\lambda I)^{-1}U^TF_2^T.$$

Step 1(a) and $F^T\mathbf{Y}$ in step 1(b) are implemented in the LINPACK routines `dqrdc` and `dqrsl`; see Dongarra et al. (1979). An implementation of $Q = F^T Q F$ in step 1(b), which uses the output of `dqrdc` in a similar manner as `dqrsl` does, is implemented in RKPACK; see Gu (1989). Golub and Van Loan (1989, §§5.1–5.2) and Dongarra et al. (1979) are good places to read about the details of these calculations. The execution of step 1 takes $O(n^2)$ flops.

Step 2(a) via Householder tridiagonalization is the most time-consuming step in Algorithm 3.1, which usually takes $4n^3/3$ flops; see, e.g., Golub and Van Loan (1989, §8.2.1). With a numerically singular Q^*, however, it is possible to speed up the process by employing a certain truncation scheme in the algorithm; see Gu et al. (1989). Step 2(b) is simply another application of the LINPACK routine `dqrsl`.

The crux of Algorithm 3.1 is in step 2(c), where one has to evaluate $U(\lambda)$, $V(\lambda)$, or $M(\lambda)$ at multiple λ values. The band Cholesky decomposition $T + n\lambda I = C^T C$ for T tridiagonal can be computed in $O(n)$ flops, where

$$C = \begin{pmatrix} a_1 & b_1 & & \\ & \ddots & \ddots & \\ & & a_{n_1-1} & b_{n_1-1} \\ & & & a_{n_1} \end{pmatrix}$$

for $n_1 = n - m$; see Golub and Van Loan (1989, §4.3.6). Through a band back substitution followed by a band forward substitution, $(T + n\lambda I)^{-1}\mathbf{x}$ is now available in $O(n)$ flops; see Golub and Van Loan (1989, §4.3.2). For $M(\lambda)$ in (3.56), $|T + n\lambda I| = \prod_{i=1}^{n_1} a_i^2$ is straightforward. The nontrivial part of this step is the efficient evaluation of the term $\mathrm{tr}(T + n\lambda I)^{-1} = \mathrm{tr}(C^{-1}C^{-T})$ in $U^*(\lambda)$ of (3.54) and $V(\lambda)$ of (3.55).

Write $C^{-T} = (\mathbf{c}_1, \dots, \mathbf{c}_{n_1})$; it is clear that $\mathrm{tr}(C^{-1}C^{-T}) = \sum_{i=1}^{n_1} \mathbf{c}_i^T \mathbf{c}_i$. From

$$C^{-T}C^T = (\mathbf{c}_1, \mathbf{c}_2, \dots, \mathbf{c}_{n_1}) \begin{pmatrix} a_1 & & & \\ b_1 & \ddots & & \\ & \ddots & a_{n_1-1} & \\ & & b_{n_1-1} & a_{n_1} \end{pmatrix} = I,$$

one has

$$\begin{aligned} a_{n_1}\mathbf{c}_{n_1} &= \mathbf{e}_{n_1}, \\ a_i \mathbf{c}_i &= \mathbf{e}_i - b_i \mathbf{c}_{i+1}, \qquad i = n_1 - 1, \dots, 1, \end{aligned}$$

where $\mathbf{e}_i$ is the ith unit vector. Because C^{-T} is lower-triangular (Problem 3.13), $\mathbf{c}_{i+1}$ is orthogonal to $\mathbf{e}_i$. Thus, one has recursive formulas

$$\begin{aligned} \mathbf{c}_{n_1}^T \mathbf{c}_{n_1} &= a_{n_1}^{-2}, \\ \mathbf{c}_i^T \mathbf{c}_i &= (1 + b_i^2 \mathbf{c}_{i+1}^T \mathbf{c}_{i+1}) a_i^{-2}, \qquad i = n_1 - 1, \dots, 1. \end{aligned} \tag{3.57}$$

The calculation in (3.57) is clearly of order $O(n)$. This technique for the efficient calculation of $\operatorname{tr}(I - A(\lambda))$ is due to Elden (1984).

At the selected λ, one has

$$\mathbf{c} = F_2 U(T + n\lambda I)^{-1}\mathbf{x},$$
$$\mathbf{d} = R^{-1}(F_1^T\mathbf{Y} - (F_1^T Q F_2)U(T + n\lambda I)^{-1}\mathbf{x}),$$

which are available in $O(n)$ flops. Also available in $O(n)$ flops are

$$\hat{\sigma}_v^2 = \frac{(n\lambda_v)\mathbf{x}(T + n\lambda_v I)^{-2}\mathbf{x}}{\operatorname{tr}(T + n\lambda_v I)^{-1}},$$
$$\hat{\sigma}_m^2 = \frac{(n\lambda_m)\mathbf{x}(T + n\lambda_m I)^{-1}\mathbf{x}}{n - M}.$$

Overall, Algorithm 3.1 takes $4n^3/3 + O(n^2)$ flops to execute, about four times what is needed for the calculation of $\mathbf{c}$ and $\mathbf{d}$ with a fixed λ.

3.4.3 Algorithm for Multiple Smoothing Parameters

We now briefly describe an algorithm for the minimization of $U(\lambda; \boldsymbol{\theta})$, $V(\lambda; \boldsymbol{\theta})$, or $M(\lambda; \boldsymbol{\theta})$ as functions of smoothing parameters λ and θ_β hidden in $Q = \sum_{\beta=1}^p \theta_\beta Q_\beta$, where Q_β has the (i,j)th entry $R_\beta(x_i, x_j)$. The algorithm operates on λ and $\vartheta_\beta = \log\theta_\beta$. We state the algorithm in terms of $V(\lambda; \boldsymbol{\theta})$, but the same procedures readily apply to $U(\lambda; \boldsymbol{\theta})$ and $M(\lambda; \boldsymbol{\theta})$.

Algorithm 3.2
Given S, Q_β, $\beta = 1, \ldots, p$, $\mathbf{Y}$, starting values $\boldsymbol{\vartheta}_0$, and possibly σ^2 as inputs, perform the following steps to minimize $V(\lambda; \boldsymbol{\theta})$ and return the associated coefficients $\mathbf{c}$, $\mathbf{d}$:

1. Initialization:
   (a) Compute the QR-decomposition $S = FR^* = (F_1, F_2)\binom{R}{O}$.
   (b) Compute $F^T\mathbf{Y}$ and $F^T Q_\beta F$, from which $\mathbf{z} = F_2^T\mathbf{Y}$, $Q_\beta^* = F_2^T Q F_2$, $F_1^T\mathbf{Y}$, and $F_1^T Q_\beta F_2$ can be extracted.
   (c) Set $\Delta\boldsymbol{\vartheta} = 0$, $\boldsymbol{\vartheta}_- = \boldsymbol{\vartheta}_0$, and $V_- = \infty$.
2. Iteration:
   (a) For the trial value $\boldsymbol{\vartheta} = \boldsymbol{\vartheta}_- + \Delta\boldsymbol{\vartheta}$, collect $Q^* = \sum_{\beta=1}^p \theta_\beta Q_\beta^*$ and scale it to have a fixed trace.
   (b) Compute $Q^* = UTU^T$, where U is orthogonal and T is tridiagonal. Compute $\mathbf{x} = U^T\mathbf{z}$.
   (c) Minimize $V(\lambda; \boldsymbol{\theta})$ with respect to λ. If $V > V_-$, set $\Delta\boldsymbol{\vartheta} = \Delta\boldsymbol{\vartheta}/2$, go to (a); else proceed.

(d) Evaluate the gradient $\mathbf{g} = (\partial/\partial\boldsymbol{\vartheta})V(\lambda;\boldsymbol{\theta})$ and the Hessian $H = (\partial^2/\partial\boldsymbol{\vartheta}\partial\boldsymbol{\vartheta}^T)V(\lambda;\boldsymbol{\theta})$.

(e) Calculate the increment $\Delta\boldsymbol{\vartheta} = -\tilde{H}^{-1}\mathbf{g}$, where $\tilde{H} = H + \text{diag}(\mathbf{e})$ is positive definite. If H itself is positive definite "enough," $\mathbf{e}$ is simply set to 0.

(f) Check convergence conditions. If the conditions fail, set $\boldsymbol{\vartheta}_- = \boldsymbol{\vartheta}$, $V_- = V$, go to (a).

3. Compute return values:

(a) Compute $\mathbf{v} = U(T + n\lambda I)^{-1}\mathbf{x}$ at the converged λ and $\boldsymbol{\vartheta}$.

(b) Return $\mathbf{c} = F_2\mathbf{v}$ and $\mathbf{d} = R^{-1}(F_1^T\mathbf{Y} - F_1^T Q F_2\mathbf{v})$, with $Q = \sum_{\beta=1}^{p} Q_\beta$.

The calculations in step 1 of Algorithm 3.2 are the same as those in step 1 of Algorithm 3.1 and can be executed in $O(n^2)$ flops. Steps 2(a) through 2(c) with fixed θ_β virtually duplicate step 2 of Algorithm 3.1, which takes $4n^3/3 + O(n^2)$ flops to execute. The calculation of gradient and Hessian in step 2(d) takes an extra $4(p-1)n^3/3 + O(n^2)$ flops; see Gu and Wahba (1991b). Each iteration of step 2 takes altogether $4pn^3/3 + O(n^2)$ flops.

The scores $U(\lambda;\boldsymbol{\theta})$, $V(\lambda;\boldsymbol{\theta})$, or $M(\lambda;\boldsymbol{\theta})$ are fully parameterized by

$$(\lambda_1, \ldots, \lambda_p) = (\lambda\theta_1^{-1}, \ldots, \lambda\theta_p^{-1}),$$

so $(\lambda, \theta_1, \ldots, \theta_p)$ form an overparameterization. This is the reason for the scaling in step 2(a). One may directly employ the Newton iteration with respect to the parameters $\log \lambda_\beta$ to minimize the scores, but the calculation of the gradient and the Hessian would take $4pn^3/3 + O(n^2)$ flops anyway. In this sense, the extra gain through step 2(c) is virtually free.

Step 2(e) returns a descent direction even when the Hessian H is not positive definite. The algorithm to use here is the modified Cholesky decomposition as described in Gill et al. (1981, §4.4.2.2), which adds positive mass to the diagonal elements of H, if necessary, to produce a factorization $\tilde{H} = G^T G$, where G is upper-triangular.

Starting from some heuristic starting values $\boldsymbol{\vartheta}_0$ as suggested in Gu and Wahba (1991b), the algorithm was found to converge in about 5 to 10 iterations on most examples. Further details are to be found in Gu and Wahba (1991b).

3.4.4 Calculation of Posterior Variances

From (3.46)—(3.48) in Theorem 3.8, one needs $(S^T M^{-1} S)^{-1}$, $\mathbf{c}_\beta$, and $\mathbf{d}_\beta$ to construct the Bayesian confidence intervals. At the converged λ and θ_β, it is easy to calculate

$$\begin{aligned} \mathbf{c}_\beta &= F_2 U(T + n\lambda I)^{-1} U^T F_2^T \boldsymbol{\xi}_\beta, \\ \mathbf{d}_\beta &= R^{-1}(F_1^T \boldsymbol{\xi}_\beta - (F_1^T Q F_2) U (T + n\lambda I)^{-1} U^T F_2^T \boldsymbol{\xi}_\beta) \end{aligned} \tag{3.58}$$

in $O(n)$ extra flops. The remaining task is the calculation of $(S^T M^{-1} S)^{-1}$. Using an elementary matrix identity (Problem 3.14), one has

$$\begin{aligned}
S^T M^{-1} S &= R^T F_1^T (Q + n\lambda I)^{-1} F_1 R \\
&= R^T (I, O) F^T (Q + n\lambda I)^{-1} F \left(\begin{smallmatrix} I \\ O \end{smallmatrix}\right) R \\
&= R^T (I, O)(F^T Q F + n\lambda I)^{-1} \left(\begin{smallmatrix} I \\ O \end{smallmatrix}\right) R \\
&= R^T ((F_1^T Q F_1 + n\lambda I) \\
&\quad - (F_1^T Q F_2)(Q^* + n\lambda I)^{-1} (F_2^T Q F_1))^{-1} R \\
&= R^T ((F_1^T Q F_1 + n\lambda I) \\
&\quad - (F_1^T Q F_2) U (T + n\lambda I)^{-1} U^T (F_2^T Q F_1))^{-1} R;
\end{aligned}$$

hence,

$$\begin{aligned}
(S^T M^{-1} S)^{-1} &= R^{-1} ((F_1^T Q F_1 + n\lambda I) \\
&\quad - (F_1^T Q F_2) U (T + n\lambda I)^{-1} U^T (F_2^T Q F_1)) R^{-T}, \qquad (3.59)
\end{aligned}$$

which is available in $O(n)$ extra flops.

3.5 Software

To place the techniques developed into the practitioner's toolbox, the algorithms of §3.4 have been implemented in open-source computer programs downloadable through the internet. The floating point calculations are done through a suite of FORTRAN compatible routines. An user-friendly front end is available in R.

3.5.1 RKPACK

The algorithms of §3.4 have been implemented in a collection of public domain RATFOR (Rational FORTRAN (Kernighan 1975)) routines collectively known as RKPACK, first released in 1989 (Gu 1989). Routines from public domain linear algebra libraries BLAS and LINPACK have been used extensively in RKPACK routines as building blocks; see Dongarra et al. (1979) for descriptions of BLAS and LINPACK. The user interface of RKPACK is through four routines, `dsidr`, `dmudr`, `dsms`, and `dcrdr`, which implement Algorithm 3.1, Algorithm 3.2, (3.59), and (3.58), respectively. A few sample application programs in RATFOR are also included in the package. RKPACK has been deposited to Netlib and StatLib. The latest version can be found at

`http://www.stat.purdue.edu/~chong/software.html`

RATFOR is a dialect of FORTRAN with a structural syntax similar to that of the popular S language (Becker et al. 1988). Most UNIX systems understand RATFOR. In compilation, RATFOR routines are translated by a RATFOR preprocessor into standard FORTRAN routines, transparent of the user, which are then sent to the compiler. For those without access to a RATFOR preprocessor, the FORTRAN translation of the routines are included in the package, but in-line comments are lost in the translation.

3.5.2 R Package `gss`

To further facilitate the application of the techniques developed, a suite of R functions have been developed to serve as a front end to RKPACK. R, an open-source clone of the popular S/Splus language (Becker et al. 1988, Chambers and Hastie 1992), was originally created by Ihaka and Gentleman (1996). At this writing, R is being developed and maintained by a core group of more than a dozen prominent statisticians/programmers stationed over several continents and is distributed under the GNU General Public License. R practices active garbage collecting, which makes it feasible to load and execute the memory hungry RKPACK routines. Add-on modules in R are known as packages, as in S/Splus. The facilities to be described below comprise part of the `gss` package, a package for general smoothing splines. The latest public release of the `gss` package can be found on any one of the CRAN (Comprehensive R Archive Network) sites, along with other R resources. The master CRAN site is at

`http://cran.r-project.org`

The installations of R and add-on packages on all major platforms are clearly explained in the R FAQ (Frequently Asked Questions on R) by Kurt Hornik, to be found on CRAN.

Some working knowledge is assumed of the modeling facilities in R, which have syntax nearly identical to those in S/Splus. A good reference on the subject is Venables and Ripley (1999). The `gss` facilities for penalized least squares regression has two layers of user interface. Most users will access the facilities through the first layer, which consists of the fitting function `ssanova` and the associated methods `predict`, `summary`, `residuals`, and `fitted`. The usage of these functions is very similar to that of the `lm` suite for linear models, as can be seen in the following examples.

Example 3.1 (Cubic spline)
Assume that the `gss` package is installed. At the R prompt, the following command loads the `gss` package into R:

```
library(gss)
```

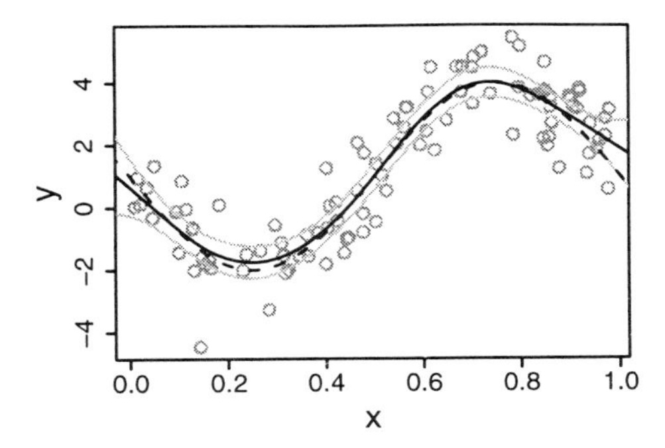

FIGURE 3.3. Cubic Spline Fit. The test function is indicated by the dashed line, the fit is indicated by the solid line, and the 95% Bayesian confidence intervals are indicated by faded solid lines. The data are superimposed as circles.

The following sequence generates some synthetic data and fits a cubic spline to the data, with the smoothing parameter selected by minimizing the GCV score $V(\lambda)$:

```
x <- runif(100)
y <- 1 + 3 * sin(2*pi*x-pi) + rnorm(x)
cubic.fit <- ssanova(y~x,type="cubic",method="v")
```

The arguments `type="cubic"` and `method="v"` are the defaults and can be omitted. The results assigned to `cubic.fit` is a list object of class `"ssanova"`. To evaluate the fit on a grid for plotting purposes, one may try the following:

```
range <- max(x) - min(x)
grid <- seq(min(x)-.05*range,max(x)+.05*range,len=51)
est <- predict(cubic.fit,data.frame(x=grid),se.fit=TRUE)
```

The flag `se.fit=TRUE` requests the calculation of the posterior standard deviation corresponding to the evaluated posterior mean. Now, `est$fit` and `est$se.fit` contain the posterior mean and the posterior standard deviation on the grid, respectively. Figure 3.3 displays a plot with the data, the test function, the cross-validated fit, and the 95% Bayesian confidence intervals, which is produced by the following commands:

```
plot(x,y); lines(grid,est$fit)
lines(grid,est$fit+1.96*est$se.fit,col=5)
lines(grid,est$fit-1.96*est$se.fit,col=5)
lines(grid,1+3*sin(2*pi*grid-pi),lty=2)
```

When executing these commands in front of a color monitor, one may want to replace the `lty` options above by preferred `col` options. To make the plot reproducible by the reader, the seed for the random number generator was set by `set.seed(5732)` before `x` and `y` were generated. □

Example 3.2 (Tensor product cubic spline)
The following sequence generates some synthetic data and fits a tensor product cubic spline to the data, with the smoothing parameters selected by minimizing the GML score $M(\lambda)$:

```
set.seed(5732)
x1 <- rnorm(100); x2 <- rnorm(100)
y <- 1 + 3 * sin(2*pi*(x1-x2)-pi) + rnorm(x1)
tpcubic.fit <- ssanova(y~x1*x2,method="m")
```

Note that `type="cubic"` is omitted in the call to `ssanova`. The model has four terms, labeled `"1"`, `"x1"`, `"x2"`, and `"x1:x2"` representing $f_\emptyset$, f_1, f_2, and $f_{1,2}$, respectively. To evaluate the fit on a grid, one may try the following:

```
range1 <- max(x1) - min(x1)
grid1 <- seq(min(x1)-.05*range1,
             max(x1)+.05*range1,length=25)
range2 <- max(x2) - min(x2)
grid2 <- seq(min(x2)-.05*range2,
             max(x2)+.05*range2,length=25)
new <- data.frame(x1=rep(grid1,25),
                  x2=rep(grid2,rep(25,25)))
est <- predict(tpcubic.fit,newdata=new,se.fit=TRUE)
post.mean <- matrix(est$fit,25,25)
post.stdev <- matrix(est$se.fit,25,25)
```

Now, let us plot the contours of the posterior mean and the posterior standard deviation, with the data superimposed:

```
contour(grid1,grid2,post.mean,sub="REML Fit")
points(x1,x2)
contour(grid1,grid2,post.stdev,sub="Standard Deviation")
points(x1,x2)
```

The plots are given in Figure 3.4. Note how quickly the posterior standard deviation takes off beyond the data range. When sitting in front of a color monitor, one may want to replace `contour` by `filled.contour`.

By default, `predict` evaluates the overall function, but a partial sum of selected model terms can also be obtained via the specification of an optional argument `include`. For example, the following command returns the interaction on the grid:

```
est.int <- predict(tpcubic.fit,grid,
                   se=TRUE,include="x1:x2")
```

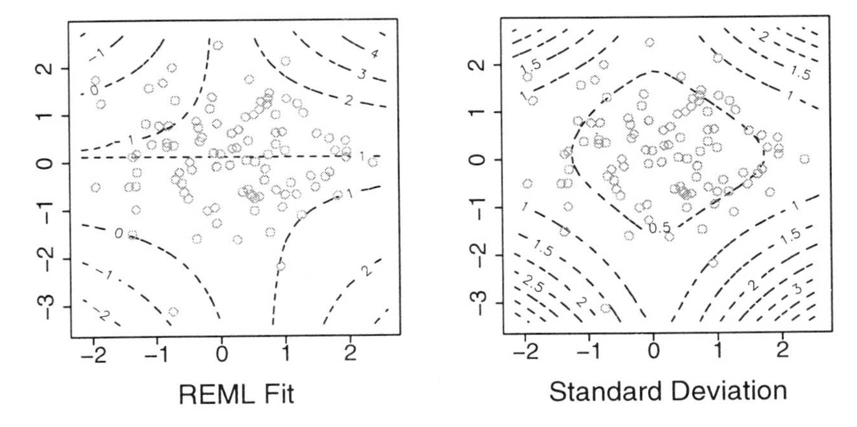

FIGURE 3.4. Tensor Product Cubic Spline Fit. Left: Fit with smoothing parameters selected by $M(\lambda)$. Right: Standard errors of the fit. The data are superimposed as circles.

One can now plot the contours of the interaction and compare with those of the overall function. □

The careful reader may have noticed that the data ranges in the above examples are rather arbitrary, whereas the formulas of §2.3 are defined on $[0, 1]$. It should be noted that the program simply maps the marginal domains onto $[0, 1]$ and uses the formulas of §2.3.3 and §2.4.3 to calculate the reproducing kernels. The marginal domains are specified via the extension of the data ranges by a certain percentage on both ends; the default is 5%, as hinted by the grid span used in the examples, resulting in marginal domains of lengths 110% of the data ranges. By construction, the averaging operators defining the ANOVA decomposition are integrations over the marginal domains scaled by the lengths of the domains.

Besides `type="cubic"` for cubic splines, other types supported in the `ssanova` suite are `type="linear"` for linear splines using the $m = 1$ formulas of §2.3.3, and `type="tp"` for the thin-plate splines, to be discussed in §4.4. Further details concerning `ssanova` and the affiliated methods can be found in the on-line documentation distributed with the code.

The generic algorithms implemented in RKPACK provide an universal numerical engine for all models sharing the structure of §2.4.5. For any specific model, all one needs are the null space basis ϕ_ν that generate the S matrix with the (i, ν)th entry $\phi_\nu(x_i)$ and the reproducing kernels R_β that generate the Q_β matrix with the (i, j)th entry $R_\beta(x_i, x_j)$, $\beta = 1, \dots, p$. All that `ssanova` does is generate the ϕ_ν and R_β functions based on the model formula for certain types of commonly used spline, calculate the matrices by evaluating the functions at the data points, and pass the matrices along with other inputs to the second layer of user interface for numerical processing. Clearly, there is no way to round up all useful models in utilities like `ssanova`, and if one desires to fit a model that is not "canned" in `ssanova` or the like, the second layer of user interface will be needed.

The second layer of user interface consists of four functions, `sspreg`, `mspreg`, `getcrdr`, and `getsms`. For single smoothing parameter models, a call to `sspreg` fits the model using Algorithm 3.1:

```
fit <- sspreg(s,q,y,method)
```

The `method` supported are `method="v"`, `method="m"`, and `method="u"`, referring to $V(\lambda)$, $M(\lambda)$, and $U(\lambda)$, respectively. If `method="u"` is specified, an external variance estimate should be provided through an extra argument `varht`. The function returns a list with the estimated $\mathbf{c}$ (`fit$c`), $\mathbf{d}$ (`fit$d`), $\log_{10} n\lambda$ (`fit$nlambda`), and σ^2 (`fit$varht`) among its components. To get $\mathbf{c}_\beta$, $\mathbf{d}_\beta$, and $(S^T MS)^{-1}$ for the posterior variances given in Theorem 3.8, one passes the fit to `getcrdr` and `getsms`:

```
crdr <- getcrdr(fit,r)
cr <- crdr$cr
dr <- crdr$dr
sms <- getsms(fit)
```

The argument `r` in the call to `getcrdr` is a matrix with the $\boldsymbol{\xi}_\beta$ of (3.49) and (3.58) in its columns, and the corresponding $\mathbf{c}_\beta$ and $\mathbf{d}_\beta$ are returned in the columns of `cr` and `dr`, respectively. For multiple smoothing parameter models, one calls `mspreg` for the fitting instead of `sspreg`:

```
fit <- mspreg(s,q,y,method,prec=1e-7,maxiter=30)
```

The extra arguments `prec` and `maxiter` control the iteration of Algorithm 3.2, and an extra component $\log_{10} \theta_\beta$ (`fit$theta`) is included in the results. The follow-up calls to `getcrdr` and `getsms` remain the same.

3.6 Model Checking Tools

Two phases of statistical modeling are model fitting and model checking. For parametric models, model checking tools include diagnostics for the lack of fit, diagnostics for the identifiability of model terms such as the collinearity in linear models, and diagnostics for the practical significance of model terms through various tests. For nonparametric models, the lack of fit is no longer a main concern, but the danger of overfitting and overinterpreting makes the other two issues ever more important.

With respect to function decompositions such as the ANOVA decomposition of §1.3.2, we introduce some geometric diagnostics for the identifiability and the practical significance of the fitted terms. The use and effectiveness of the diagnostics are illustrated through simple simulations. Also presented are some heuristic arguments and related conceptual discussion concerning the diagnostics.

3.6.1 Cosine Diagnostics

Consider $\eta = \sum_{\beta=0}^{p} f_\beta$, where $f_0 \propto 1$ and f_β, $\beta > 0$ are terms in a function decomposition such as the ANOVA decomposition of §1.3.2. Evaluating a fit at the sampling points x_i, one obtains a retrospective linear model

$$\mathbf{Y} = \mathbf{f}_0 + \mathbf{f}_1 + \cdots + \mathbf{f}_p + \mathbf{e}, \tag{3.60}$$

where $\mathbf{f}_\beta = (f_\beta(x_1), \ldots, f_\beta(x_n))^T$. Projecting (3.60) onto $\{\mathbf{1}\}^\perp = \{\mathbf{f} : \mathbf{f}^T\mathbf{1} = 0\}$ to remove the constant term, one gets

$$\mathbf{Y}^* = \mathbf{f}_1^* + \cdots + \mathbf{f}_p^* + \mathbf{e}^*. \tag{3.61}$$

The collinearity indices κ_β of $(\mathbf{f}_1^*, \ldots, \mathbf{f}_p^*)$ (Stewart 1987), which equal the square roots of the variance inflation factors, measure the identifiability of the f_β's in the fit. Denoting by C the $p \times p$ matrix with the (β, γ)th entry $\cos(\mathbf{f}_\beta^*, \mathbf{f}_\gamma^*)$, the κ_β^2's are given by the diagonals of C^{-1}. Write $\hat{\mathbf{Y}}^* = \mathbf{f}_1^* + \cdots + \mathbf{f}_p^*$. The scaled dot products $\pi_\beta = (\mathbf{f}_\beta^*)^T\hat{\mathbf{Y}}^*/\|\hat{\mathbf{Y}}^*\|^2$ provide a "decomposition" of unity, $\sum_{\beta=1}^{p} \pi_\beta = 1$, although π_β can be negative. When $\mathbf{f}_\beta^*$ are nearly orthogonal to each other, the π_β's come close to form a percentage decomposition of the sum of squares of $\hat{\mathbf{Y}}^*$ into those of its components.

The $\mathbf{f}_\beta^*$'s are supposed to predict the response $\mathbf{Y}^*$, so a near-orthogonal angle between an $\mathbf{f}_\beta^*$ and $\mathbf{Y}^*$ indicates a noise term. Signal terms should be reasonably orthogonal to the residuals, so a large cosine between an $\mathbf{f}_\beta^*$ and $\mathbf{e}^*$ makes a term suspect. Among informative measures for the signal-to-noise ratio are $\cos(\mathbf{Y}^*, \mathbf{e}^*)$ and $R^2 = \|\mathbf{Y}^* - \mathbf{e}^*\|^2/\|\mathbf{Y}^*\|^2$. Finally, a very small Euclidean norm of an $\mathbf{f}_\beta^*$ as compared to that of $\mathbf{Y}^*$ also indicates a negligible term.

These geometric diagnostics will be collectively referred to as the cosine diagnostics, as they are largely based on the cosines among the vectors appearing in (3.61)

For weighted data, one may simply premultiply (3.60) by $W^{1/2}$, project the terms onto $\{W^{1/2}\mathbf{1}\}^\perp$, and operate from the resulting vectors. For replicated data, κ_β and π_β remain the same regardless of whether the retrospective linear model is based on (3.35) (unweighted) or (3.36) (weighted), but entities involving $\mathbf{Y}^*$ or $\mathbf{e}^*$ do vary; see Problem 3.16.

3.6.2 Examples

As illustrations of the use and effectiveness of the cosine diagnostics, we now analyze a few simple synthetic examples on $[0, 1]^3$ using the `ssanova` facilities in the R package `gss`.

Example 3.3 (Independent design)
First, generate some synthetic data as follows:

```
set.seed(5732)
x1 <- runif(100)
x2 <- runif(100)
x3 <- runif(100)
y <- 10*sin(pi*x2)+exp(3*x3)
     +5*cos(2*pi*(x1-x2))+3*rnorm(x1)
```

A tensor product linear spline model is then fitted to the data:

```
fit <- ssanova(y~x1*x2*x3-x1:x2:x3,type="linear")
```

Note that a tensor product cubic spline is infeasible here, as it has way too many smoothing parameters (twelve, to be exact). The diagnostics for the fit can be requested from the method `summary`:

```
sum.fit <- summary(fit,diagnostics=TRUE)
```

A look at the κ_β's confirms that there is no identifiability problem with this fit; the pound sign # is added in front of each line of the computer printout to distinguish it from the command one types in:

```
round(sum.fit$kappa,2)
#    x1    x2    x3 x1:x2 x1:x3 x2:x3
#  1.04  1.01  1.09  1.07  1.08  1.08
```

Given below are the π_β's, the cosines between $\mathbf{Y}^*$, $\mathbf{e}^*$ and the $\mathbf{f}_\beta$'s, and the norms of the vectors, where the `cos.y` line gives $\cos(\mathbf{Y}^*, \cdot)$ and the `cos.e` line gives $\cos(\mathbf{e}^*, \cdot)$:

```
round(sum.fit$pi,2)
#    x1    x2    x3 x1:x2 x1:x3 x2:x3
#  0.04  0.17  0.48  0.29  0.00  0.02
round(sum.fit$cosines,2)
#          x1    x2    x3 x1:x2 x1:x3 x2:x3
# cos.y  0.35  0.42  0.71  0.61  0.32  0.35
# cos.e  0.26  0.14  0.14  0.35  0.20  0.23
# norm   8.79 26.70 42.48 32.78  0.00  3.09
#        yhat     y     e
# cos.y  0.97  1.00  0.57
# cos.e  0.36  0.57  1.00
# norm  67.47 76.55 19.36
```

The interaction "`x1:x3`" was apparently eliminated by the fitting procedure. The terms "`x1`" and "`x2:x3`" appear very weak, both from the π_β's and from their "high" correlations with the residuals relative to their correlations with the response. Eliminating "`x1:x3`" and "`x2:x3`" but keeping "`x1`" due to the presence of "`x1:x2`", a new model is fitted to the data:

```
fit.new <- ssanova(y~x1*x2+x3,type="linear")
```

A quick check shows that there is virtually no change in the diagnostics associated with the remaining terms. The elimination of "x1:x3" and "x2:x3" knocks out six smoothing parameters from the cubic spline model, so we are tempted to give it a try:

```
fit.new.c <- ssanova(y~x1*x2+x3,type="cubic")
```

The diagnostics for the cubic spline fit agree well with those for the linear spline fit:

```
sum.new.c<-summary(fit.new.c,TRUE)
round(sum.new.c$pi,2)
#    x1    x2    x3 x1:x2
#  0.01  0.14  0.55  0.30
round(sum.new.c$cos,2)
#          x1    x2    x3 x1:x2  yhat     y     e
# cos.y 0.14  0.38  0.70  0.55  0.92  1.00  0.48
# cos.e 0.07  0.07  0.03  0.10  0.10  0.48  1.00
# norm  3.43 23.87 47.14 34.92 67.31 76.55 30.28
```

Note that the residuals generally are larger in the cubic spline fit, as the fit does not bend as easily as the linear spline fit. □

Example 3.4 (Simple aliasing design)
Instead of an independent design, we now put $x_{i\langle 1\rangle}$ and $x_{i\langle 2\rangle}$ on a curve to create some identifiability problem:

```
set.seed(5732)
x2 <- runif(100)
x3 <- runif(100)
x1 <- sqrt(x2)
y <- 10*sin(pi*x2)+exp(3*x3)
     +5*cos(2*pi*(x1-x2))+3*rnorm(x1)
```

Fit a tensor product linear spline model to the data and obtain the diagnostics. The κ_β's indicate severe concurvity, but all interactions plus the "x2" term were effectively eliminated:

```
fit <- ssanova(y~x1*x2*x3-x1:x2:x3,type="linear")
sum.fit <- summary(fit,TRUE)
round(sum.fit$kappa,2)
#    x1    x2    x3 x1:x2 x1:x3 x2:x3
#  7.61  9.15  1.70  3.04  7.34  7.05
round(sum.fit$pi,2)
#    x1    x2    x3 x1:x2 x1:x3 x2:x3
#  0.16  0.00  0.84  0.00  0.00  0.00
```

```
round(sum.fit$cos,2)
#           x1   x2    x3 x1:x2 x1:x3 x2:x3
# cos.y   0.35 0.35  0.79  0.24  0.38  0.24
# cos.e   0.14 0.15  0.10  0.14  0.27  0.26
# norm   23.32 0.00 47.38  0.00  0.00  0.20
#         yhat     y     e
# cos.y   0.91  1.00  0.56
# cos.e   0.16  0.56  1.00
# norm   50.23 59.76 25.26
```

Formally removing the terms and refit, no change is to be found in the diagnostics but in the κ_β's. We shall skip ahead to the cubic spline fit and get its diagnostics:

```
fit.new.c <- ssanova(y~x1+x3,type="cubic")
sum.new.c <- summary(fit.new.c,TRUE)
round(sum.new.c$pi,2)
#   x1   x3
# 0.16 0.84
round(sum.new.c$cos,2)
#           x1    x3  yhat     y     e
# cos.y   0.33  0.78  0.89  1.00  0.50
# cos.e   0.05  0.02  0.04  0.50  1.00
# norm   23.82 48.96 51.88 59.76 27.67
```

Once again, the cubic spline fit has larger residuals. □

Example 3.5 (Complex aliasing design)
We now change the aliasing pattern to $x_{i\langle 1\rangle} = (x^2_{i\langle 2\rangle} + x^2_{i\langle 3\rangle})/2$ and obtain the tensor product linear spline fit and its diagnostics:

```
set.seed(5732)
x2 <- runif(100)
x3 <- runif(100)
x1 <- (x2^2+x3^2)/2
y <- 10*sin(pi*x2)+exp(3*x3)
     +5*cos(2*pi*(x1-x2))+3*rnorm(x1)
fit <- ssanova(y~x1*x2*x3-x1:x2:x3,type="linear")
sum.fit <- summary(fit,TRUE)
```

The κ_β's indicate mild concurvity. All terms involving "x1" were weak, if not eliminated outright:

```
round(sum.fit$kappa,2)
#   x1   x2   x3 x1:x2 x1:x3 x2:x3
# 1.80 1.39 1.47  1.82  1.73  1.54
round(sum.fit$pi,2)
```

```
#     x1    x2    x3 x1:x2 x1:x3 x2:x3
#   0.00  0.21  0.68  0.00  0.02  0.09
round(sum.fit$cosines,2)
#          x1    x2    x3 x1:x2 x1:x3 x2:x3
# cos.y  0.37  0.45  0.78  0.76  0.38  0.42
# cos.e  0.19  0.14  0.10  0.34  0.28  0.22
# norm   0.00 33.41 60.92  0.00  4.56 16.66
#        yhat     y     e
# cos.y  0.96  1.00  0.47
# cos.e  0.21  0.47  1.00
# norm  74.32 82.45 23.27
```

Formally removing the terms involving "x1" and refit a cubic spline model, the diagnostics point to no problems:

```
fit.new.c <- ssanova(y~x2*x3,type="cubic")
sum.new.c <- summary(fit.new.c,TRUE)
round(sum.new.c$kappa,2)
#    x2    x3 x2:x3
#  1.02  1.01  1.03
round(sum.new.c$pi,2)
#    x2    x3 x2:x3
#  0.21  0.67  0.13
round(sum.fit.new$cosines,2)
#          x2    x3 x2:x3  yhat     y     e
# cos.y  0.45  0.77  0.42  0.95  1.00  0.37
# cos.e  0.05  0.01  0.07  0.05  0.37  1.00
# norm  34.27 61.78 22.64 76.68 82.45 26.45
```

The residuals are, again, larger for the cubic spline fit. □

3.6.3 Concepts and Heuristics

We now briefly discuss the heuristics behind the cosine diagnostics and some related concepts. The primary issues are the identifiability of the f_β's, which the κ_β's are designed to diagnose, and the practical significance of individual terms, which the $\cos(\mathbf{Y}^*, \mathbf{f}^*_\beta)$'s are designed to diagnose.

First, consider the identifiability. By construction, the decomposition $\eta = \sum_{\beta=0}^{p} f_\beta$ is well defined on its domain, say $\mathcal{X}$. When the function is being estimated from the data, however, information only comes from the sampling points $\mathcal{X}_0 = \{x_i\}_{i=1}^{n}$, and the identifiability of the terms in the decomposition depends on how well the decomposition is supported on the restricted domain $\mathcal{X}_0$. Parallel to collinearity, such an identifiability problem is called concurvity by Buja et al. (1989).

There exist two kinds of concurvity: the retrospective, or observed, concurvity, and the prospective concurvity. The observed concurvity can be defined as the collinearity of the restrictions of the estimated f_β's to $\mathcal{X}_0$, which the κ_β's are designed to diagnose. Prospective concurvity, the same in spirit as what was under discussion in Buja et al. (1989), is a (undesirable) property of the model and the design $\mathcal{X}_0$ based on preobservation analysis. For a parametric linear model, concurvity reduces to collinearity, the form of the fit is fully predictable from the model and the design, so there is no distinction between prospective and retrospective collinearity.

What is so bad about concurvity? One calculates an estimate $f = \sum_{\beta=0}^{p} f_\beta$ based on information from $\mathcal{X}_0$, but its restriction to $\mathcal{X}_0$, say $f^0 = \sum_{\beta=0}^{p} f_\beta^0$, is not well defined. If there is an alternative breakup $f^0 = \sum_{\beta=0}^{p} \alpha_\beta f_\beta^0$, then one could have used an alternative estimate $g = \sum_{\beta=0}^{p} \alpha_\beta f_\beta$ instead of $f = \sum_{\beta=0}^{p} f_\beta$. For this to be of serious concern to us, however, the difference $(\alpha_\beta - 1) f_\beta$ would have to be practically meaningful, and $J(f-g) = \sum_\beta (\alpha_\beta - 1)^2 J_\beta(f_\beta)$ would have to be negligible, where $J_\beta(f_\beta)$ is the roughness contribution of f_β to $J(f)$. This pretty much rules out the participation of "nonparametric" components in serious concurvity: For $(\alpha_\beta - 1) f_\beta$ to be practically significant, one must have negligible $J_\beta(f_\beta)$; hence, f_β would be primarily a parametric component in $\mathcal{N}_J$. The main concern of Buja et al. (1989), the numerical instability caused by concurvity to their back-fitting algorithm, is, however, not an issue here, as all terms are estimated simultaneously via the stable linear system (3.4).

Now, consider the practical significance of individual terms. Recall that in a parametric regression model, insignificant terms are often detected using various F-statistics. Consider a linear model $\mathbf{Y} = \alpha\mathbf{1} + \beta\mathbf{x} + \boldsymbol{\epsilon}$, where $\mathbf{1}^T\mathbf{x} = 0$; if $\mathbf{1}^T\mathbf{x} \neq 0$, replace $\mathbf{x}$ by $(I - \mathbf{1}\mathbf{1}^T/n)\mathbf{x}$. Write $\mathbf{f}_0 = \hat{\alpha}\mathbf{1}$ and $\mathbf{f}_1 = \hat{\beta}\mathbf{x} = \mathbf{x}(\mathbf{x}^T\mathbf{x})^{-1}\mathbf{x}^T\mathbf{Y}$. The standard F-statistic for testing $\beta = 0$, or $\mathbf{f}_1 = 0$, is

$$F = \frac{\mathbf{Y}^T\mathbf{x}(\mathbf{x}^T\mathbf{x})^{-1}\mathbf{x}^T\mathbf{Y}}{\mathbf{Y}^T(I - \mathbf{1}\mathbf{1}^T/n - \mathbf{x}(\mathbf{x}^T\mathbf{x})^{-1}\mathbf{x}^T)\mathbf{Y}} = \frac{\cos^2(\mathbf{Y}^*, \mathbf{f}_1^*)}{1 - \cos^2(\mathbf{Y}^*, \mathbf{f}_1^*)}, \tag{3.62}$$

which is monotone in

$$\cos^2(\mathbf{Y}^*, \mathbf{f}_1^*) = \frac{\mathbf{Y}^T\mathbf{x}(\mathbf{x}^T\mathbf{x})^{-1}\mathbf{x}^T\mathbf{Y}}{\mathbf{Y}^T(I - \mathbf{1}\mathbf{1}^T/n)\mathbf{Y}}; \tag{3.63}$$

see Problem 3.17. Hence, $\cos(\mathbf{Y}^*, \mathbf{f}_\beta^*)$ coincide with the classical measures in a specific simple parametric setting.

We suggest that $\cos(\mathbf{Y}^*, \mathbf{f}_\beta^*)$ be taken as absolute measures when the smoothing parameters are selected using a data-adaptive criterion such as $V(\lambda)$, for, in such a circumstance, different terms are allowed to compete with each other and with the residual term for shares of resources based on their qualifications as predictors of $\mathbf{Y}$. These diagnostics are objective quantities, but their calibration has to be subjective in the lack of sampling distributions. Our limited experience seems to suggest that a term with $\cos(\mathbf{Y}^*, \mathbf{f}_\beta^*) > 0.4$ shall not be overlooked and a term with

$\cos(\mathbf{Y}^*, \mathbf{f}_\beta^*) < 0.25$ may be safely suppressed. The calibration of $\|\mathbf{f}_\beta\|$ (an analog of χ^2-statistics) is much more difficult, so their use is limited and is of secondary importance. The π_β's provide reasonable measures for the relative strengths of the fitted terms, especially when the terms $\mathbf{f}_\beta^*$ are nearly orthogonal.

3.7 Case Studies

We now apply the techniques developed so far to analyze a few real-life data sets. As for all data analysis exercises, subjective choices will have to be made along the way, and the author's preferences by no means represent the only "correct" solutions.

3.7.1 Nitrogen Oxides in Engine Exhaust

In an experiment reported by Brinkman (1981), a single-cylinder engine was run with ethanol to see how the NO_x concentration in the exhaust depended on the compression ratio and the equivalence ratio. There were 88 measurements made, and the data were analyzed by Cleveland and Devlin (1988) and Breiman (1991), among others, using other smoothing methods.

The data were read into R as a data frame `nox` with components `nox`, `comp`, and `equi`. A tensor product cubic spline was fitted to the data and the diagnostics obtained:

```
fit.nox<-ssanova(log10(nox)~comp*equi,data=nox)
sum.nox<-summary(fit.nox,TRUE)
round(sum.nox$kappa,2)
#       comp       equi comp:equi
#       1.08       1.04      1.05
round(sum.nox$pi,2)
#       comp       equi comp:equi
#      -0.02       1.00      0.02
round(sum.nox$cos,2)
#         comp equi comp:equi yhat    y    e
# cos.y -0.08 0.95       0.08 0.99 1.00 0.20
# cos.e  0.02 0.02       0.03 0.04 0.20 1.00
# norm   1.91 8.33       1.60 8.02 8.18 1.37
```

The NO_x concentrations are positive with some near-zero readings, so a log transform was made. The effect of equivalence ratio was dominant, but the compression ratio had little impact. Eliminating terms involving `comp`, one can fit a cubic spline in `equi` and plot the data, the fit, and the 95% Bayesian confidence intervals, as in Figure 3.5:

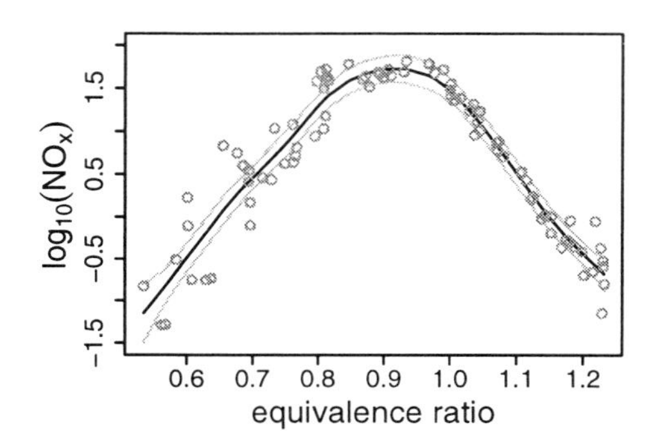

FIGURE 3.5. Cubic Spline Fit to NO_x Data. The 95% Bayesian confidence intervals are indicated by the faded lines and the data by the circles.

```
fit.nox<-ssanova(log10(nox)~equi,data=nox,met="m")
grid<-sort(nox$equi)
est<-predict(fit.nox,data.frame(equi=grid),se=TRUE)
plot(nox$equi,log10(nox$nox),xlab="equivalence ratio",
     ylab=expression(log[10](NO[x])))
lines(grid,est$fit)
lines(grid,est$fit+1.96*est$se,col=5)
lines(grid,est$fit-1.96*est$se,col=5)
```

We switched to $M(\lambda)$ for smoothing parameter selection after seeing a rough cross-validation fit.

The compression ratio had only five distinctive values, so it could have been treated as an ordinal discrete variable; it would not make a difference though, as `nox` is plain flat on the `comp` axis. Cleveland and Devlin (1988) and Breiman (1991) both used the cubic root transform for `nox` instead of the log transform; parallel analysis using the cubic root transform yield essentially the same results.

3.7.2 Ozone Concentration in Los Angeles Basin

Daily measurements of ozone concentration and eight meteorological quantities in the Los Angeles basin were recorded for 330 days of 1976. The data were used by Breiman and Friedman (1985) to illustrate their ACE algorithm (alternating conditional expectation) and by Buja et al. (1989) to illustrate nonparametric additive models through the back-fitting algorithm. The data are read into an R data frame `ozone` with the following components:

`upo3` Upland ozone concentration (ppm).

`vdht` Vandenberg 500 millibar height (m).

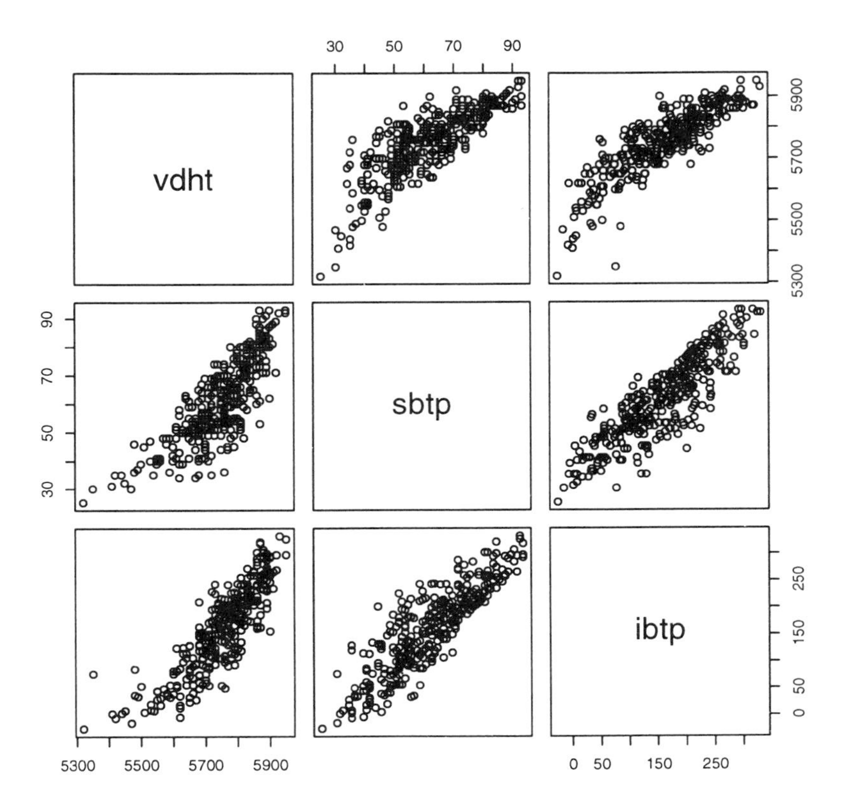

FIGURE 3.6. Scatter Plot Matrix of Ozone Data: A Correlated Group.

`wdsp` Wind speed (mph).

`hmdt` Humidity (%).

`sbtp` Sandburg Air Base temperature (^oC).

`ibht` Inversion base height (ft).

`dgpg` Dagget pressure gradient (mmHg).

`ibtp` Inversion base temperature (^oF).

`vsty` Visibility (miles).

From the scatter plot matrix, the three variables `vdht`, `sbtp`, and `ibtp` appeared to be highly correlated; see Figure 3.6. We decided not to include these variables simultaneously in our preliminary analysis. We also decided not to include the variable `wdsp`, which showed little relation with any of the other variables.

Our first attempt was to fit tensor product linear spline models on five variables: one of **vdht**, **sbtp**, or **ibtp**, plus four others, **hmdt**, **ibht**, **dgpg**, and **vsty**. Included in the models were 5 main effects and 10 pairwise interactions. The log transform was applied to the response since it is positive with some readings near zero. The measure $R^2 = \|\mathbf{Y}^* - \mathbf{e}^*\|^2/\|\mathbf{Y}^*\|^2$ was calculated to be 0.74, 0.71, and 0.82 for the three fits. The fit involving **ibtp** and its diagnostics are illustrated below, which yields $R^2 = 0.82$:

```
fit.ozone0<-ssanova(log10(upo3)~
                    (ibtp+hmdt+ibht+dgpg+vsty)^2,
                    data=ozone,type="linear")
sum.ozone0<-summary(fit.ozone0,TRUE)
round(sum.ozone0$kappa,2)
#        ibtp       hmdt       ibht       dgpg       vsty
#        2.05       2.07       2.21       1.72       1.21
#  ibtp:hmdt  ibtp:ibht  ibtp:dgpg  ibtp:vsty  hmdt:ibht
#        1.30       1.36       1.16       1.39       1.25
#  hmdt:dgpg  hmdt:vsty  ibht:dgpg  ibht:vsty  dgpg:vsty
#        1.67       1.23       1.16       1.24       1.33
round(sum.ozone0$cos,2)
#          ibtp  hmdt  ibht  dgpg  vsty  ibtp:hmdt
# cos.y  0.77  0.34 -0.56  0.38  0.45       0.60
# cos.e  0.05  0.07  0.06  0.06  0.08       0.31
# norm   3.71  0.00  0.00  1.59  0.45       1.37
#        ibtp:ibht ibtp:dgpg ibtp:vsty hmdt:ibht
# cos.y       0.14      0.34      0.54      0.14
# cos.e       0.18      0.21      0.13      0.13
# norm        0.00      0.00      0.52      0.35
#        hmdt:dgpg hmdt:vsty ibht:dgpg ibht:vsty
# cos.y       0.40     -0.27      0.29      0.23
# cos.e       0.11      0.12      0.21      0.19
# norm        0.23      0.00      0.75      0.00
#        dgpg:vsty  yhat     y     e
# cos.y       0.20  0.96  1.00  0.46
# cos.e       0.15  0.20  0.46  1.00
# norm        0.07  5.34  5.90  1.66
```

Modest concurvity appeared to exist, but seven interactions were negligible, among which the terms **ibtp:ibht**, **ibtp:dgpg**, **hmdt:vsty**, and **ibht:vsty** were not showing any sign of life and **hmdt:ibht**, **ibht:dgpg**, and **dgpg:vsty** had small cosines with the response. This left us with only three interactions, **ibtp:hmdt**, **ibtp:vsty**, and **hmdt:dgpg**. The nil main effect **ibht** would be removed, as all interactions involving it were negligible, but **hmdt** would stay for now. Fitting a tensor product cubic spline to the data with four main effects and three pairwise interactions, one had the fit below with $R^2 = 0.77$:

```
fit.ozone1<-ssanova(log10(upo3)~ibtp+hmdt+dgpg+vsty
                    +ibtp:(hmdt+vsty)+hmdt:dgpg,
                    data=ozone)
sum.ozone1<-summary(fit.ozone1,TRUE)
round(sum.ozone1$kappa,2)
#       ibtp       hmdt       dgpg       vsty
#       1.08       1.95       1.53       1.05
# ibtp:hmdt ibtp:vsty hmdt:dgpg
#       1.04       1.05       1.43
round(sum.ozone2$cos,2)
#         ibtp  hmdt dgpg vsty ibtp:hmdt ibtp:vsty
# cos.y 0.74 -0.43 0.33 0.31      0.23      0.34
# cos.e 0.00  0.00 0.01 0.04      0.05      0.08
# norm  4.09  0.50 1.95 0.55      1.21      0.82
#          hmdt:dgpg yhat    y    e
# cos.y         0.41 0.90 1.00 0.48
# cos.e         0.03 0.04 0.48 1.00
# norm          1.07 5.18 5.90 2.62
```

There appeared to be concurvity among hmdt, dgpg, and hmdt:dgpg, and the negative cosine between hmdt and the response might suggest that it was offsetting the effect of hmdt:dgpg, so hmdt:dgpg looked like a candidate for removal. Another term to be removed was ibtp:hmdt because of its small cosine with the response. Removing the two interactions, and adding as main effects the previously excluded variables vdht, sbtp, wdsp to double check their effects, one had the following fit with $R^2 = 0.75$:

```
fit.ozone2<-update(fit.ozone1,.~.+vdht+sbtp+wdsp
                   -(ibtp+dgpg):hmdt)
sum.ozone2<-summary(fit.ozone2,TRUE)
round(sum.ozone2$kappa,2)
#       ibtp       hmdt       dgpg       vsty
#       1.89       1.12       1.25       1.10
#       vdht       sbtp       wdsp ibtp:vsty
#       1.09       1.98       1.07       1.23
round(sum.ozone2$cos,2)
#         ibtp hmdt dgpg vsty  vdht sbtp wdsp
# cos.y 0.73 0.11 0.43 0.46 -0.08 0.79 0.27
# cos.e 0.01 0.05 0.02 0.03  0.03 0.01 0.05
# norm  1.95 0.33 1.33 1.09  0.43 1.92 0.32
#         ibtp:vsty yhat   y    e
# cos.y        0.51 0.88 1.0 0.50
# cos.e        0.03 0.03 0.5 1.00
# norm         0.80 5.12 5.9 2.75
```

The terms hmdt, vdht, and wdsp were negligible. Eliminating these terms, one had the following fit with $R^2 = 0.75$:

```
fit.ozone3<-update(fit.ozone2,.~.-hmdt-vdht-wdsp)
sum.ozone3<-summary(fit.ozone3,TRUE)
round(sum.ozone3$kappa,2)
#        ibtp        dgpg        vsty        sbtp ibtp:vsty
#        2.16        1.12        1.06        2.20      1.05
round(sum.ozone3$cos,2)
#         ibtp dgpg vsty sbtp ibtp:vsty yhat   y    e
# cos.y 0.74 0.43 0.39 0.78        0.41 0.88 1.0 0.50
# cos.e 0.00 0.02 0.04 0.00        0.08 0.03 0.5 1.00
# norm  2.38 1.61 0.88 1.66        0.92 5.10 5.9 2.82
round(sum.ozone3$pi,2)
#        ibtp        dgpg        vsty        sbtp ibtp:vsty
#        0.40        0.15        0.07        0.29      0.08
```

The term sbtp is concurve with ibtp, and its removal reduced the R^2 from 0.75 to 0.74:

```
fit.ozone4<-update(fit.ozone3,.~.-sbtp)
sum.ozone4<-summary(fit.ozone4,TRUE)
round(sum.ozone4$kappa,2)
#        ibtp        dgpg        vsty ibtp:vsty
#        1.09        1.03        1.07      1.06
round(sum.ozone4$cos,2)
#         ibtp dgpg vsty ibtp:vsty yhat    y    e
# cos.y 0.74 0.41 0.42        0.42 0.87 1.00 0.51
# cos.e 0.00 0.02 0.03        0.08 0.03 0.51 1.00
# norm  3.79 1.93 0.96        0.94 5.06 5.90 2.90
round(sum.ozone4$pi,2)
#        ibtp        dgpg        vsty ibtp:vsty
#        0.65        0.18        0.09      0.08
```

We now plot the main effects of ibtp, dgpg, and vsty in fit.ozone3 and fit.ozone4. To obtain the fit and standard error of ibtp at the data points in fit.ozone3, one can use

```
est3.ibtp<-predict(fit.ozone3,ozone,
                   se=TRUE,include="ibtp")
```

Other terms can be similarly obtained. The plots are given in Figure 3.7, with the rugs on the bottom of the plots marking the sampling points. It is easily seen that fit.ozone3 has a weaker ibtp effect with a larger standard error. The ibtp effect in fit.ozone4 is split between ibtp and sbtp in fit.ozone3 and the split suffers some identifiability problem due to concurvity. The vsty effect is plotted on a grid instead of the data points, as the sampling points are discrete beyond 150 with gaps of 50.

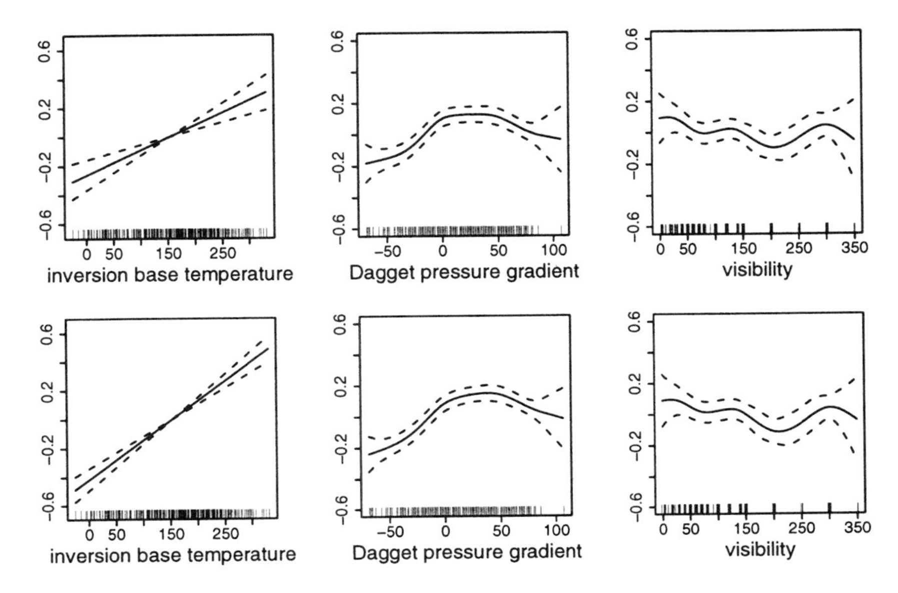

FIGURE 3.7. Three Main Effects of Ozone Fits with 95% Bayesian Confidence Intervals. Top: `fit.ozone3` with concurvity. Bottom: `fit.ozone4` without concurvity. The rugs on the bottom of the plots mark the sampling points, with the visibility jittered.

3.8 Computation: Special Algorithms

The generic algorithms of §3.4 are of order $O(n^3)$, which makes them impractical for very large data sets. For some problems, however, structures can be introduced through alternative formulations, yielding much faster algorithms for calculations with fixed smoothing parameter. To select the smoothing parameter using $U(\lambda)$ or $V(\lambda)$, one needs algorithms of comparable speed for the evaluation of $\mathrm{tr}A(\lambda)$, which is the focus of this section. According to current knowledge, the score $M(\lambda)$ is largely beyond reach with the alternative formulations, so are the posterior variances which one would need for the construction of Bayesian confidence intervals.

For polynomial splines on $[0,1]$, bandedness can be introduced into the matrices involved through the use of ordered local-support basis, and $O(n)$ algorithms are available for both $\hat{\mathbf{Y}}$ and $\mathrm{tr}A(\lambda)$ (§3.8.1). For problems such as tomographical reconstruction and the smoothing of digital images, one usually solves sparse or highly structured linear systems through iterative procedures, and the term $\mathrm{tr}A(\lambda)$ can be estimated through a parallel run with some $\mathbf{w} \sim N(0, I)$ replacing $\mathbf{Y}$ (§3.8.2).

3.8.1 Fast Algorithm for Polynomial Splines

A polynomial smoothing spline on $[0,1]$, the minimizer of

$$\frac{1}{n}\sum_{i=1}^{n}(Y_i-\eta(x_i))^2+\lambda\int_0^1(\eta^{(m)})^2dx, \tag{3.64}$$

is called a natural spline in the numerical analysis literature. It is a piecewise polynomial of order $2m-1$, with up to the $(2m-2)$nd derivatives continuous and the $(2m-1)$st derivative jumping at the knots $\xi_1<\cdots<\xi_q$, the ordered distinctive sampling points x_i. On $[0,\xi_1]$ and $[\xi_q,1]$, it is a polynomial of order $m-1$. See, e.g., de Boor (1978).

The natural splines with a given set of knots $\xi_1<\cdots<\xi_q$ form a linear space of dimension q; see Problem 3.22. There exists a local-support basis $\{B_j(x), j=1,\dots,q\}$ for these natural splines, with each of the B_j's supported on at most $2m$ of the adjacent intervals $[0,\xi_1]$, $[\xi_1,\xi_2]$, ..., $[\xi_q,1]$, and at most $2m$ of the B_j's are nonzero at any $x\in[0,1]$; see Schumaker (1981, §8.2). Plugging the expression $\eta(x)=\sum_{j=1}^{q}c_jB_j(x)$ into (3.64), one has

$$(\mathbf{Y}-X\mathbf{c})^T(\mathbf{Y}-X\mathbf{c})+n\lambda\mathbf{c}^TJ\mathbf{c}, \tag{3.65}$$

where X is $n\times q$ with the (i,j)th entry $B_j(x_i)$ and J is $q\times q$ with the (i,j)th entry $\int_0^1 B_i^{(m)}B_j^{(m)}dx$. Minimizing (3.65) with respect to $\mathbf{c}$, one gets $\mathbf{c}=(X^TX+n\lambda J)^{-1}X^T\mathbf{Y}$ and $\hat{\mathbf{Y}}=X(X^TX+n\lambda J)^{-1}X^T\mathbf{Y}$.

Ordering the basis functions B_j increasingly by their supports, one has

$$B_i(x)B_j(x)=B_i^{(m)}(x)B_j^{(m)}(x)=0$$

for $|i-j|\geq 2m$. It is clear that X^TX and J are both banded with bandwidth $4m-1$. The band Cholesky decomposition $(X^TX+n\lambda J)=C^TC$ takes $O(q)$ flops, with the upper-triangular C banded with bandwidth $2m$; see Golub and Van Loan (1989, §4.3.6). The coefficients $\mathbf{c}$ then are available in $O(q)$ extra flops through a band back substitution followed by a band forward substitution; see Golub and Van Loan (1989, §4.3.2).

The nontrivial part of the algorithm is the fast evaluation of $\mathrm{tr}A(\lambda)=\mathrm{tr}[(X^TX+n\lambda J)^{-1}(X^TX)]$. Write $B=X^TX$ and $C^{-T}=(\mathbf{c}_1,\dots,\mathbf{c}_q)$. It follows that $\mathrm{tr}A(\lambda)=\sum_{i,j}b_{i,j}\mathbf{c}_i^T\mathbf{c}_j$. Since B is symmetric and banded with bandwidth $4m-1$, only $\mathbf{c}_i^T\mathbf{c}_j$ for $0\leq i-j<2m$ need to be computed. From $C^{-T}C^T=I$, one has

$$\mathbf{e}_i=\sum_{j=1}^{q}d_{i,j}\mathbf{c}_j=\sum_{j=i}^{q\wedge(i+2m-1)}d_{i,j}\mathbf{c}_j,$$

where $\mathbf{e}_i$ is the ith unit vector and $d_{i,j}$ is the (i,j)th entry of C with $d_{i,j}=0$ for $j<i$ and $j\geq i+2m$. Write $n(i)=q\wedge(i+2m-1)$. From

$$d_{i,i}\mathbf{c}_i=\mathbf{e}_i-\sum_{j=i+1}^{n(i)}d_{i.j}\mathbf{c}_j,$$

one has, recursively,

$$\begin{aligned}\mathbf{c}_q^T\mathbf{c}_q&=d_q^{-2},\\ \mathbf{c}_i^T\mathbf{c}_k&=-d_{i,i}^{-1}\sum_{j=i+1}^{n(i)}d_{i,j}\mathbf{c}_j^T\mathbf{c}_k,\quad i<k,\\ \mathbf{c}_i^T\mathbf{c}_i&=d_{i,i}^{-2}\left(1+\sum_{j=i+1}^{n(i)}\sum_{l=i+1}^{n(i)}d_{i,j}d_{i,l}\mathbf{c}_j^T\mathbf{c}_l\right),\end{aligned}\tag{3.66}$$

where the fact that $\mathbf{e}_i^T\mathbf{c}_k=0$ for $i<k$ is used. These formulas are immediate extensions of (3.57) on page 78. Using (3.66), one can fill $\mathbf{c}_i^T\mathbf{c}_j$ in the band $0\leq i-j<2m$, from the bottom row up, backward within each row, without any reference to entries outside the band. The calculations take $O(q)$ flops.

The key to this algorithm is the band structure made available by the ordered local-support basis. Many authors use the popular B-spline basis as the $B_j(x)$ in the above formulation, which makes no difference in computation and performance, but, technically, B-splines are not natural splines, as they have different boundary conditions; see de Boor (1978) and Schumaker (1981) for details. The algorithm has been implemented for B-splines independently by Finbarr O'Sullivan and by H. J. Woltring, with code available from the NETLIB at

```
http://www.netlib.org/gcv
```

3.8.2 Iterative Algorithms and Monte Carlo Cross-Validation

Smoothing with a quadratic penalty is a special case of generalized ridge regression and can often be formulated in the form of (3.65) for some X. Fixing the smoothing parameter, one solves the linear system

$$(X^TX+n\lambda J)\mathbf{c}=X^T\mathbf{Y}\tag{3.67}$$

for $\mathbf{c}$ and calculates $\hat{\mathbf{Y}}=X\mathbf{c}=A(\lambda)\mathbf{Y}$ and $\mathbf{e}=\mathbf{Y}-\hat{\mathbf{Y}}=(I-A(\lambda))\mathbf{Y}$. In many applications, the matrix $X^TX+n\lambda J$ is sparse or highly structured, although not necessarily banded as in §3.8.1, which allows for the fast calculation of the matrix-vector multiplication $(X^TX+n\lambda J)\mathbf{c}$. Iterative procedures such as the conjugate gradient method are often the most efficient for solving such linear systems; see, e.g., Golub and Van Loan (1989, Chapter 10).

Examples of such formulation can be found in, e.g., Girard (1989). Detailed algorithmic specifications, which vary from problem to problem, are not directly relevant to our discussion. Our primary concern here is the implementation of automatic smoothing parameter selection through scores like $U(\lambda)$ or $V(\lambda)$ when iterative procedures are used to solve (3.67).

When the linear system (3.67) is solved iteratively, one has no direct access to the structure of the smoothing matrix $A(\lambda)$ and its trace. To use $U(\lambda)$ or $V(\lambda)$ for the selection of the smoothing parameter in such a circumstance, a Monte Carlo approximation of $\operatorname{tr}A(\lambda)$ was proposed by Girard (1989). The idea is simple and easy to implement. Let $\mathbf{w}$ be a vector of n independent standard normal deviates. Passing $\mathbf{w}$ through the same iterative procedures that produce $\hat{\mathbf{Y}} = A(\lambda)\mathbf{Y}$, one obtains $A(\lambda)\mathbf{w}$. One then can use $\mathbf{w}^T A(\lambda)\mathbf{w}$ to approximate $\operatorname{tr}A(\lambda)$ and select the smoothing parameter by minimizing

$$\tilde{U}(\lambda) = \frac{1}{n}\mathbf{Y}^T(I - A(\lambda))^2\mathbf{Y} + 2\frac{\sigma^2}{n}\mathbf{w}^T A(\lambda)\mathbf{w}$$

for σ^2 known, or by minimizing

$$\tilde{V}(\lambda) = \frac{n^{-1}\mathbf{Y}^T(I - A(\lambda))^2\mathbf{Y}}{[1 - n^{-1}\mathbf{w}^T A(\lambda)\mathbf{w}]^2}$$

for σ^2 unknown. The justification of the approximation is through the following theorem.

Theorem 3.9

Assume independent noise ϵ_i with mean zero, a common variance σ^2, and uniformly bounded fourth moments. If Condition 3.2.1 of §3.2.1 holds, then

$$\tilde{U}(\lambda) - L(\lambda) - n^{-1}\boldsymbol{\epsilon}^T\boldsymbol{\epsilon} = o_p(L(\lambda)). \tag{3.68}$$

If, in addition, Condition 3.2.2 of §3.2.2 also holds, then

$$\tilde{V}(\lambda) - L(\lambda) - n^{-1}\boldsymbol{\epsilon}^T\boldsymbol{\epsilon} = o_p(L(\lambda)). \tag{3.69}$$

Proof: Recalling (3.17) and (3.19) in the proof of Theorem 3.1, one has

$$\tilde{U}(\lambda) - U(\lambda) = 2\frac{\sigma^2}{n}(\mathbf{w}^T A(\lambda)\mathbf{w} - \operatorname{tr}A(\lambda)) = o_p(L(\lambda)),$$

which together with Theorem 3.1 yields (3.68). To prove (3.69) with Theorem 3.3 in mind, it suffices to show that $\tilde{V}(\lambda) - V(\lambda) = o_p(L(\lambda))$. Write $\mu = n^{-1}\operatorname{tr}A(\lambda)$ and $\tilde{\mu} = n^{-1}\mathbf{w}^T A(\lambda)\mathbf{w}$. Simple algebra yields

$$\tilde{V}(\lambda) - V(\lambda) = V(\lambda)\left[\frac{(1-\mu)^2}{(1-\tilde{\mu})^2} - 1\right] = V(\lambda)\frac{2-\mu-\tilde{\mu}}{(1-\tilde{\mu})^2}(\tilde{\mu} - \mu),$$

which is $o_p(L(\lambda))$ since $\tilde{\mu} - \mu = o_p(L(\lambda))$, $V(\lambda) = O_p(1)$, and $\mu = o(1)$. This completes the proof. □

It is clear that each evaluation of $\tilde{U}(\lambda)$ or $\tilde{V}(\lambda)$ takes about twice as many flops as the calculation of $\hat{\mathbf{Y}}$ alone. In practice, it is advisable to generate a single $\mathbf{w}$ for use in $\tilde{U}(\lambda)$ or $\tilde{V}(\lambda)$ for all evaluations. One benefit of this is the continuity of the resulting score, and the other benefit is possible faster convergence of the iteration when $A(\lambda)\mathbf{w}$ at some nearby λ is used as the starting value. The approximation may be improved a little by averaging $\mathbf{w}^T A(\lambda)\mathbf{w}$ over a few replicates of $\mathbf{w}$ at further computing cost. Since n is usually very large when $\tilde{U}(\lambda)$ or $\tilde{V}(\lambda)$ is used, however, any benefit from such practice, if any, may not be worth the extra cost.

Compared to $\tilde{\mu} = n^{-1}\mathbf{w}^T A(\lambda)\mathbf{w}$, $\mu^* = \mathbf{w}^T A(\lambda)\mathbf{w}/\mathbf{w}^T\mathbf{w}$ provides a better estimator of $\mu = n^{-1}\text{tr}A(\lambda)$ that one may use in practice; see Problem 3.23. Theorem 3.9 remains valid when $\tilde{\mu}$ is replaced by μ^*.

3.9 Bibliographic Notes

Section 3.1

The general problem of penalized least squares regression with multiple penalty terms was formulated by Wahba (1986) and studied numerically by Gu, Bates, Chen, and Wahba (1989) and Gu and Wahba (1991b). The linear system (3.4) as the basis for computation first appeared in Wahba and Wendelberger (1980). The smoothing matrix in the form of (3.7) was given by Wahba (1978).

Section 3.2

The score $U(\lambda)$, originally proposed by Mallows (1973) for use in ridge regression, is usually referenced as Mallows' C_L. An equivalent form of Theorem 3.1 was proved by Li (1986). Cross-validation is a classical technique for model selection in a variety of parametric and nonparametric problems. The generalized cross-validation score $V(\lambda)$ was due to Craven and Wahba (1979). An equivalent form of Theorem 3.3 was proved by Li (1986) using decision-theoretical techniques. The simple direct proof presented here appears to be new, which is largely adapted from related arguments in Craven and Wahba (1979).

The score $M(\lambda)$ was proposed and studied in the context by Wahba (1985). Restricted maximum likelihood (REML) has been widely used in the literature on variance components and mixed effect models; see, e.g., Harville (1977) and Robinson (1991). In Bayesian statistics, such an approach to the estimation of prior parameters is known as the type-II maximum likelihood; see, e.g., Berger (1985, §3.5.4).

The variance estimate $\hat{\sigma}_v^2$ was proposed by Wahba (1983) based on heuristic arguments and excellent simulation results. The motivation by equating λ_u and λ_v represents an alternative interpretation of the arguments developed by Gu, Heckman, and Wahba (1992) for smoothing parameter selection with replicated data. The primary result of Gu, Heckman, and Wahba (1992) was the calculus leading to (3.37)—(3.39).

Section 3.3

The Bayesian confidence intervals were proposed by Wahba (1983), with the across-the-function coverage property suggested through heuristic arguments and demonstrated via empirical simulations. A more rigorous treatment of the across-the-function coverage property for univariate polynomial splines can be found in Nychka (1988). The componentwise intervals derived through Theorem 3.8 were explored in Gu and Wahba (1993a).

Section 3.4

The developments in this section draw heavily on some standard numerical linear algebra results, for which Golub and Van Loan (1989) and Dongarra, Moler, Bunch, and Stewart (1979) are excellent references. Algorithm 3.1 was proposed by Gu, Bates, Chen, and Wahba (1989), with important ideas borrowed from earlier work by Elden (1984) and Bates, Lindstrom, Wahba, and Yandell (1987). Algorithm 3.2 was developed by Gu and Wahba (1991b), where further details are to be found.

Section 3.5

RKPACK was first released to the public in 1989, with the two drivers `dsidr` and `dmudr` each having two options for smoothing parameter selection, $V(\lambda)$ or $M(\lambda)$. The option $U(\lambda)$ and the two utility routines `dcrdr` and `dsms` were added in 1992. The R package `gss` was first released to the public in 1999. It is still under active development.

Section 3.6

An excellent review of diagnostics for collinearity can be found in Stewart (1987), where the collinearity indices are introduced. Earlier discussion of concurvity and its numerical ramifications can be found in Buja, Hastie, and Tibshirani (1989). This section draws heavily on the materials of Gu (1992b), where more examples and further discussion are to be found. The values of κ_β^2 were mistakenly reported as κ_β in the examples of Gu (1992b), although the mistake was inconsequential.

Section 3.7

In earlier analyses of the NO_x data, Cleveland and Devlin (1988) used multivariate local weighted regression and Breiman (1991) used his $\prod$ method, and both concluded that the interaction between the compression ratio and the equivalence ratio was significant. The analysis presented in §3.7.1 concludes otherwise.

In Breiman and Friedman (1985), an additive model in `sbtp`, `ibht`, `dgpg`, and `vsty` was fitted to the Los Angeles ozone data using alternating conditional expectation (ACE). Buja, Hastie, and Tibshirani (1989) used the data as a running example in the discussion of additive models and backfitting algorithm. A slew of analyses of the ozone data using a variety of techniques were compared in Hastie and Tibshirani (1990, §10.3), where a scatter plot matrix of all the variables can be found.

Section 3.8

A comprehensive treatment of natural splines can be found in Schumaker (1981, Chapter 8). The $O(n)$ evaluation of $\mathrm{tr}A(\lambda)$ was proposed by Hutchinson and de Hoog (1985); see also O'Sullivan (1985). The distinction between the B-splines and the natural splines is discussed in de Boor (1978) and Schumaker (1981).

The Monte Carlo approximation of the trace term $\mathrm{tr}A(\lambda)$ was proposed by Girard (1989); see also Hutchinson (1989). Theorem 3.9 largely represents restatements of some of the results found in Girard (1991), but with greatly simplified proofs.

3.10 Problems

Section 3.1

3.1 Consider the least squares functional $L(f) = \sum_{i=1}^{n}(Y_i - f(x_i))^2$ in a reproducing kernel Hilbert space $\mathcal{H}$ with a square seminorm $J(f)$.

(a) Prove that $L(f)$ is continuous, convex, and Fréchet differentiable.

(b) Let $\{\phi_\nu, \nu = 1, \ldots, m\}$ be a basis of $\mathcal{N}_J = \{f : J(f) = 0\}$ and S be the $n \times m$ matrix with the (i, ν)th entry $\phi_\nu(x_i)$. Prove that if S is of full column rank, then $L(f)$ is strictly convex in $\mathcal{N}_J$.

(c) Prove that if S is of full column rank, then $L(f) + \lambda J(f)$ is strictly convex in $\mathcal{H}$.

3.2 Prove that the linear system

$$\begin{aligned}(Q + n\lambda I)\mathbf{c} + S\mathbf{d} &= \mathbf{Y},\\ S^T\mathbf{c} &= 0,\end{aligned}$$

where S is of full column rank, Q non-negative definite, and $\lambda > 0$, has a unique solution that satisfies

$$Q\{(Q + n\lambda I)\mathbf{c} + S\mathbf{d} - \mathbf{Y}\} = 0,$$
$$S^T\{Q\mathbf{c} + S\mathbf{d} - \mathbf{Y}\} = 0.$$

3.3 Prove that the eigenvalues of the smoothing matrix $A(\lambda)$ as defined in (3.7) are all in the range $[0, 1]$.

3.4 Show that the solution of (3.10) minimizes

$$(\mathbf{Y} - S\mathbf{d} - Q\mathbf{c})^T W(\mathbf{Y} - S\mathbf{d} - Q\mathbf{c}) + n\lambda\mathbf{c}^T Q\mathbf{c}.$$

Section 3.2

3.5 Prove Theorem 3.1 under the general moment conditions on ϵ_i as stated in the theorem.

(a) Let B and C be $n \times n$ matrices, where B is symmetric. Show that

$$\text{Var}[\boldsymbol{\epsilon}^T B\boldsymbol{\epsilon}] \leq 2\sigma^4 \text{tr}B^2 + \sum_{i=1}^n b_{ii}^2(K - 3\sigma^4), \tag{3.70}$$
$$\text{Var}[\boldsymbol{\eta}^T C\boldsymbol{\epsilon}] = \sigma^2\boldsymbol{\eta}^T CC^T\boldsymbol{\eta}, \tag{3.71}$$

where K bounds $E[\epsilon_i^4]$ uniformly.

(b) Prove (3.17) by applying (3.70) with $B = A^2(\lambda)$ and applying (3.71) with $C = (I - A(\lambda))A(\lambda)$. Note that the Cauchy-Schwarz inequality can be used to bound $\text{Cov}[\boldsymbol{\epsilon}^T B\boldsymbol{\epsilon}, \boldsymbol{\eta}^T C\boldsymbol{\epsilon}]$.

(c) Prove (3.18) by applying (3.71) with $C = I - A(\lambda)$.

(d) Prove (3.19) by applying (3.70) with $B = A(\lambda)$.

3.6 Show that (3.27) is the minus log likelihood of $\mathbf{Z} = F_2^T\mathbf{Y}$.

3.7 Prove Theorem 3.5.

3.8 Consider replicated data $Y_{i,j} = \eta(x_i) + \epsilon_{i,j}$, where $j = 1, \ldots, w_i$, $i = 1, \ldots, n$. Denote the total sample size by $N = \sum_{i=1}^n w_i$ and the response vector of length N by $\tilde{\mathbf{Y}}$. Let S be $n \times m$ with entries $\phi_\nu(x_i)$, Q be $n \times n$ with entries $R_J(x_i, x_j)$, and $P = \text{diag}(\mathbf{1}_{w_i})$ of size $N \times n$.

(a) Write $\bar{Y}_i = \sum_{j=1}^{w_i} Y_{i,j}/w_i$. Show that

$$\sum_{i=1}^n \sum_{j=1}^{w_i} (Y_{i,j} - \eta(x_i))^2 = \sum_{i=1}^n w_i(\bar{Y}_i - \eta(x_i))^2 + \sum_{i=1}^n \sum_{j=1}^{w_i} (Y_{i,j} - \bar{Y}_i)^2.$$

(b) Solving (3.35) directly through (3.3) with Y, S, Q replaced by $\tilde{Y}$, $\tilde{S}$, $\tilde{Q}$, respectively; verify that $\tilde{S} = PS$ and $\tilde{Q} = PQP^T$.

(c) Let $\mathbf{Y}_w$ be of length n with the ith entry $\sqrt{w_i}\bar{Y}_i$. Verify that $\mathbf{Y}_w = W^{-1/2}P^T\tilde{Y}$, where $W = P^TP = \text{diag}(w_i)$.

(d) Consider F_2 orthogonal of size $n \times (n-m)$ satisfying $F_2^TW^{1/2}S = O$, and F_3 orthogonal of size $N \times (N-n)$ satisfying $F_3^TP = O$. Verify that $\tilde{F}_2 = (PW^{-1/2}F_2, F_3)$ is orthogonal and satisfies $\tilde{F}_2^T\tilde{S} = O$.

(e) The smoothing matrix for (3.35) is given by

$$\tilde{A}(\lambda) = I - n\lambda\tilde{F}_2(\tilde{F}_2^T\tilde{Q}\tilde{F}_2 + n\lambda I)^{-1}\tilde{F}_2^T,$$

and that for (3.36) is given by

$$A_w(\lambda) = I - n\lambda F_2(F_2^TW^{1/2}QW^{1/2}F_2 + n\lambda I)^{-1}F_2^T;$$

see (3.7) and (3.11). Show that

$$I - \tilde{A}(\lambda) = PW^{-1/2}(I - A_w(\lambda))W^{-1/2}P^T + F_3F_3^T.$$

Section 3.3

3.9 Prove Theorem 3.8. Similar to the proofs of Theorems 2.8 and 3.6, first consider independent proper priors for $\psi_\nu = d_\nu\phi_\nu$, $d_\nu \sim N(0, \tau^2)$, then let $\tau^2 \to \infty$.

(a) Find the covariance matrix of $\mathbf{Y}$, $\psi_\nu(x)$, and $\psi_\mu(x)$ and use it to prove (3.44) and (3.46).

(b) Find the covariance matrix of $\mathbf{Y}$, $\eta_\beta(x)$, and $\eta_\gamma(x)$ and use it to prove (3.45) and (3.48).

(c) Find the covariance matrix of $\mathbf{Y}$, ψ_ν and $\eta_\beta(x)$ and use it to prove (3.47).

3.10 Derive the results of Theorems 3.6 and 3.8 for weighted data with $\epsilon_i \sim N(0, \sigma^2/w_i)$.

3.11 Verify that (3.50) simplifies to $n\lambda A(\lambda)$.

3.12 Show that for weighted data with weights w_i, $b^{-1}\text{Var}[\eta(x_i)|\mathbf{Y}]$ is the (i,i)th entry of $n\lambda W^{-1/2}A_w(\lambda)W^{-1/2}$.

Section 3.4

3.13 For L lower-triangular, prove that L^{-1} is also lower-triangular.

3.14 For an invertible block matrix $M = \begin{pmatrix} A & B \\ C & D \end{pmatrix}$, show that

$$M^{-1} = \begin{pmatrix} E^{-1} & -E^{-1}BD^{-1} \\ -D^{-1}CE^{-1} & D^{-1} + D^{-1}CE^{-1}BD^{-1} \end{pmatrix},$$

where $E = A - BD^{-1}C$.

Section 3.5

3.15 Suppose $Y_i = \eta(x_i) + \epsilon_i$, where $\eta = \sum_{\nu=1}^{4} \psi_\nu + \sum_{\beta=1}^{5} f_\beta$ with fixed effects ψ_ν and random effects f_β, as in Theorem 3.8.

(a) Derive $E[\psi_3(x) + f_2(x)|\mathbf{Y}]$ and $b^{-1}\text{Var}[\psi_3(x) + f_2(x)|\mathbf{Y}]$.

(b) Derive $E[\psi_4(x)+f_3(x)+f_4(x)+f_5(x)|\mathbf{Y}]$ and $b^{-1}\text{Var}[\psi_4(x)+f_3(x)+f_4(x) + f_5(x)|\mathbf{Y}]$.

Section 3.6

3.16 Consider the replicated data of Problem 3.8 and keep all the notation and definitions. Write the retrospective linear model corresponding to (3.35) as

$$\tilde{\mathbf{Y}} = \tilde{\mathbf{f}}_0 + \tilde{\mathbf{f}}_1 + \cdots + \tilde{\mathbf{f}}_p + \tilde{\mathbf{e}} \tag{3.72}$$

and that corresponding to (3.36) as

$$\bar{\mathbf{Y}} = \mathbf{f}_0 + \mathbf{f}_1 + \cdots + \mathbf{f}_p + \mathbf{e}, \tag{3.73}$$

where $\bar{\mathbf{Y}} = W^{-1}P^T\tilde{\mathbf{Y}}$ has the ith entry $\bar{Y}_i$. It is easy to see that $\tilde{\mathbf{f}}_\beta = P\mathbf{f}_\beta$.

(a) Verify that $PW^{-1}P^T$ is a projection matrix and $I - PW^{-1}P^T = F_3F_3^T$.

(b) Show that $\tilde{\mathbf{Y}} = F_3F_3^T\tilde{\mathbf{Y}} + P\bar{\mathbf{Y}}$, $F_3F_3^T\tilde{\mathbf{Y}} = F_3F_3^T\tilde{\mathbf{e}}$, and $W^{-1}P^T\tilde{\mathbf{e}} = \mathbf{e}$.

(c) Projecting (3.72) onto $\{\mathbf{1}_N\}^\perp$, where the subscript N indicates the length of the vector, one gets $\tilde{\mathbf{Y}}^* = \tilde{\mathbf{f}}_1^* + \cdots + \tilde{\mathbf{f}}_p^* + \tilde{\mathbf{e}}^*$. Show that

$$\begin{aligned} \tilde{\mathbf{f}}_\beta^* &= P(I - \mathbf{1}_n\mathbf{1}_n^TW/N)\mathbf{f}_\beta, \\ \tilde{\mathbf{Y}}^* &= P(I - \mathbf{1}_n\mathbf{1}_n^TW/N)\bar{\mathbf{Y}} + F_3F_3^T\tilde{\mathbf{Y}}, \\ \tilde{\mathbf{e}}^* &= P(I - \mathbf{1}_n\mathbf{1}_n^TW/N)\mathbf{e} + F_3F_3^T\tilde{\mathbf{Y}}. \end{aligned}$$

(d) Verify that $I - W^{1/2}\mathbf{1}_n\mathbf{1}_n^TW^{1/2}/N$ is the projection matrix onto $\{W^{1/2}\mathbf{1}\}^\perp$.

(e) For $(\tilde{\mathbf{a}}, \mathbf{a}) = (\tilde{\mathbf{f}}^*_\gamma, \mathbf{f}_\gamma), (\tilde{\mathbf{Y}}^*, \bar{\mathbf{Y}}), (\tilde{\mathbf{e}}^*, \mathbf{e})$, show that

$$\tilde{\mathbf{a}}^T \tilde{\mathbf{f}}^*_\beta = (W^{1/2}\mathbf{a})^T (I - W^{1/2}\mathbf{1}_n\mathbf{1}_n^T W^{1/2}/N)(W^{1/2}\mathbf{f}_\beta)$$

(f) For $(\tilde{\mathbf{a}}, \mathbf{a}), (\tilde{\mathbf{b}}, \mathbf{b}) = (\tilde{\mathbf{Y}}^*, \bar{\mathbf{Y}}), (\tilde{\mathbf{e}}^*, \mathbf{e})$, show that

$$\tilde{\mathbf{a}}^T \tilde{\mathbf{b}} = (W^{1/2}\mathbf{a})^T (I - W^{1/2}\mathbf{1}_n\mathbf{1}_n^T W^{1/2}/N)(W^{1/2}\mathbf{b}) + \tilde{\mathbf{Y}}^T F_3 F_3^T \tilde{\mathbf{Y}}.$$

3.17 Verify (3.62) and (3.63).

Section 3.7

3.18 Analyze the NO_x data of §3.7.1, with the cubic root of NO_x concentration as the response.

3.19 Analyze the NO_x data of §3.7.1, with the compression ratio treated as an ordinal factor.

3.20 Consider the ozone data of §3.7.2.

(a) Fit a tensor product linear spline in the variables `vdht`, `hmdt`, `ibht`, `dgpg`, and `vsty`, with all pairwise interactions included.

(b) Simplify the model with the help of cosine diagnostics and refit with cubic splines; iterate the process if necessary.

(c) Obtain selected main effects from the final model and compare with those illustrated in Figure 3.7.

3.21 Consider the ozone data of §3.7.2.

(a) Fit a cubic spline additive model in all variables.

(b) Simplify the model with the help of cosine diagnostics and refit; iterate the process if necessary.

(c) Obtain selected main effects from the final model and compare with those illustrated in Figure 3.7.

Section 3.8

3.22 Given a set of knots $0 < \xi_1 < \cdots < \xi_q < 1$, a natural spline is a piecewise polynomial of order $2m-1$ on $[\xi_1, \xi_q]$, $m-1$ on $[0, \xi_1]$ and $[\xi_q, 1]$, with up to the $(2m-2)$nd derivatives continuous and the $(2m-1)$st derivative jumping at the knots. Verify that a natural spline has q free parameters.

3.23 Prove the inequality $E[\mu^* - \mu]^2 < E[\tilde{\mu} - \mu]^2$, where $\mu = n^{-1}\text{tr}A(\lambda)$, $\mu^* = \mathbf{w}^T A(\lambda)\mathbf{w}/\mathbf{w}^T\mathbf{w}$, and $\tilde{\mu} = n^{-1}\mathbf{w}^T A(\lambda)\mathbf{w}$, for $\mathbf{w} \sim N(0, I)$.

(a) Show that without loss of generality, one may assume $A(\lambda)$ to be diagonal.

(b) Show that $\mathbf{w}/\sqrt{\mathbf{w}^T\mathbf{w}}$ and $\mathbf{w}^T\mathbf{w}$ are independent.

(c) For $A(\lambda) = \text{diag}(d_i)$, calculate

$$E[\mu^* - \mu]^2 = \frac{E[n^{-1}\sum d_i w_i^2 - (n^{-1}\sum d_i)(n^{-1}\sum w_i^2)]^2}{E[n^{-1}\sum w_i^2]^2},$$

and compare with $E[\tilde{\mu} - \mu]^2 = E[n^{-1}\sum d_i w_i^2 - n^{-1}\sum d_i]^2$.

4
More Splines

The framework for model construction as laid out in Chapter 2 takes as building blocks any reproducing kernel. The polynomial splines of §2.3 are the standard choices on continuous domains, but generalizations or restrictions are sometimes called for by the nature of the applications. The technical underpinnings of the variants are generally different from that of polynomial splines, but once the reproducing kernels are specified, everything else remains largely intact.

In this chapter, we present several variants of polynomial splines that have a broad range of applications. Discussed in §4.2 are splines on the circle, or periodic polynomial splines, which are often used to model periodic phenomena as well as to showcase asymptotic calculations. L-Splines are discussed in §4.3, where the null space $\mathcal{N}_J$ of the roughness penalty $J(f)$ is not restricted to lower-order polynomials. To model spatial data in a natural manner, one has at his disposal the isotropically invariant thin-plate splines on the domain $\mathcal{X} = (-\infty, \infty)^d$ (§4.4). The derivation of the reproducing kernels is the main focus of the discussion, although some advanced mathematical background is relegated to the literature.

The simple but useful idea of partial splines is also briefly discussed and illustrated (§4.1).

4.1 Partial Splines

From time to time, one may need to include certain parametric terms in a model that are not in the span of the null space $\mathcal{N}_J$ of $J(f)$. Such a need can be easily met by using the partial spline technique, in which one simply adds the needed terms to the list of fixed effect terms; see §2.5 for a discussion concerning the fixed and random effects in a spline model.

Let $\{\psi_\mu\}_{\mu=1}^q$ be a set of known functions on the domain $\mathcal{X}$. The minimizer of the functional

$$\frac{1}{n}\sum_{i=1}^n \Big(Y_i - \sum_{\mu=1}^q \beta_\mu \psi_\mu(x_i) - \eta(x_i)\Big)^2 + \lambda J(\eta) \tag{4.1}$$

with respect to β_μ and $\eta \in \mathcal{H} = \{f : J(f) < \infty\}$ is called a partial spline. Write $\mathcal{H} = \mathcal{H}_0 \oplus \mathcal{H}_J$ as in §3.1, where $\mathcal{H}_0 = \{\phi_\nu\}_{\nu=1}^m$. A partial spline is simply a regular spline that minimizes

$$\frac{1}{n}\sum_{i=1}^n (Y_i - \tilde{\eta}_0(x_i) - \eta_1(x_i))^2 + \lambda J(\eta_1) \tag{4.2}$$

for $\tilde{\eta}_0 \in \tilde{\mathcal{H}}_0$ and $\eta_1 \in \mathcal{H}_J$, where

$$\tilde{\mathcal{H}}_0 = \text{span}\{\phi_\nu, \nu = 1, \ldots, m; \psi_\mu, \mu = 1, \ldots, q\}.$$

Theorem 2.9, the solution expression (3.2) on page 60, and the computation of the fit stay intact with the augmented null space $\tilde{\mathcal{H}}_0$.

Partial spline models can be fitted using the `ssanova` facilities in the R package `gss` (§3.5.2) through the specification of an optional argument `partial`.

Example 4.1 (Cubic spline with jump)
To estimate a function $\tilde{\eta}$ that has a possible jump at a known location $x = 0.7$ but otherwise believed to be smooth on $[0, 1]$, one may minimize

$$\frac{1}{n}\sum_{i=1}^n (Y_i - \beta I_{[x_i > 0.7]} - \eta(x_i))^2 + \lambda \int_0^1 \ddot{\eta}^2 dx$$

with respect to β and $\eta \in \{f : \int_0^1 \ddot{f}^2 dx < \infty\}$. The solution has an expression

$$\tilde{\eta}(x) = \sum_{\nu=1}^3 d_\nu \phi_\nu(x) + \sum_{i=1}^n c_i R_J(x_i, x),$$

where $\phi_1(x) = 1$, $\phi_2(x) = (x - 0.5)$, $\phi_3(x) = I_{[x>0.7]}$, and $R_J(x, y) = k_2(x)k_2(y) - k_4(x - y)$; see (2.27) on page 37 for the expressions of $k_2(x)$ and $k_4(x)$.

The following R code generates some synthetic data and fits a cubic spline with jump:

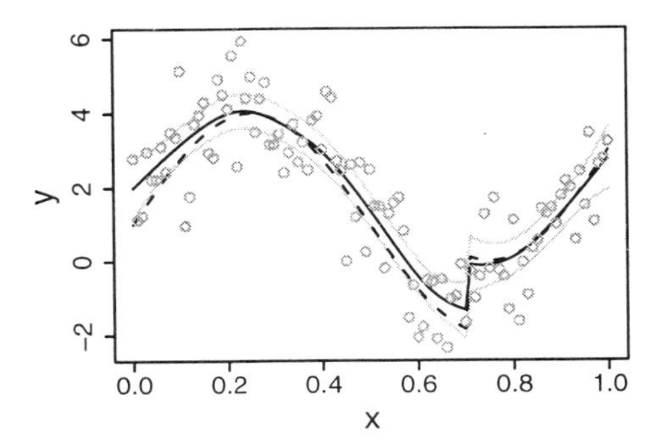

FIGURE 4.1. Cubic Spline Fit with Jump. The test function is indicated by the dashed line, the fit is indicated by the solid line, and the 95% confidence intervals are indicated by faded solid lines. The data are superimposed as circles.

```
set.seed(5732)
x <- (0:100)/100
y <- 1 + 3 * sin(2*pi*x) + 2 * (x>.7) + rnorm(x)
part.fit <- ssanova(y~x,partial=as.numeric(x>.7),ext=0)
```

The argument `ext=0` sets the domain to be the data range $[0, 1]$. The following commands evaluate the fit and plot it as shown in Figure 4.1:

```
new <- data.frame(x=x,partial=as.numeric(x>.7))
est <- predict(part.fit,new,se=TRUE)
plot(x,y); lines(x,est$fit)
lines(x,est$fit+1.96*est$se,col=5)
lines(x,est$fit-1.96*est$se,col=5)
lines(x,1+3*sin(2*pi*x)+2*(x>.7),lty=2)
```

Obviously, one should avoid using the variable name `"partial"` for anything appearing in the model formula. □

4.2 Splines on the Circle

Functions on the circle are isomorphic to periodic functions on $[0, 1]$. A periodic function $f(x)$ on $[0, 1]$ can usually be expressed in the form of a Fourier series expansion

$$f(x) = a_0 + \sum_{\mu=1}^{\infty} (a_\mu \cos 2\pi\mu x + b_\mu \sin 2\pi\mu x), \tag{4.3}$$

where $\sum_{\mu=1}^{\infty}(a_\mu^2 + b_\mu^2) < \infty$. Denote by $\mathcal{P}[0, 1]$ the linear space of all functions on $[0, 1]$ permitting the Fourier series expansion (4.3); all continuous periodic functions belong to $\mathcal{P}[0, 1]$.

In parallel to §2.3.3, we present a family of reproducing kernels on $[0,1]$ for periodic polynomial splines. With equally spaced data, a periodic polynomial spline is shown to be equivalent to a low-pass filter through an analytical spectral decomposition of the matrix Q appearing in (3.4). Assisted by such an analytical spectral decomposition, it is also possible to illustrate further details of the asymptotics of §3.2 concerning smoothing parameter selection.

4.2.1 Periodic Reproducing Kernel Hilbert Spaces on [0, 1]

Consider the space $\mathcal{H} = \{f : f \in \mathcal{P}[0,1], f^{(m)} \in \mathcal{L}_2[0,1]\}$. By the orthogonality of the trigonometric basis, it is easy to calculate

$$\int_0^1 (f^{(m)})^2 dx = \frac{1}{2}\sum_{\mu=1}^{\infty}(a_\mu^2 + b_\mu^2)(2\pi\mu)^{2m} \tag{4.4}$$

for $f \in \mathcal{P}[0,1]$, noting that $\int_0^1 \sin^2 2\pi\mu x\, dx = \int_0^1 \cos^2 2\pi\mu x\, dx = 1/2$; see Problem 4.1. Hence, $\mathcal{H} = \{f : f \in \mathcal{P}[0,1], \sum_{\mu=1}^{\infty}(a_\mu^2 + b_\mu^2)\mu^{2m} < \infty\}$. With an inner product

$$(f,g) = \left(\int_0^1 f dx\right)\left(\int_0^1 g dx\right) + \int_0^1 f^{(m)} g^{(m)} dx,$$

the reproducing kernel is seen to be

$$\begin{aligned} R(x,y) &= 1 + \sum_{\mu=1}^{\infty} \frac{2}{(2\pi\mu)^{2m}}(\cos 2\pi\mu x \cos 2\pi\mu y + \sin 2\pi\mu x \sin 2\pi\mu y) \\ &= 1 + \sum_{\mu=1}^{\infty} \frac{2\cos 2\pi\mu(x-y)}{(2\pi\mu)^{2m}}; \end{aligned} \tag{4.5}$$

see Problem 4.2. Comparing this with (2.18) on page 35, it is easy to verify that $R(x,y) = 1 + (-1)^{m-1}k_{2m}(x-y)$; see Problem 4.3. A one-way ANOVA decomposition with the averaging operator $Af = \int_0^1 f dx$ is built in with $R_0 = 1$ generating the "mean" space and $R_1 = (-1)^{m-1}k_{2m}(x-y)$ generating the "contrast" space.

4.2.2 Splines as Low-Pass Filters

Consider $Y_i = \eta(x_i) + \epsilon_i$, where $\epsilon_i \sim N(0,\sigma^2)$. The minimizer η_λ of

$$\frac{1}{n}\sum_{i=1}^{n}(Y_i - \eta(x_i))^2 + \lambda\int_0^1 (\eta^{(m)})^2 dx, \tag{4.6}$$

for $\eta \in \mathcal{H} \subset \mathcal{P}[0,1]$, is called a periodic polynomial spline. In the notation of §3.1, one has $\mathcal{N}_J = \text{span}\{1\}$ and $R_J = (-1)^{m-1}k_{2m}(x-y)$. To compute the minimizer η_λ of (4.6) via (3.4) on page 61, one has $S = \mathbf{1}$ and Q with the (i,j)th entry $(-1)^{m-1}k_{2m}(x_i - x_j)$.

Now, consider equally spaced data with $x_i = (i-1)/n$. The (i,j)th entry of Q is then $(-1)^{m-1}k_{2m}((i-j)/n)$. Substituting in the expression (2.18), straightforward algebra yields

$$\begin{aligned}(-1)^{m-1}k_{2m}((i-j)/n) &= \left(\sum_{\mu=-\infty}^{-1} + \sum_{\mu=1}^{\infty}\right)\frac{\exp(2\pi\mathbf{i}\mu(i-j)/n)}{(2\pi\mu)^{2m}} \\ &= \left(\sum_{\xi=-\infty}^{-1} + \sum_{\xi=1}^{\infty}\right)\frac{\exp(2\pi\mathbf{i}(n\xi)(i-j)/n)}{(2\pi n\xi)^{2m}} \\ &\quad + \sum_{\nu=1}^{n-1}\sum_{\xi=-\infty}^{\infty}\frac{\exp(2\pi\mathbf{i}(\nu+n\xi)(i-j)/n)}{(2\pi(\nu+n\xi))^{2m}} \\ &= \sum_{\nu=0}^{n-1}\lambda_\nu\frac{\exp(2\pi\mathbf{i}\nu(i-j)/n)}{n},\end{aligned} \tag{4.7}$$

where $\mathbf{i} = \sqrt{-1}$ and

$$\begin{aligned}\lambda_0 &= 2n(2\pi)^{-2m}\sum_{\xi=1}^{\infty}(n\xi)^{-2m} \\ \lambda_\nu &= n(2\pi)^{-2m}\sum_{\xi=-\infty}^{\infty}(\nu+n\xi)^{-2m}, \quad \nu = 1,\dots,n-1.\end{aligned} \tag{4.8}$$

Hence, one has the spectral decomposition $Q = \Gamma\Lambda\Gamma^H$, where Γ is the Fourier matrix with the (i,j)th entry $n^{-1/2}\exp(2\pi\mathbf{i}(i-1)(j-1)/n)$, Γ^H the conjugate transpose of Γ, $\Gamma^H\Gamma = \Gamma\Gamma^H = I$, and $\Lambda = \mathrm{diag}(\lambda_0,\dots,\lambda_{n-1})$; see Problem 4.4. Note that $\lambda_\nu = \lambda_{n-\nu}$, $\nu = 1,\dots,n-1$.

The operation $\tilde{\mathbf{z}} = \Gamma^H\mathbf{z}$ defines the discrete Fourier transform of $\mathbf{z}$ and $\mathbf{z} = \Gamma\tilde{\mathbf{z}}$ defines its inverse. It is easy to see that $\Gamma^H\mathbf{1} = \sqrt{n}\mathbf{e}_1$, where $\mathbf{e}_1$ is the first unit vector. Let $\tilde{Y}$ be the discrete Fourier transform of $\mathbf{Y}$ and $\tilde{\mathbf{c}}$ be that of $\mathbf{c}$. The linear system (3.4) reduces to

$$\begin{aligned}(\Lambda + n\lambda I)\tilde{\mathbf{c}} + \sqrt{n}\mathbf{e}_1 d &= \tilde{\mathbf{Y}}, \\ \tilde{c}_1 &= 0.\end{aligned}$$

Hence, one has $\tilde{c}_\nu = \tilde{Y}_\nu/(\lambda_{\nu-1} + n\lambda)$, $\nu = 2,\dots,n$. Remember that $\hat{\mathbf{Y}} = \mathbf{Y} - n\lambda\mathbf{c}$, so $\tilde{\hat{Y}}_\nu = w_\nu\tilde{Y}_\nu$, where $w_1 = 1$, $w_\nu = \lambda_{\nu-1}/(\lambda_{\nu-1} + n\lambda)$, $\nu = 2,\dots,n$. The eigenvalues λ_ν of Q monotonically decreases up to $\nu = n/2$, so a periodic spline with equally spaced data is virtually a low-pass filter.

For $n = 128$ and $m = 1,2,3$, $\log_{10}\lambda_\nu$, $\nu = 1,\dots,64$, are plotted in the left frame of Figure 4.2, and w_ν with $n\lambda = \lambda_9$, $\nu = 1,\dots,65$, are plotted in the right frame. The order m controls the shape of the filter, and the smoothing parameter λ determines the half-power frequency.

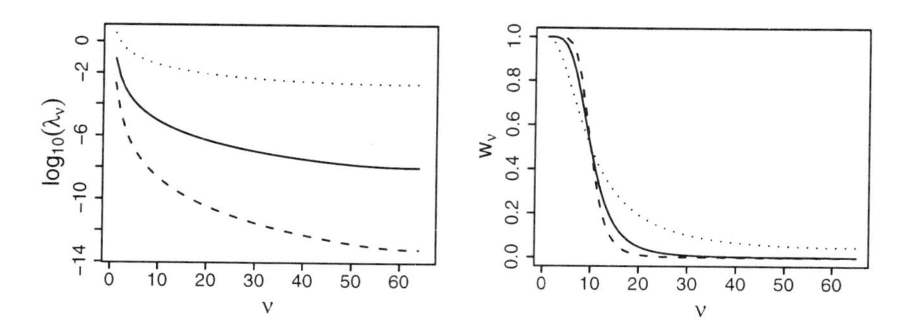

FIGURE 4.2. Splines as Low-Pass Filters. Left: Eigenvalues λ_ν of Q. Right: Damping factors w_ν with $w_{10} = .5$. The dotted lines are for $m = 1$, the solid lines for $m = 2$, and the dashed lines for $m = 3$. The sample size is $n = 128$.

4.2.3 More on Asymptotics of §3.2

Assisted by the analytical spectral decomposition $Q = \Gamma\Lambda\Gamma^H$ for periodic splines with equally spaced data, we can now look into further details of the asymptotics of §3.2 concerning smoothing parameter selection.

Write $W = \text{diag}(w_1, \dots, w_n)$, where $w_1 = 1$ and $w_\nu = \lambda_{\nu-1}/(\lambda_{\nu-1} + n\lambda)$, $\nu = 2, \dots, n$. From $\hat{\tilde{\mathbf{Y}}} = W\tilde{\mathbf{Y}}$, one has $A(\lambda) = \Gamma W \Gamma^H$. It follows that

$$\text{tr}A(\lambda) = 1 + \sum_{\nu=1}^{n-1} \frac{\lambda_\nu}{\lambda_\nu + n\lambda} = 1 + \sum_{\nu=1}^{n-1} \frac{1}{1 + \lambda\rho_\nu}, \tag{4.9}$$

where $\rho_\nu = n/\lambda_\nu$, and

$$\text{tr}A^2(\lambda) = 1 + \sum_{\nu=1}^{n-1} \frac{1}{(1 + \lambda\rho_\nu)^2}. \tag{4.10}$$

For $\nu \leq n/2$, it follows from (4.8) that $\rho_\nu = n/\lambda_\nu = h_\nu \nu^{2m}$, where $h_\nu \in (\beta_1, \beta_2)$ for some constants $0 < \beta_1 < \beta_2 < \infty$. As $\lambda \to 0$ and $n\lambda^{1/2m} \to \infty$,

$$\begin{aligned}
\text{tr}A(\lambda) &= K_1 + 2\Big(\sum_{\nu \leq \lambda^{-1/2m}} + \sum_{\lambda^{-1/2m} < \nu < n/2} \Big) \frac{1}{1 + \lambda\rho_\nu} \\
&= K_1 + K_2\lambda^{-1/2m} + K_3 \int_{\lambda^{-1/2m}}^{n/2} \frac{1}{1 + \lambda x^{2m}} dx \\
&= K_1 + K_2\lambda^{-1/2m} + K_3\lambda^{-1/2m} \int_1^\infty \frac{1}{1 + x^{2m}} dx \\
&= K_4\lambda^{-1/2m},
\end{aligned}$$

where the K_i's are constants bounded away from 0 and ∞. Similarly, one has $\text{tr}A^2(\lambda) = K_5\lambda^{-1/2m}$ for some $K_5 \in (0, \infty)$. Condition 3.2.2 of §3.2.2, that $\{n^{-1}\text{tr}A(\lambda)\}^2/n^{-1}\text{tr}A^2(\lambda) \to 0$, follows when $\lambda \to 0$, $n\lambda^{1/2m} \to \infty$.

We now calculate the risk $R(\lambda) = E[n^{-1}\sum_{i=1}^n(\eta_\lambda(x_i) - \eta(x_i))^2]$ and verify Condition 3.2.1 of §3.2.1, that $nR(\lambda) \to \infty$. From (3.16) on page 63, one has

$$R(\lambda) = \frac{1}{n}\boldsymbol{\eta}^T(I - A(\lambda))^2\boldsymbol{\eta} + \frac{\sigma^2}{n}\text{tr}A^2(\lambda) = B(\lambda) + O(n^{-1}\lambda^{-1/2m}), \quad (4.11)$$

where $\boldsymbol{\eta} = (\eta(0), \eta(1/n), \ldots, \eta((n-1)/n))^T$, with the bias term

$$\begin{aligned} B(\lambda) = \frac{1}{n}\boldsymbol{\eta}^T(I - A(\lambda))^2\boldsymbol{\eta} &= \frac{1}{n}\sum_{\nu=1}^{n-1}\frac{(n\lambda)^2}{(\lambda_\nu + n\lambda)^2}|\tilde{\eta}_{\nu+1}|^2 \\ &= \lambda\sum_{\nu=1}^{n-1}\frac{\lambda\rho_\nu}{(1+\lambda\rho_\nu)^2}\frac{\rho_\nu}{n}|\tilde{\eta}_{\nu+1}|^2, \end{aligned} \quad (4.12)$$

where $\tilde{\eta}_{\nu+1}$ is the $(\nu+1)$st entry of $\Gamma^H\boldsymbol{\eta}$. It will be shown that $B(\lambda) = O(\lambda^p)$ for some $p \in [1,2]$, so Condition 3.2.1 follows when $n\lambda^p \to \infty$ and $\lambda \to 0$. The optimal λ is of the order $O_p(n^{-2m/(2pm+1)})$, which satisfies these conditions.

We now show that $B(\lambda) = O(\lambda^p)$ for some $p \in [1,2]$. For $\eta \in \mathcal{P}[0,1]$,

$$\eta(i/n) = \tilde{a}_0 + \sum_{\mu=1}^{n-1}(\tilde{a}_\mu\cos(2\pi\mu i/n) + \tilde{b}_\mu\sin(2\pi\mu i/n)),$$

where $\tilde{a}_0 = \sum_{\xi=0}^\infty a_{n\xi}$, $\tilde{a}_\nu = \sum_{\xi=0}^\infty a_{\nu+n\xi}$, and $\tilde{b}_\nu = \sum_{\xi=0}^\infty b_{\nu+n\xi}$, $\nu = 1, \ldots, n-1$. For integers ν and μ, one has the orthogonality relations

$$\begin{aligned} &\sum_{i=1}^n\cos(2\pi\nu i/n)\cos(2\pi\mu i/n) = \frac{n}{2}\delta_{\nu,\mu}, && \nu,\mu \in [1, n/2), \\ &\sum_{i=1}^n\cos^2(2\pi\nu i/n) = n, && \nu = n/2, \\ &\sum_{i=1}^n\sin(2\pi\nu i/n)\sin(2\pi\mu i/n) = \frac{n}{2}\delta_{\nu,\mu}, && \nu,\mu \in [1, n/2), \\ &\sum_{i=1}^n\cos(2\pi\nu i/n)\sin(2\pi\mu i/n) = 0, \end{aligned} \quad (4.13)$$

where $\delta_{\nu,\mu}$ is the Kronecker delta. It follows that

$$\tilde{\eta}_{\nu+1} = \frac{\sqrt{n}}{2}\{(\tilde{a}_\nu + \tilde{a}_{n-\nu}) + \mathbf{i}(b_\nu - b_{n-\nu})\}, \quad \nu = 1, \ldots, n-1, \quad (4.14)$$

so

$$|\tilde{\eta}_{\nu+1}|^2 = \frac{n}{4}\{(\tilde{a}_\nu + \tilde{a}_{n-\nu})^2 + (\tilde{b}_\nu - \tilde{b}_{n-\nu})^2\}, \quad \nu = 1, \ldots, n-1;$$

see Problem 4.5. For $\nu > 0$, by the Cauchy-Schwarz inequality,

$$\tilde{a}_\nu^2 \le \left(\sum_{\xi=0}^{\infty} a_{\nu+n\xi}^2 (\nu + n\xi)^{2m} \right) \left(\sum_{\xi=0}^{\infty} (\nu + n\xi)^{-2m} \right),$$
$$\tilde{b}_\nu^2 \le \left(\sum_{\xi=0}^{\infty} b_{\nu+n\xi}^2 (\nu + n\xi)^{2m} \right) \left(\sum_{\xi=0}^{\infty} (\nu + n\xi)^{-2m} \right),$$

where $\sum_{\xi=0}^{\infty} (\nu+n\xi)^{-2m} = O(\lambda_\nu/n) = O(\rho_\nu^{-1})$. Since $\sum_{\mu=1}^{\infty} (a_\mu^2 + b_\mu^2)\mu^{2m} < \infty$, one has $\sum_{\nu=1}^{n-1} \rho_\nu \tilde{a}_\nu^2 < \infty$ and $\sum_{\nu=1}^{n-1} \rho_\nu \tilde{b}_\nu^2 < \infty$. It follows that

$$\sum_{\nu=1}^{n-1} \frac{\rho_\nu}{n} |\tilde{\eta}_{\nu+1}|^2 \le \frac{1}{2} \sum_{\nu=1}^{n/2} \rho_\nu \{(\tilde{a}_\nu + \tilde{a}_{n-\nu})^2 + (\tilde{b}_\nu - \tilde{b}_{n-\nu})^2\} = O(1).$$

Plugging this into (4.12) and noting that $\lambda\rho_\nu/(1 + \lambda\rho_\nu)^2 < 1$, one has $B(\lambda) = O(\lambda)$. When η is "supersmooth," in the sense that

$$\sum_{\mu=1}^{\infty} (a_\mu^2 + b_\mu^2)\mu^{2pm} < \infty \tag{4.15}$$

holds for some $p > 1$, similar calculation yields $B(\lambda) = O(\lambda^p)$, for p up to 2. When (4.15) holds for $p > 2$ but $B_2 = \lambda^{-2}B(\lambda)|_{\lambda=0} > 0$, it can be shown that $\lambda^{-2}B(\lambda) - B_2 = o(1)$ for $\lambda = o(1)$ (Problem 4.6), so $O(\lambda^2)$ is the best attainable rate for $B(\lambda)$.

Finally, let us see how the minimizer λ_m of the score $M(\lambda)$ undersmoothes when η is "supersmooth." Plugging the spectral decomposition $A(\lambda) = \Gamma W \Gamma^H$ into (3.29) on page 68, after some algebra one gets

$$M(\lambda) = \frac{\frac{1}{n} \sum_{\nu=1}^{n-1} \frac{\lambda\rho_\nu}{1 + \lambda\rho_\nu} |\tilde{Y}_{\nu+1}|^2}{\left(\prod_{\nu=1}^{n-1} \frac{\lambda\rho_\nu}{1 + \lambda\rho_\nu} \right)^{1/(n-1)}}.$$

Straightforward calculation yields

$$\begin{aligned} \frac{d \log M(\lambda)}{d \log \lambda} &= \frac{\frac{1}{n} \sum_{\nu=1}^{n-1} \frac{\lambda\rho_\nu}{(1 + \lambda\rho_\nu)^2} |\tilde{Y}_{\nu+1}|^2}{\frac{1}{n} \sum_{\nu=1}^{n-1} \frac{\lambda\rho_\nu}{1 + \lambda\rho_\nu} |\tilde{Y}_{\nu+1}|^2} - \frac{1}{n-1} \sum_{\nu=1}^{n-1} \frac{1}{1 + \lambda\rho_\nu} \\ &= \frac{N(\lambda)}{D(\lambda)} - \frac{1}{n-1} \mathrm{tr} A(\lambda), \end{aligned} \tag{4.16}$$

say; see Problem 4.7. As shown above, $(n-1)^{-1}\text{tr}A(\lambda) = K_6 n^{-1}\lambda^{-1/2m}$ for some $K_6 \in (0,\infty)$. Now

$$|\tilde{Y}_{\nu+1}|^2 = |\tilde{\eta}_{\nu+1}|^2 + |\tilde{\epsilon}_{\nu+1}|^2 + (\tilde{\eta}_{\nu+1}\bar{\tilde{\epsilon}}_{\nu+1} + \bar{\tilde{\eta}}_{\nu+1}\tilde{\epsilon}_{\nu+1}),$$

where $\bar{z}$ denotes the conjugate of complex number z, and, correspondingly, $N(\lambda)$ and $D(\lambda)$ can each be decomposed into three terms. We shall calculate the rates for the terms corresponding to $|\tilde{\eta}_{\nu+1}|^2$ and $|\tilde{\epsilon}_{\nu+1}|^2$, which control the rate of the cross-term through the Cauchy-Schwarz inequality; see Problem 4.8.

It is easy to verify that

$$\frac{1}{n}\sum_{\nu=1}^{n-1}\frac{\lambda\rho_\nu}{1+\lambda\rho_\nu}|\tilde{\eta}_{\nu+1}|^2 = O(\lambda)$$

and that

$$\frac{1}{n}\sum_{\nu=1}^{n-1}\frac{\lambda\rho_\nu}{1+\lambda\rho_\nu}|\tilde{\epsilon}_{\nu+1}|^2 = \frac{1}{n}\boldsymbol{\epsilon}^T(I-A(\lambda))\boldsymbol{\epsilon} = \sigma^2(1-\mu_1) + o_p(R(\lambda)+n^{-1}),$$

where $\mu_1 = n^{-1}\text{tr}A(\lambda)$ and (3.19) on page 64 is used. It follows that $D(\lambda) = \sigma^2(1+o_p(1))$. Similarly,

$$N_1(\lambda) = \frac{1}{n}\sum_{\nu=1}^{n-1}\frac{\lambda\rho_\nu}{(1+\lambda\rho_\nu)^2}|\tilde{\eta}_{\nu+1}|^2 = O(\lambda)$$

and

$$\frac{1}{n}\sum_{\nu=1}^{n-1}\frac{\lambda\rho_\nu}{(1+\lambda\rho_\nu)^2}|\tilde{\epsilon}_{\nu+1}|^2 = \frac{1}{n}\boldsymbol{\epsilon}^T(A(\lambda)-A^2(\lambda))\boldsymbol{\epsilon} = O_p(n^{-1}\lambda^{-1/2m}),$$

so $N(\lambda) = O_p(\lambda + n^{-1}\lambda^{-1/2m})$.

When η is "supersmooth" but $\lambda^{-1}N_1(\lambda)|_{\lambda=0} > 0$, one has $N_1(\lambda) = K_7\lambda$ for some $K_7 \in (0,\infty)$; the proof is similar to Problem 4.6. Hence, $O(\lambda)$ is the best attainable rate for $N_1(\lambda)$. Putting things together, it follows that λ cannot exceed the order of $n^{-1}\lambda^{-1/2m}$ for (4.16) to evaluate to zero. This leads to $\lambda_m = O(n^{-2m/(2m+1)})$, which is smaller than the optimal λ of order $O_p(n^{-2m/(2pm+1)})$ when $p > 1$.

4.3 L-Splines

Consider functions on $[0,1]$. Given a general differential operator L and a weight function $h(x) > 0$, the minimizer of

$$\frac{1}{n}\sum_{i=1}^{n}(Y_i - \eta(x_i))^2 + \lambda\int_0^1 (L\eta)^2(x)h(x)dx \tag{4.17}$$

is called an L-spline. The polynomial splines of §2.3 are special cases of L-splines. In applications where $\mathcal{N}_L = \{f : Lf = 0\}$ provides a more natural parametric model than a low-order polynomial, an L-spline other than a polynomial spline often provides a better estimate.

Popular examples of L-splines include trigonometric splines and Chebyshev splines, which we will discuss in §4.3.1 and §4.3.2, respectively; of interest are the characterization of the null space of L and the derivation of the reproducing kernels. A general approach to the construction of reproducing kernels for L-splines is described next (§4.3.3), and data analysis with L-splines is illustrated through a real-data example (§4.3.4). Based on a special structure in the reproducing kernel from the general construction of §4.3.3, a fast algorithm similar to that of §3.8.1 is also described for the computation of L-splines (§4.3.5).

4.3.1 Trigonometric Splines

Consider $f \in \mathcal{P}[0,1]$ periodic with $\int_0^1 f dx = a_0 = 0$. The differential operator

$$L_2 = D^2 + (2\pi)^2 \tag{4.18}$$

has a null space $\mathcal{N}_L = \text{span}\{\cos 2\pi x, \sin 2\pi x\}$. To the inner product

$$\left(\int_0^1 f(x) \cos 2\pi x dx\right) \left(\int_0^1 g(x) \cos 2\pi x dx\right) + \left(\int_0^1 f(x) \sin 2\pi x dx\right) \left(\int_0^1 g(x) \sin 2\pi x dx\right)$$

in $\mathcal{N}_L$ corresponds the reproducing kernel

$$\cos 2\pi x \cos 2\pi y + \sin 2\pi x \sin 2\pi y.$$

Take $h(x) = 1$ and define $\mathcal{H} = \{f : f \in \mathcal{P}[0,1], a_0 = 0, \int_0^2 (L_2 f)^2 dx < \infty\}$, and consider $\mathcal{H}_L = \mathcal{H} \ominus \mathcal{N}_L$ with the inner product $\int_0^1 (L_2 f)(L_2 g) dx$. Since

$$f(x) = \sum_{\mu=2}^{\infty} (a_\mu \cos 2\pi\mu x + b_\mu \sin 2\pi\mu x)$$

for $f \in \mathcal{H}_L$, the reproducing kernel of $\mathcal{H}_L$ is easily seen to be

$$\begin{aligned} R_2(x,y) &= \sum_{\mu=2}^{\infty} \frac{2}{(2\pi)^4(\mu^2-1)^2} (\cos 2\pi\mu x \cos 2\pi\mu y + \sin 2\pi\mu x \sin 2\pi\mu y) \\ &= \sum_{\mu=2}^{\infty} \frac{2 \cos 2\pi\mu(x-y)}{(2\pi)^4(\mu^2-1)^2}; \end{aligned} \tag{4.19}$$

see Problem 4.9. Note that for $f \in \mathcal{P}[0,1]$,

$$\int_0^1 (L_2 f)^2 dx = (2\pi)^4 \sum_{\mu=2}^{\infty} (a_\mu^2 + b_\mu^2)(\mu^2 - 1)^2, \tag{4.20}$$

so $\int_0^1 (L_2 f)^2 dx < \infty$ is equivalent to $\int_0^1 \ddot{f}^2 dx < \infty$; compare (4.20) with (4.4) of §4.2 for $m = 2$. Naturally, one would like to add the constant term a_0 back in as a fixed effect, which can be achieved by using $\lambda \sum_{\mu=2}^{\infty} (a_\mu^2 + b_\mu^2)(\mu^2 - 1)^2$ as the penalty term instead of $\lambda \int_0^1 (L_2 f)^2 dx$. This procedure is technically an application of the partial spline technique discussed in §4.1.

More generally, the differential operator

$$L_{2r} = (D^2 + (2\pi)^2) \cdots (D^2 + (2\pi r)^2) \tag{4.21}$$

has a null space $\mathcal{N}_L = \operatorname{span}\{\cos 2\pi\nu x, \sin 2\pi\nu x, \nu = 1, \ldots, r\}$. In the space

$$\mathcal{H}_L = \{f : f = \textstyle\sum_{\mu=r+1}^{\infty} (a_\mu \cos 2\pi\mu x + b_\mu \sin 2\pi\mu x), \int_0^1 (L_{2r} f)^2 dx < \infty\}$$

with the inner product $\int_0^1 (L_{2r} f)(L_{2r} g) dx$, the reproducing kernel is seen to be

$$R_{2r}(x, y) = \sum_{\mu=r+1}^{\infty} \frac{2 \cos 2\pi\mu(x - y)}{(2\pi)^{4r} (\mu^2 - 1)^2 \cdots (\mu^2 - r^2)^2}; \tag{4.22}$$

see Problem 4.10.

With the differential operator

$$L_3 = D(D^2 + (2\pi)^2), \tag{4.23}$$

the null space $\mathcal{N}_L = \operatorname{span}\{1, \cos 2\pi x, \sin 2\pi x\}$ automatically contains the constant term. To the inner product

$$\left(\int_0^1 f dx\right)\left(\int_0^1 g dx\right) + \left(\int_0^1 f(x) \cos 2\pi x dx\right)\left(\int_0^1 g(x) \cos 2\pi x dx\right) + \left(\int_0^1 f(x) \sin 2\pi x dx\right)\left(\int_0^1 g(x) \sin 2\pi x dx\right)$$

in $\mathcal{N}_L$ corresponds the reproducing kernel

$$1 + \cos 2\pi x \cos 2\pi y + \sin 2\pi x \sin 2\pi y.$$

Take $h(x) = 1$ and define $\mathcal{H} = \{f : f \in \mathcal{P}[0,1], \int_0^1 (L_3 f)^2 dx < \infty\}$. Corresponding to the inner product $\int_0^1 (L_3 f)(L_3 g) dx$, the reproducing kernel of $\mathcal{H}_L = \mathcal{H} \ominus \mathcal{N}_L$ is seen to be

$$R_3(x, y) = \sum_{\mu=2}^{\infty} \frac{2 \cos 2\pi\mu(x - y)}{(2\pi)^6 \mu^2 (\mu^2 - 1)^2}; \tag{4.24}$$

see Problem 4.11. For $f \in \mathcal{P}[0,1]$,

$$\int_0^1 (L_3 f)^2 dx = (2\pi)^6 \sum_{\mu=2}^{\infty} (a_\mu^2 + b_\mu^2)\mu^2(\mu^2 - 1)^2, \tag{4.25}$$

so $\int_0^1 (L_3 f)^2 dx < \infty$ is equivalent to $\int_0^1 (f^{(3)})^2 dx < \infty$; compare (4.25) with (4.4) of §4.2 for $m = 3$.

In general, the differential operator

$$L_{2r+1} = D(D^2 + (2\pi)^2) \cdots (D^2 + (2\pi r)^2) \tag{4.26}$$

has a null space $\mathcal{N}_L = \text{span}\{1, \cos 2\pi\nu x, \sin 2\pi\nu x, \nu = 1, \dots, r\}$. In the space

$$\mathcal{H}_L = \{f : f = \textstyle\sum_{\mu=r+1}^{\infty} (a_\mu \cos 2\pi\mu x + b_\mu \sin 2\pi\mu x), \int_0^1 (L_{2r+1} f)^2 dx < \infty\}$$

with the inner product $\int_0^1 (L_{2r+1} f)(L_{2r+1} g) dx$, the reproducing kernel is given by

$$R_{2r+1}(x, y) = \sum_{\mu=r+1}^{\infty} \frac{2\cos 2\pi\mu(x-y)}{(2\pi)^{4r+2}\mu^2(\mu^2-1)^2 \cdots (\mu^2 - r^2)^2}; \tag{4.27}$$

see Problem 4.12.

4.3.2 Chebyshev Splines

Let $w_i(x) \in \mathcal{C}^{(m-i+1)}[0,1]$, $i = 1, \dots, m$, be strictly positive functions with $w_i(0) = 1$. Consider the differential operator

$$L_m = D_m \cdots D_1, \tag{4.28}$$

where $D_i f = D(f/w_i)$.

The null space $\mathcal{N}_L$ of L_m is spanned by

$$\begin{aligned}
\phi_1(x) &= w_1(x) \\
\phi_2(x) &= w_1(x) \int_0^x w_2(t_2) dt_2 \\
&\vdots \\
\phi_m(x) &= w_1(x) \int_0^x w_2(t_2) dt_2 \cdots \int_0^{t_{m-1}} w_m(t_m) dt_m,
\end{aligned} \tag{4.29}$$

which form a so-called Chebyshev system on $[0,1]$, in the sense that

$$\det[\phi_j(x_i)]_{i,j=1}^m > 0 \quad \text{for all } x_1 < x_2 < \cdots < x_m,\ [x_1, x_m] \subseteq [0,1];$$

see Schumaker (1981, §2.5, Theorem 9.2). The functions ϕ_ν in (4.29) also form an extended Chebyshev system, in the sense that

$$\det[\phi_j^{(i-1)}(x)]_{i,j=1}^m > 0, \quad \forall x \in [0,1];$$

see Karlin and Studden (1966, §1.2, Theorem 1.2 on page 379). The matrix

$$[\phi_j^{(i-1)}(x)]_{i,j=1}^m = \begin{pmatrix} \phi_1(x) & \phi_2(x) & \cdots & \phi_m(x) \\ \dot{\phi}_1(x) & \dot{\phi}_2(x) & \cdots & \dot{\phi}_m(x) \\ \vdots & \vdots & & \vdots \\ \phi_1^{(m-1)}(x) & \phi_2^{(m-1)}(x) & \cdots & \phi_m^{(m-1)}(x) \end{pmatrix},$$

is known as the Wronskian matrix of $\phi = (\phi_1, \ldots, \phi_m)^T$. Write $L_0 = I$, $L_1 = D_1$, …, $L_{m-1} = D_{m-1} \cdots D_1$. One has $(L_\mu \phi_\nu)(0) = \delta_{\mu+1,\nu}$, $\mu = 0, \ldots, m-1$, $\nu = 1, \ldots, m$, where $\delta_{\mu,\nu}$ is the Kronecker delta. It follows that $\sum_{\nu=1}^m \phi_\nu(x)\phi_\nu(y)$ is the reproducing kernel of $\mathcal{N}_L$ corresponding to the inner product

$$\sum_{\nu=1}^m (L_{\nu-1}f)(0)(L_{\nu-1}g)(0).$$

Actually, $\{\phi_\nu\}_{\nu=1}^m$ form an orthonormal basis of $\mathcal{N}_L$ under the given inner product.

Define $\mathcal{H} = \{f : \int_0^1 (L_m f)^2 h dx < \infty\}$ and denote $\mathcal{H}_L = \mathcal{H} \ominus \mathcal{N}_L$. For $f \in \mathcal{H}_L$, noting that $(L_\nu f)(0) = 0$, $\nu = 0, \ldots, m-1$, it is straightforward to verify that

$$\begin{aligned} f(x) &= w_1(x) \int_0^x w_2(t_2)dt_2 \cdots \int_0^{t_{m-1}} w_m(t_m)dt_m \int_0^{t_m} (L_m f)(u)du \\ &= \int_0^x (L_m f)(u)du \left\{ w_1(x) \int_u^x w_2(t_2)dt_2 \cdots \int_u^{t_{m-1}} w_m(t_m)dt_m \right\} \\ &= \int_0^x G(x;u)(L_m f)(u)du, \end{aligned} \tag{4.30}$$

where

$$G(x;u) = \begin{cases} w_1(x) \int_u^x w_2(t_2)dt_2 \cdots \int_u^{t_{m-1}} w_m(t_m)dt_m, & u \le x, \\ 0, & u > x; \end{cases} \tag{4.31}$$

see Problem 4.13. The function $G(x;u)$ is called a Green's function associated with the differential operator L_m. After some algebra, one has the expression

$$G(x;u) = \begin{cases} \sum_{\nu=1}^m \phi_\nu(x)\psi_\nu(u), & u \le x, \\ 0, & u > x, \end{cases} \tag{4.32}$$

where

$$\psi_\nu(u) = -\int_0^u w_{\nu+1}(t_{\nu+1})dt_{\nu+1} \times \int_u^{t_{\nu+1}} w_{\nu+2}(t_{\nu+2})dt_{\nu+2} \cdots \int_u^{t_{m-1}} w_m(t_m)dt_m,$$

$\nu = 1, \dots, m-2$, $\psi_{m-1}(u) = -\int_0^u w_m(t_m)dt_m$, and $\psi_m(u) = 1$ (Problem 4.14). Write

$$R_x(y) = \int_0^1 G(x;u)G(y;u)(h(u))^{-1}du.$$

It is straightforward to verify that $(L_\nu R_x)(0) = 0$, $\nu = 0, \dots, m-1$, and that $(L_m R_x)(y) = G(x;y)/h(y)$; see Problem 4.15. Hence, by (4.30), the reproducing kernel in $\mathcal{H}_L$ corresponding to an inner product $\int_0^1 (L_m f)(L_m g)h dx$ is given by

$$R_L(x,y) = \int_0^1 G(x;u)G(y;u)(h(u))^{-1}du. \tag{4.33}$$

By Theorem 2.5, the reproducing kernel of $\mathcal{H}$ under the inner product

$$\sum_{\nu=1}^m (L_{\nu-1}f)(0)(L_{\nu-1}g)(0) + \int_0^1 (L_m f)(L_m g)h dx$$

is seen to be

$$R(x,y) = \sum_{\nu=1}^m \phi_\nu(x)\phi_\nu(y) + \int_0^1 G(x;u)G(y;u)(h(u))^{-1}du.$$

Parallel to (2.6) on page 32, one has, for $f \in \mathcal{H}$, the generalized Taylor expansion,

$$f(x) = \sum_{\nu=1}^m (L_{\nu-1}f)(0)\phi_\nu(x) + \int_0^x G(x;u)(L_m f)(u)du.$$

Since $G(x;u) = 0$, $u > x$, one may rewrite (4.33) as

$$R_L(x,y) = \int_0^{x \wedge y} G(x;u)G(y;u)(h(u))^{-1}du.$$

It is easy to see that the calculus of this section applies on any domain of the form $[0,a]$, where a is not necessarily scaled to 1.

Example 4.2 (Polynomial splines)
Setting $w_i(x) = 1$, $i = 1, \dots, m$, and $h(x) = 1$, one gets the polynomial splines of §2.3.1; see Problem 4.16. □

Example 4.3 (Exponential splines)
Setting $w_i(x) = e^{\beta_i x}$, $i = 1, \ldots, m$, where $\beta_i \geq 0$ with the strict inequality holding for $i > 1$, one gets the so-called exponential splines; see, e.g., Schumaker (1981, §9.9). Denote $\alpha_i = \sum_{j=1}^{i} \beta_j$. It is easy to verify that

$$L_\nu = e^{-\alpha_\nu x}(D - \alpha_\nu) \cdots (D - \alpha_1), \quad \nu = 1, \ldots, m,$$

and that L_m has the null space $\mathcal{N}_L = \text{span}\{e^{\alpha_i x}, i = 1, \ldots, m\}$.

As a specific case, consider $m = 2$, $\beta_1 = 0$, and $\beta_2 = \theta$. One has $L_2 = e^{-\theta x}(D - \theta)D$ with the null space $\mathcal{N}_L = \text{span}\{1, e^{\theta x}\}$. The orthonormal basis of $\mathcal{N}_L$ consists of $\phi_1 = 1$ and $\phi_2 = (e^{\theta x} - 1)/\theta$. Now,

$$G(x; u) = \int_u^x e^{\theta t} dt = \theta^{-1}(e^{\theta x} - e^{\theta u}) = \phi_2(x) - \phi_2(u), \quad u \leq x,$$

so

$$\begin{aligned} R_L(x, y) &= \int_0^{x \wedge y} G(x; u) G(y; u) (h(u))^{-1} du \\ &= \int_0^{x \wedge y} (\phi_2(x) - \phi_2(u))(\phi_2(y) - \phi_2(u))(h(u))^{-1} du \end{aligned}$$

The generalized Taylor expansion is seen to be

$$f(x) = f(0) + \dot{f}(0)\phi_2(x) + \int_0^x (\phi_2(x) - \phi_2(u)) e^{-\theta u}(\ddot{f}(u) - \theta \dot{f}(u)) du, \quad (4.34)$$

which, after a change of variable $\tilde{x} = \phi_2(x)$, reduces to

$$g(\tilde{x}) = g(0) + \dot{g}(0)\tilde{x} + \int_0^{\tilde{x}} (\tilde{x} - \tilde{u}) \ddot{g}(\tilde{u}) d\tilde{u}, \quad (4.35)$$

where $g(\tilde{x}) = f(\phi_2^{-1}(\tilde{x}))$ for ϕ_2^{-1} the inverse of ϕ_2; see Problem 4.17. With $1/h(x) = d\phi_2/dx = e^{\theta x}$,

$$\begin{aligned} R_L(x, y) &= \int_0^{x \wedge y} G(x; u) G(y; u) \frac{d\phi_2(u)}{du} du \\ &= \int_0^{x \wedge y} (\phi_2(x) - \phi_2(u))(\phi_2(y) - \phi_2(u)) d\phi_2(u), \end{aligned} \quad (4.36)$$

so the formulation virtually yields a cubic spline in $\tilde{x} = \phi_2(x)$; compare (4.36) with (2.10) on page 33 for $m = 2$.

More generally, an exponential spline on $[0, a]$ with $\beta_1 = 0$, $\beta_i = \theta$, $i = 2, \ldots, m$, and $h(x) = e^{-\theta x}$ reduces to a polynomial spline in $\tilde{x} = \phi_2(x)$ with a penalty proportional to $\int_0^{\phi_2(a)} (g^{(m)}(\tilde{x}))^2 d\tilde{x}$; see Problem 4.18. □

Example 4.4 (Hyperbolic splines)
For $m = 2r$, let $\beta_1 = 0$, $\beta_i > 0$, $i = 2, \ldots, r$, and denote $\alpha_i = \sum_{j=1}^{i} \beta_j$, $i = 1, \ldots, r$. Setting $w_i(x) = e^{\beta_i x}$, $i = 1, \ldots, r$, $w_{r+1}(x) = e^{-2\alpha_r x}$, and $w_{r+i}(x) = w_{r-i+2}(x)$, $i = 2, \ldots, r$, one gets the so-called hyperbolic splines; see Schumaker (1981, §9.9). It is straightforward to verify that

$$\begin{aligned} L_\nu &= e^{-\alpha_\nu x}(D - \alpha_\nu) \cdots (D - \alpha_1), && \nu = 1, \ldots, r, \\ L_{2r-\nu+1} &= e^{\alpha_\nu x}(D + \alpha_\nu) \cdots (D + \alpha_r) \\ &\qquad \times (D - \alpha_r) \cdots (D - \alpha_1), && \nu = r, \ldots, 1. \end{aligned}$$

The differential operator

$$L_m = D(D + \alpha_2) \cdots (D + \alpha_r)(D - \alpha_r) \cdots (D - \alpha_2)D$$

has the null space $\mathcal{N}_L = \text{span}\{1, x, e^{-\alpha_\nu x}, e^{\alpha_\nu x}, \nu = 2, \ldots, r\}$.

Consider the case with $r = 2$ and $\beta_2 = \theta$. One has $L_4 = D(D+\theta)(D-\theta)D$ with the null space $\mathcal{N}_L = \text{span}\{1, x, e^{-\theta x}, e^{\theta x}\}$. The orthonormal basis of $\mathcal{N}_L$ consists of $\phi_1 = 1$, $\phi_2 = (e^{\theta x} - 1)/\theta$, $\phi_3 = (\cosh\theta x - 1)/\theta^2$, and $\phi_4 = (\sinh\theta x - \theta x)/\theta^3$. The Green's function is

$$G(x, u) = (\sinh\theta(x - u) - \theta(x - u))/\theta^3,$$

for $u \le x$; see Problem 4.19.

More generally, with $\beta_i = \theta$, $i = 2, \ldots, r$, one can show that, for $\phi_2 = (e^{\theta x} - 1)/\theta$,

$$\begin{aligned} \phi_\nu &= \frac{\phi_2^{\nu-1}(x)}{(\nu - 1)!}, \\ \phi_{r+\nu} &= \int_0^x \frac{\phi_2^{\nu-1}(v)}{(\nu - 1)!} \frac{(\phi_2(x) - \phi_2(v))^{r-1}}{(r - 1)!} \frac{d\phi_2(v)}{(1 + \theta\phi_2(v))^{2r-1}}, \end{aligned} \tag{4.37}$$

$\nu = 1, \ldots, r$, and that

$$G(x; u) = \int_u^x \frac{(\phi_2(v) - \phi_2(u))^{r-1}}{(r - 1)!} \frac{(\phi_2(x) - \phi_2(v))^{r-1}}{(r - 1)!} \frac{d\phi_2(v)}{(1 + \theta\phi_2(v))^{2r-1}}, \tag{4.38}$$

for $u \le x$; see Problem 4.20. □

4.3.3 General Construction

Consider a differential operator of the form

$$L = D^m + \sum_{j=0}^{m-1} a_j(x) D^j. \tag{4.39}$$

This effectively covers the operator L_m of (4.28) as a special case, which can be written as

$$L_m = \big\{ \prod_{i=1}^m w_i(x) \big\}^{-1} \big(D^m + \sum_{j=0}^{m-1} a_j(x) D^j \big),$$

since the factor $\{\prod_{i=1}^m w_i(x)\}^{-1}$ can be absorbed into the weight function $h(x)$. When $a_j \in \mathcal{C}^{(m-j)}[0,1]$, it is known that the null space of L, $\mathcal{N}_L = \{f : Lf = 0\}$, is an m-dimensional linear subspace of infinitely differentiable functions; see Schumaker (1981, §10.1). Let ϕ_ν, $\nu = 1, \ldots, m$, be a basis of such an $\mathcal{N}_L$. The Wronskian matrix of $\boldsymbol{\phi} = (\phi_1, \ldots, \phi_m)^T$,

$$W(\boldsymbol{\phi})(x) = \begin{pmatrix} \phi_1(x) & \phi_2(x) & \cdots & \phi_m(x) \\ \dot{\phi}_1(x) & \dot{\phi}_2(x) & \cdots & \dot{\phi}_m(x) \\ \vdots & \vdots & & \vdots \\ \phi_1^{(m-1)}(x) & \phi_2^{(m-1)}(x) & \cdots & \phi_m^{(m-1)}(x) \end{pmatrix},$$

is known to be nonsingular, $\forall x \in [0,1]$; see Schumaker (1981, §10.1). Since $W(\boldsymbol{\phi})(0)$ is invertible, $\sum_{\nu=1}^m f^{(\nu-1)}(0) g^{(\nu-1)}(0)$ forms an inner product in $\mathcal{N}_L$ (Problem 4.21). Define $\tilde{\boldsymbol{\phi}} = [W(\boldsymbol{\phi})(0)]^{-T} \boldsymbol{\phi}$. It is easy to verify that $\tilde{\phi}_\nu^{(\mu-1)}(0) = \delta_{\mu,\nu}$, $\mu, \nu = 1, \ldots, m$, so $\tilde{\phi}_\nu$, $\nu = 1, \ldots, m$, form an orthonormal basis of $\mathcal{N}_L$ and $\sum_{\nu=1}^m \tilde{\phi}_\nu(x) \tilde{\phi}_\nu(y)$ is its reproducing kernel.

An m-dimensional function space on an interval is called a Chebyshev space if it has a basis that is a Chebyshev system on the interval; see Schumaker (1981, §2.5). A function in an m-dimensional Chebyshev space is uniquely determined by its values on m distinctive points on the interval. The space $\mathcal{N}_L$ may not be a Chebyshev space on $[0,1]$, but for some $\delta > 0$, it is always a Chebyshev space on intervals shorter than δ; see Schumaker (1981, Theorem 10.5).

Define $\mathcal{H} = \{f : \int_0^1 (Lf)^2 h dx < \infty\}$ and $\mathcal{H}_L = \mathcal{H} \ominus \mathcal{N}_L$. Let $\psi_\nu(x)$, $\nu = 1, \ldots, m$, be the entries of the last column of $[W(\boldsymbol{\phi})(x)]^{-1}$. It is easy to see that

$$\begin{aligned} &\sum_{\nu=1}^m \phi_\nu^{(j)}(x) \psi_\nu(x) = 0, \quad j = 0, \ldots, m-2, \\ &\sum_{\nu=1}^m \phi_\nu^{(m-1)}(x) \psi_\nu(x) = 1. \end{aligned} \tag{4.40}$$

Write

$$G(x;u) = \begin{cases} \sum_{\nu=1}^m \phi_\nu(x) \psi_\nu(u), & u \le x, \\ 0, & u > x; \end{cases} \tag{4.41}$$

we show that $G(x;u)$ is a Green's function associated with L in (4.39). For $g \in \mathcal{L}_2[0,1]$, define

$$\tilde{g}(x) = \int_0^1 G(x;u) g(u) du.$$

Using (4.40), it is easy to calculate

$$\begin{aligned}\tilde{g}^{(j)}(x) &= \sum_{\nu=1}^{m} \phi_\nu^{(j)}(x) \int_0^x \psi_\nu(u) g(u) du, \quad j = 0, \ldots, m-1, \\ \tilde{g}^{(m)}(x) &= \sum_{\nu=1}^{m} \phi_\nu^{(m)}(x) \int_0^x \psi_\nu(u) g(u) du + g(x);\end{aligned} \tag{4.42}$$

see Problem 4.22. Hence, $\tilde{g}^{(j)}(0) = 0$, $j = 0, \ldots, m-1$, and since $\phi_\nu^{(m)}(x) + \sum_{j=0}^{m-1} a_j(x)\phi_\nu^{(j)}(x) = 0$ as $\phi_\nu \in \mathcal{N}_L$, $L\tilde{g} = g$. It follows that for $f \in \mathcal{H}_L$,

$$f(x) = \int_0^x G(x;u)(Lf)(u)du,$$

and corresponding to the inner product $\int_0^1 (Lf)(Lg)hdx$, one has the reproducing kernel

$$R_L(x,y) = \int_0^{x \wedge y} G(x;u)G(y;u)(h(u))^{-1}du. \tag{4.43}$$

For $f \in \mathcal{H}$, one has the generalized Taylor expansion

$$f(x) = \sum_{\nu=1}^{m} f^{(\nu-1)}(0)\tilde{\phi}_\nu(x) + \int_0^x G(x;u)(Lf)(u)du.$$

Example 4.5 (Cubic spline)
Consider $L = D^2$ with $\phi_1(x) = 1$ and $\phi_2(x) = x$. The Wronskian matrix and its inverse are respectively

$$W(\phi)(x) = \begin{pmatrix} 1 & x \\ 0 & 1 \end{pmatrix} \quad \text{and} \quad [W(\phi)(x)]^{-1} = \begin{pmatrix} 1 & -x \\ 0 & 1 \end{pmatrix}.$$

One has $\tilde{\phi}_1 = \phi_1$, $\tilde{\phi}_2 = \phi_2$, and $G(x;u) = x - u$ for $u \leq x$. The results coincide with those derived in §2.3.1 and Example 4.2. □

Example 4.6 (Exponential spline)
Consider $L = (D - \theta)D$ for $\theta > 0$ with $\phi_1(x) = 1$ and $\phi_2(x) = e^{\theta x}$. The Wronskian matrix and its inverse are respectively

$$W(\phi)(x) = \begin{pmatrix} 1 & e^{\theta x} \\ 0 & \theta e^{\theta x} \end{pmatrix} \quad \text{and} \quad [W(\phi)(x)]^{-1} = \begin{pmatrix} 1 & -\theta^{-1} \\ 0 & \theta^{-1}e^{-\theta x} \end{pmatrix}.$$

One has $\tilde{\phi}_1(x) = 1$, $\tilde{\phi}_2(x) = (e^{\theta x} - 1)/\theta$, and

$$G(x;u) = e^{-\theta u}(\tilde{\phi}_2(x) - \tilde{\phi}_2(u))$$

for $u \leq x$. The results agree with those of Example 4.3 for $m = 2$, after adjusting for the factor $e^{-\theta x}$ appearing in the operator $L_2 = e^{-\theta x}(D - \theta)D$ of Example 4.3. □

Example 4.7
Consider $L = (D+\theta)D$ for $\theta > 0$ with $\phi_1(x) = 1$ and $\phi_2(x) = e^{-\theta x}$. Substituting $-\theta$ for θ in Example 4.6, one has $\tilde{\phi}_1(x) = 1$, $\tilde{\phi}_2(x) = (1 - e^{-\theta x})/\theta$, and

$$G(x;u) = e^{\theta u}(\tilde{\phi}_2(x) - \tilde{\phi}_2(u))$$

for $u \le x$. With a weight function $h(x) = e^{3\theta x}$, one obtains a cubic spline in $\tilde{\phi}_2(x)$. □

Example 4.8 (Trigonometric splines)
Consider $L = D^2 + (2\pi)^2$ with $\phi_1(x) = \sin 2\pi x$ and $\phi_2(x) = \cos 2\pi x$. The Wronskian matrix and its inverse are respectively

$$W(\boldsymbol{\phi})(x) = \begin{pmatrix} \sin 2\pi x & \cos 2\pi x \\ (2\pi)\cos 2\pi x & -(2\pi)\sin 2\pi x \end{pmatrix}$$

and

$$[W(\boldsymbol{\phi})(x)]^{-1} = \begin{pmatrix} \sin 2\pi x & (2\pi)^{-1}\cos 2\pi x \\ \cos 2\pi x & -(2\pi)^{-1}\sin 2\pi x \end{pmatrix}.$$

One has $\tilde{\phi}_1 = \cos 2\pi x$, $\tilde{\phi}_2(x) = (2\pi)^{-1}\sin 2\pi x$, and

$$G(x;u) = \frac{1}{2\pi}\sin 2\pi(x-u)$$

for $u \le x$. The reproducing kernel of $\mathcal{H}_L$ corresponding to the inner product $\int_0^1 (Lf)(Lg)dx$ is thus

$$\begin{aligned} R_L &= \frac{1}{(2\pi)^2}\int_0^{x\wedge y} \sin 2\pi(x-u)\sin 2\pi(y-u)du \\ &= \frac{(x\wedge y)\cos 2\pi(x-y)}{2(2\pi)^2} - \frac{\sin 2\pi(x+y) - \sin 2\pi|x-y|}{4(2\pi)^3}. \end{aligned} \tag{4.44}$$

This reproducing kernel is different from the one given in (4.19) of §4.3.1, where the constant and the nonperiodic functions are excluded.

Now, consider $L = D(D^2 + (2\pi)^2)$ with $\phi_1(x) = 1$, $\phi_2(x) = \sin 2\pi x$, and $\phi_3(x) = \cos 2\pi x$. The Wronskian matrix and its inverse are respectively

$$W(\boldsymbol{\phi})(x) = \begin{pmatrix} 1 & \sin 2\pi x & \cos 2\pi x \\ 0 & (2\pi)\cos 2\pi x & -(2\pi)\sin 2\pi x \\ 0 & -(2\pi)^2\sin 2\pi x & -(2\pi)^2\cos 2\pi x \end{pmatrix}$$

and

$$[W(\boldsymbol{\phi})(x)]^{-1} = \begin{pmatrix} 1 & 0 & (2\pi)^{-2} \\ 0 & (2\pi)^{-1}\cos 2\pi x & -(2\pi)^{-2}\sin 2\pi x \\ 0 & -(2\pi)^{-1}\sin 2\pi x & -(2\pi)^{-2}\cos 2\pi x \end{pmatrix}.$$

One has $\tilde{\phi}_1(x) = 1$, $\tilde{\phi}_2(x) = (2\pi)^{-1}\sin 2\pi x$, $\tilde{\phi}_3(x) = (2\pi)^{-2}(1 - \cos 2\pi x)$, and

$$G(x;u) = \frac{1}{(2\pi)^2}(1 - \cos 2\pi(x-u))$$

for $u \leq x$. The reproducing kernel of $\mathcal{H}_L$ corresponding to the inner product $\int_0^1 (Lf)(Lg)dx$ is thus

$$\begin{aligned} R_L &= \frac{1}{(2\pi)^4}\int_0^{x\wedge y}(1-\cos 2\pi(u-x))(1-\cos 2\pi(u-y))du \\ &= \frac{x\wedge y}{(2\pi)^4} - \frac{\sin 2\pi x + \sin 2\pi y - \sin 2\pi|x-y|}{(2\pi)^5} \\ &\quad + \frac{(x\wedge y)\cos 2\pi(x-y)}{2(2\pi)^4} + \frac{\sin 2\pi(x+y) - \sin 2\pi|x-y|}{4(2\pi)^5} \end{aligned} \tag{4.45}$$

This reproducing kernel is different from the one given in (4.24) of §4.3.1, where the nonperiodic functions are excluded. □

Example 4.9 (Logistic spline)
Consider $D(D - \gamma\theta e^{-\theta x}/(1+\gamma e^{-\theta x}))$ for $\theta, \gamma > 0$, with $\phi_1(x) = 1$ and $\phi_2(x) = 1/(1+\gamma e^{-\theta x})$. The Wronskian matrix and its inverse are respectively

$$W(\phi)(x) = \begin{pmatrix} 1 & (1+\gamma e^{-\theta x})^{-1} \\ 0 & \gamma\theta e^{-\theta x}(1+\gamma e^{-\theta x})^{-2} \end{pmatrix}$$

and

$$[W(\phi)(x)]^{-1} = \begin{pmatrix} 1 & -(\gamma\theta)^{-1}e^{\theta x}(1+\gamma e^{-\theta x}) \\ 0 & (\gamma\theta)^{-1}e^{\theta x}(1+\gamma e^{-\theta x})^2 \end{pmatrix}.$$

One has $\tilde{\phi}_1(x) = 1$,

$$\tilde{\phi}_2(x) = \frac{(1+\gamma)^2}{\gamma\theta}\left(\frac{1}{1+\gamma e^{-\theta x}} - \frac{1}{1+\gamma}\right),$$

and

$$G(x;u) = \frac{e^{\theta u}(1+\gamma e^{-\theta u})^2}{(1+\gamma)^2}(\tilde{\phi}_2(x) - \tilde{\phi}_2(u))$$

for $u \leq x$. With a weight function $h(x) \propto e^{3\theta x}(1+\gamma e^{-\theta x})^6$, one gets a cubic spline in $\tilde{\phi}_2(x)$. □

4.3.4 Case Study: Weight Loss of Obese Patient

Obese patients on a weight rehabilitation program tend to lose adipose tissue at a diminishing rate as the treatment progresses. A data set concerning the weight loss of a male patient can be found in the R package `MASS`, as a data frame `wtloss` with two components, `Weight` and `Days`. A nonlinear regression model was considered in Venables and Ripley (1999, Chapter 8),

$$Y = \beta_0 + \beta_1 2^{-x/\theta} + \epsilon, \tag{4.46}$$

where Y was the weight at x days after the start of the rehabilitation program. The least squares estimates of the parameters were given by $\hat{\beta}_0 = 81.374$, $\hat{\beta}_1 = 102.68$, and $\hat{\theta} = 141.91$. The parameter β_0 may be interpreted as the ultimate lean weight, β_1 the total amount to be lost, and θ the "half-decay" time.

Note that $2^{-x/\theta} = e^{-\tilde{\theta}x}$ with $\tilde{\theta} = \log 2/\theta$. The nonlinear model (4.46) is in the null space of the differential operator $L = (D+\theta)D$ considered in Example 4.7. To allow for possible departures from the parametric model, we consider a cubic spline in $e^{-\tilde{\theta}x}$, which is an L-spline with $L = (D+\tilde{\theta})D$ and $h(x) = e^{3\tilde{\theta}x}$. Fixing $\tilde{\theta}$, the smoothing parameter can be selected using the GCV score $V(\lambda)$ of (3.23), and to choose $\tilde{\theta}$, one may compare the minimum $V(\lambda)$ scores obtained with different $\tilde{\theta}$. Note that Theorem 3.3 is still useful in this situation. The R code below finds the GCV estimate of the parameter $\tilde{\theta}$:

```
library(MASS); data(wtloss)
tmp.fun <- function(theta) {
    theta <- theta/100
    ssanova(Weight~exp(-theta*Days),data=wtloss)$score
}
nlm(tmp.fun,1)$estimate
# 0.4884628
```

The `tmp.fun` function returns the minimum $V(\lambda)$ score for fixed $\tilde{\theta}$. The `nlm` function finds the minimal point of `tmp.fun` using a Newton-type algorithm with numerical derivatives; see Dennis and Schnabel (1996) for algorithmic details. The scaling of `theta` in `tmp.fun` was introduced so that `nlm` would use appropriate differencing steps for the calculation of numerical derivatives. The solution corresponds to $\theta = \log(2)/0.004884628 = 141.9038$, matching the least squares estimate in the parametric model. The minimum $V(\lambda)$ for $\tilde{\theta} = 0.004885$ is 0.8166.

The fit with $\tilde{\theta} = 0.004885$ can now be calculated and plotted as the solid line in the left frame of Figure 4.3, which is indistinguishable from the parametric fit plotted as the dashed line; the data are superimposed as circles. A cubic spline in x is also calculated and superimposed as the long dashed line, which is nearly indistinguishable from the other two fits; the minimum $V(\lambda)$ for the cubic spline is 0.9283.

```
# calculate the L-spline and cubic spline fits
wtloss$dd <- exp(-.004885*wtloss$Days)
wtloss.fit1 <- ssanova(Weight~dd,data=wtloss)
wtloss.fit2 <- ssanova(Weight~Days,data=wtloss)
```

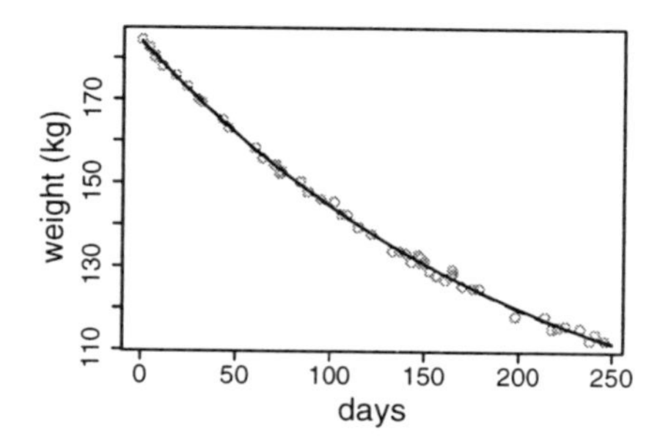

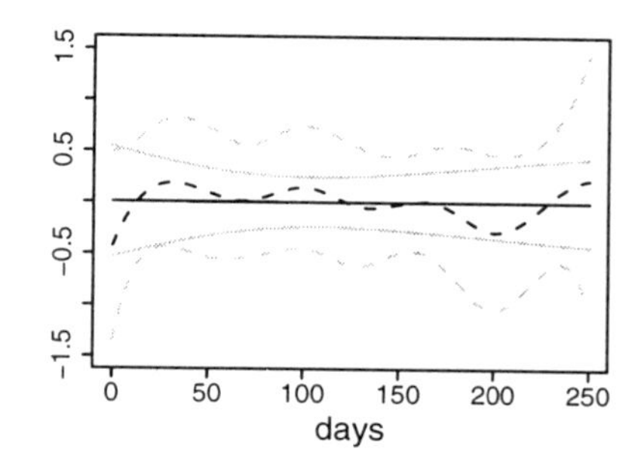

FIGURE 4.3. Weight Loss of Obese Patient. Left: The L-spline fit, the cubic spline fit, and the nonlinear parametric fit are visually indistinguishable; the data are represented by circles. Right: Spline fits and Bayesian confidence intervals minus the parametric fit; the L-spline fit is indicated by solid lines and the cubic spline fit is indicated by dashed lines.

```
tt <- seq(0,250,length=101)
est1 <- predict(wtloss.fit1,
                data.frame(dd=exp(-.004885*tt)),se=TRUE)
est2 <- predict(wtloss.fit2,data.frame(Days=tt),se=TRUE)
est0 <- 81.374+102.68*2^(-tt/141.91)
# plot the fits
plot(wtloss$Days,wtloss$Weight)
lines(tt,est1$fit)
lines(tt,est0,lty=2)
lines(tt,est2$fit,lty=5)
```

In the right frame of Figure 4.3, the L-spline and cubic spline fits and their corresponding Bayesian confidence intervals are plotted after the parametric fit is subtracted from each curve.

```
plot(tt,est1$fit-est0,type="l",ylim=c(-1.5,1.5))
lines(tt,est2$fit-est0,lty=5)
lines(tt,est1$fit-est0-1.96*est1$se,lty=2)
lines(tt,est1$fit-est0+1.96*est1$se,lty=2)
lines(tt,est2$fit-est0-1.96*est2$se,lty=3)
lines(tt,est2$fit-est0+1.96*est2$se,lty=3)
```

It is clear that the L-spline fit has smaller standard errors than the cubic spline fit.

Admittedly, the relative noise level in the weight measurements is way below what one usually sees in stochastic data, although the displayed nonlinearity might not be detectable at a higher noise level. To confirm the usefulness of the demonstrated techniques on "ordinary" data, a simple simulation is conducted below. On $x_i = (i - 0.5)/100$, $i = 1, \ldots, 100$,

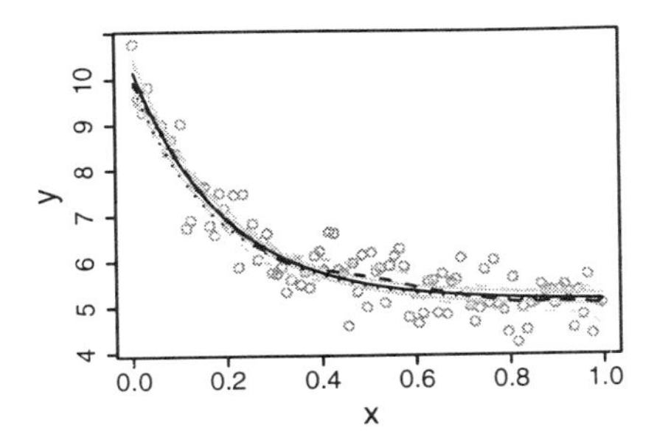

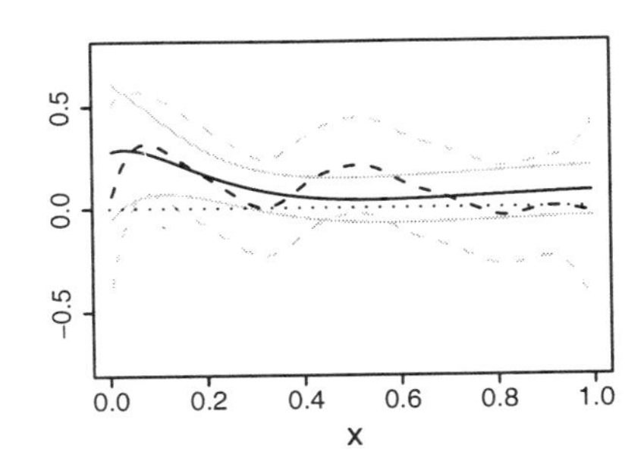

FIGURE 4.4. L-Spline Simulation. Left: The L-spline fit and the corresponding Bayesian confidence intervals are indicated by solid lines, the cubic spline fit is indicated by dashed lines, the test function is indicated by dotted line, and the data are represented by circles. Right: The left frame curves minus the test function.

responses are generated according to $Y_i = 5 + 3e^{-4x_i} + 2e^{-8x_i} + \epsilon_i$, where $\epsilon_i \sim N(0, (0.5)^2)$:

```
set.seed(5732)
tt <- ((1:100)-.5)/100
yy <- 5+3*exp(-4*tt)+2*exp(-8*tt)+.5*rnorm(tt)
```

L-Splines with $L = (D + \tilde{\theta})D$ and $h(x) = e^{3\tilde{\theta}x}$ are tried, and the $\tilde{\theta}$ that minimizes the minimum $V(\lambda)$ is obtained:

```
tmp.fun <- function(theta) {
    ssanova(yy~exp(-theta*tt))$score
}
nlm(tmp.fun,4)$estimate
# 5.255928
```

The minimum $V(\lambda)$ for $\tilde{\theta} = 5.2559$ is 0.2456, and that for a cubic spline in x is 0.2610:

```
ssanova(yy~exp(-5.2559*tt))$score
# 0.2456418
ssanova(yy~tt)$score
# 0.2610069
```

One can now calculate and plot the fits as in the left frame of Figure 4.4, where the L-spline fit and the corresponding Bayesian confidence intervals are drawn in solid and faded solid lines, the cubic spline in dashed and faded dashed lines, and the test function in the dotted line. The data are superimposed as circles.

```
ttt <- exp(-5.2559*tt)
```

```
fit.L <- ssanova(yy~ttt)
est.L <- predict(fit.L,data.frame(ttt=ttt),se=TRUE)
fit.c <- ssanova(yy~tt)
est.c <- predict(fit.c,data.frame(tt=tt),se=TRUE)
#
plot(tt,yy)
lines(tt,est.L$fit)
lines(tt,est.L$fit-1.96*est.L$se,col=5)
lines(tt,est.L$fit+1.96*est.L$se,col=5)
lines(tt,est.c$fit,lty=2)
lines(tt,est.c$fit-1.96*est.c$se,col=5,lty=2)
lines(tt,est.c$fit+1.96*est.c$se,col=5,lty=2)
lines(tt,5+3*exp(-4*tt)+2*exp(-8*tt),lty=3)
```

Subtracting the test function from each of the lines, one gets the right frame of Figure 4.4.

4.3.5 Fast Algorithm

We now describe a fast algorithm for the computation of L-splines due to Heckman and Ramsay (2000). The algorithm assumes that $x_1 < x_2 < \cdots < x_n$, that the space $\mathcal{N}_L = \text{span}\{\phi_\nu, \nu = 1, \ldots, m\}$ is Chebyshev on the intervals $[x_{i+1}, x_{i+m}]$, $i = 1, \ldots, n-m$, and that

$$R_L(x,y) = \int_0^1 G(x;u)G(y;u)(h(u))^{-1}du,$$

where $G(x;u)$ is of the form $\sum_{\nu=1}^{m} \phi_\nu(x)\psi_\nu(u)$ for $u \leq x$. For replicated data, one may work with (3.36) on page 69 and select λ using $U(\lambda)$ of (3.37) or $V(\lambda)$ of (3.38). As with the algorithms of §3.8, the score $M(\lambda)$ and the posterior variances of §3.3 are not available through the fast algorithm, according to current knowledge.

Without loss of generality, consider (3.10) on page 62. From $S_w^T \mathbf{c}_w = 0$, $\mathbf{c}_w = T\boldsymbol{\gamma}$ for some $n \times (n-m)$ matrix T of full column rank satisfying $S_w^T T = O$. Premultiplying the first equation of (3.10) by T^T and plugging in $T\boldsymbol{\gamma}$ for $\mathbf{c}_w$, one has

$$(T^T Q_w T + (n\lambda) T^T T)\boldsymbol{\gamma} = T^T \mathbf{Y}_w.$$

Now, since $\mathbf{Y}_w - \hat{\mathbf{Y}}_w = (I - A_w(\lambda))\mathbf{Y}_w = (n\lambda)\mathbf{c}_w$, one has

$$I - A_w(\lambda) = (n\lambda)T(T^T Q_w T + (n\lambda)T^T T)^{-1} T^T.$$

If T can be chosen such that both $T^T T$ and $T^T Q_w T$ are banded, then the $O(n)$ algorithm of §3.8.1 can be readily applied to calculate L-splines with λ selected by $U(\lambda)$ or $V(\lambda)$.

Let $\mathbf{t}_i$ be an n-vector with $i-1$ leading zeros, $n-m-i$ trailing zeros, and the middle $m+1$ entries $t_{j,i}$ satisfying conditions $t_{i,i} \neq 0$ and

$$\textstyle\sum_{j=i}^{i+m} t_{j,i}\sqrt{w_j}\mathbf{s}_j = 0,$$

where $\mathbf{s}_j^T = (\phi_1(x_j), \dots, \phi_m(x_j))$ is the jth row of S; the latter condition is possible because $\mathbf{s}_j$, $j = i, \dots, i+m$, are linearly dependent, and the former condition is possible because $\mathbf{s}_j$, $j = i+1, \dots, i+m$, are linearly independent since x_j's are distinctive and $\mathcal{N}_L$ is Chebyshev on $[x_{i+1}, x_{i+m}]$. Set $T = (\mathbf{t}_1, \dots, \mathbf{t}_{n-m})$. It is obvious that $S_w^T T = O$ and that T is of full column rank. It is also clear that $T^T T$ is banded with bandwidth $2m+1$. Plugging in the expression $G(x;u) = \sum_{\nu=1}^m \phi_\nu(x)\psi_\nu(u)$ for $u \leq x$, the (k,l)th entry of Q_w can be written as

$$\begin{aligned} q_{k,l} = \sqrt{w_k}\sqrt{w_l}R_L(x_k, x_l) &= \int_0^{x_k \wedge x_l} G(x_k;u)G(x_l;u)(h(u))^{-1}du \\ &= (\sqrt{w_k}\mathbf{s}_k)^T P(x_k \wedge x_l)(\sqrt{w_l}\mathbf{s}_l), \end{aligned}$$

where $P(v)$ is $m \times m$ with the (μ,ν)th entry $\int_0^v \psi_\mu(u)\psi_\nu(u)(h(u))^{-1}du$. Now, for $i < j$, consider the (i,j)th entry of $T^T Q_w T$,

$$\begin{aligned} r_{i,j} &= \sum_{k,l} t_{k,i}(\sqrt{w_k}\mathbf{s}_k)^T P(x_k \wedge x_l)(\sqrt{w_l}\mathbf{s}_l)t_{l,j} \\ &= \sum_{k \leq l} t_{k,i}(\sqrt{w_k}\mathbf{s}_k)^T P(x_k)(\sqrt{w_l}\mathbf{s}_l)t_{l,j} \\ &\quad + \sum_{k > l} t_{k,i}(\sqrt{w_k}\mathbf{s}_k)^T P(x_l)(\sqrt{w_l}\mathbf{s}_l)t_{l,j} \\ &= r'_{i,j} + r''_{i,j}, \end{aligned}$$

say. By the construction of T, $\sum_{l=k}^n t_{l,j}(\sqrt{w_l}\mathbf{s}_l) = 0$ unless $j < k \leq j+m$, and $t_{k,i} = 0$ unless $i \leq k \leq i+m$, so one must have $j < i+m$, or $j-i < m$, for $r'_{i,j} \neq 0$. Similarly, one must have $j-i < m$ for $r''_{i,j} \neq 0$. Hence, $T^T Q_w T$ is banded with bandwidth $2m-1$.

The algorithm relies on the particular form $\int_0^1 G(x;u)G(y;u)(h(u))^{-1}du$ of reproducing kernels with $G(x;u) = \sum_{\nu=1}^m \phi_\nu(x)\psi_\nu(u)$, $u \leq x$, so it does not work with the reproducing kernels of §2.3.3 and §4.3.1.

4.4 Thin-Plate Splines

A thin-plate spline is the minimizer of

$$\frac{1}{n}\sum_{i=1}^n (Y_i - \eta(x_i))^2 + \lambda J_m^d(\eta) \tag{4.47}$$

on the d-dimensional domain $\mathcal{X} = (-\infty, \infty)^d$, where

$$J_m^d(f) = \sum_{\alpha_1+\cdots+\alpha_d=m} \frac{m!}{\alpha_1!\cdots\alpha_d!} \times \int\cdots\int \left(\frac{\partial^m f}{\partial x_{\langle 1\rangle}^{\alpha_1}\cdots\partial x_{\langle d\rangle}^{\alpha_d}} \right)^2 dx_{\langle 1\rangle}\cdots dx_{\langle d\rangle}. \tag{4.48}$$

The null space of $J_m^d(f)$ consists of polynomials of up to $(m-1)$ total order, which is of dimension $M = \binom{d+m-1}{d}$; see Problem 4.23. The functional $J_m^d(f)$ is invariant under a rotation of the coordinates; see Problem 4.24. In the space $\mathcal{H} = \{f : J_m^d(f) < \infty\}$ with $J_m^d(f)$ as a square semi norm, it is necessary that $2m - d > 0$ for the evaluation functional $[x]f = f(x)$ to be continuous; see Duchon (1977), Meinguet (1979), and Wahba and Wendelberger (1980).

The derivation of reproducing kernels for thin-plate splines requires some advanced knowledge of differential equation theory; details can be found in Duchon (1977), Meinguet (1979) and references cited therein. In the sections to follow, we try to keep the exposition to an elementary level, leaving the technically more advanced discussion to the literature. For the fitting of a thin-plate spline alone, an easy-to-evaluate, conditionally non-negative definite semi-kernel is all that one would need (§4.4.1), but to compute the Bayesian confidence intervals or to use thin-plate marginals to construct tensor product splines, genuine reproducing kernels have to be constructed (§4.4.2). Tensor product thin-plate splines are briefly discussed in §4.4.3, and the case study previewed in §1.4.1 is developed in full in §4.4.4.

4.4.1 Semi-Kernels for Thin-Plate Splines

When the parametric least squares estimate in the null space of $J_m^d(f)$ uniquely exists, the minimizer η_λ of (4.47) uniquely exists; see Theorem 2.9. From Duchon (1977, Theorem 4 bis), η_λ has an expression

$$\eta_\lambda(x) = \sum_{\nu=1}^{M} d_\nu \phi_\nu(x) + \sum_{i=1}^{n} c_i E(|x_i - x|), \tag{4.49}$$

where $\{\phi_\nu\}_{\nu=1}^M$ span the null space of $J_m^d(f)$, c_i's are subject to the constraints $S^T\mathbf{c} = 0$ with S the $n \times M$ matrix with the (i,ν)th entry $\phi_\nu(x_i)$, $|x-y|$ is the Euclidean distance, and

$$E(x) = \begin{cases} \theta_{m,d} x^{2m-d} \log x, & d \text{ even, for } \\ & \quad \theta_{m,d} = \frac{(-1)^{d/2+m+1}}{2^{2m-1}\pi^{d/2}(m-1)!(m-d/2)!}, \\ \theta_{m,d} x^{2m-d}, & d \text{ odd, for } \\ & \quad \theta_{m,d} = \frac{\Gamma(d/2-m)}{2^{2m}\pi^{d/2}(m-1)!}. \end{cases} \tag{4.50}$$

The constant $\theta_{m,d}$ in (4.50) is not really needed for (4.49), as it is readily absorbed into c_i's. The reproducing kernels, however, are expressed in terms of $E(x)$ with $\theta_{m,d}$ attached, as will be seen shortly.

For c_i's satisfying $S^T\mathbf{c} = 0$, it can be shown that

$$J_m^d\big(\textstyle\sum_{i=1}^n c_i E(|x_i - x|)\big) = \sum_i \sum_j c_i c_j E(|x_i - x_j|); \tag{4.51}$$

see Meinguet (1979) and Wahba and Wendelberger (1980). Plugging (4.49) and (4.51) into (4.47), the estimation reduces to the minimization of

$$(\mathbf{Y} - S\mathbf{d} - K\mathbf{c})^T(\mathbf{Y} - S\mathbf{d} - K\mathbf{c}) + n\lambda\mathbf{c}^T K\mathbf{c} \tag{4.52}$$

with respect to $\mathbf{c}$ and $\mathbf{d}$, subject to the constraints $S^T\mathbf{c} = 0$, where K is $n \times n$ with the (i,j)th entry $E(|x_i - x_j|)$.

Compare (4.52) with (3.3) on page 60 and (3.4) on page 61. It is easily seen that the solution of the linear system

$$\begin{aligned} (K + n\lambda I)\mathbf{c} + S\mathbf{d} &= \mathbf{Y}, \\ S^T\mathbf{c} &= 0, \end{aligned} \tag{4.53}$$

is a solution of the constrained minimization problem (4.52).

To compute a thin-plate spline, one may use Algorithm 3.1 of §3.4.2, which was designed for the linear system (3.4). The only difference between (3.4) and (4.53) is that Q in (3.4) is non-negative definite, whereas K in (4.53) is not. It is easy to check, however, that one only needs $F_2^T Q F_2$ to be non-negative definite for Algorithm 3.1 to work, and indeed it is the case; check (4.51). The matrix K satisfying

$$S^T\mathbf{c} = 0 \quad \Longrightarrow \quad \mathbf{c}^T K\mathbf{c} \geq 0$$

is said to be conditionally non-negative definite.

The bivariate function $E(|x - y|)$ acts like a reproducing kernel in this approach to the computation of thin-plate splines, and hence is called a semi-kernel. Note that only the sign of $\theta_{m,d}$ matters for the calculation, as the magnitude can be absorbed into c_i's and λ.

Example 4.10 (Cubic spline)
With $d = 1$ and $m = 2$, one has $J_2^1(f) = \int_{-\infty}^{\infty} \ddot{f}^2 dx$, yielding a cubic spline on the real line. Since $\Gamma(1/2 - 2) > 0$, $E(|x - y|) \propto |x - y|^3$. One has

$$\eta_\lambda(x) = d_1 + d_2 x + \sum_{i=1}^n c_i |x_i - x|^3,$$

with $\mathbf{c}$ and $\mathbf{d}$ solving (4.53) for K with the (i,j)th entry $|x_i - x_j|^3$. Under this formulation, one does not need to map the data into $[0, 1]$. □

Example 4.11
With $d = 2$ and $m = 2$, one has $J_2^2(f) = \int\int(\ddot{f}^2_{\langle 11\rangle} + 2\ddot{f}^2_{\langle 12\rangle} + \ddot{f}^2_{\langle 22\rangle})dx_{\langle 1\rangle}dx_{\langle 2\rangle}$. Obviously, $d/2 + m + 1$ is even, so $E(|x-y|) \propto |x-y|^2 \log|x-y|$. It follows that

$$\eta_\lambda(x) = d_1 + d_2 x_{i\langle 1\rangle} + d_3 x_{i\langle 2\rangle} + \sum_{i=1}^n c_i |x_i - x|^2 \log|x_i - x|,$$

with $\mathbf{c}$ and $\mathbf{d}$ the solution of (4.53), where the matrix K has the (i,j)th entry $|x_i - x_j|^2 \log|x_i - x_j|$. □

4.4.2 Reproducing Kernels for Thin-Plate Splines

For the calculation of the fit alone, it is sufficient to know the semi-kernel. To evaluate the posterior variance for the Bayesian confidence intervals of §3.3 or to construct tensor product splines of §2.4 with thin-plate splines as building blocks on the marginal domains, one will have to calculate the genuine reproducing kernel, which is the subject of this section.

Denote by ψ_ν a set of polynomials that span $\mathcal{N}_J$, the null space of $J_m^d(f)$. Define

$$(f,g)_0 = \sum_{i=1}^N p_i f(u_i) g(u_i), \tag{4.54}$$

where $u_i \in (-\infty,\infty)^d$, $p_i > 0$, $\sum_{i=1}^N p_i = 1$ are specified such that the Gram matrix with the (ν,μ)th entry $(\psi_\nu,\psi_\mu)_0$ is nonsingular. Following some standard orthogonalization procedure, one can find an orthonormal basis ϕ_ν, $\nu = 1,\ldots,M$, for $\mathcal{N}_J$ with $\phi_1(x) = 1$ and $(\phi_\nu,\phi_\mu)_0 = \delta_{\nu,\mu}$, where $\delta_{\nu,\mu}$ is the Kronecker delta. The reproducing kernel in $\mathcal{N}_J$ is seen to be

$$R_0(x,y) = \sum_{\nu=1}^M \phi_\nu(x)\phi_\nu(y). \tag{4.55}$$

The projection of f onto $\mathcal{N}_J$ is defined by the operator P through

$$(Pf)(x) = \sum_{\nu=1}^M (f,\phi_\nu)_0 \phi_\nu(x). \tag{4.56}$$

Define

$$R_1(x,y) = (I - P_{(x)})(I - P_{(y)})E(|x-y|), \tag{4.57}$$

where I is the identity operator and $P_{(x)}$ and $P_{(y)}$ are the projection operators of (4.56) applied to the arguments x and y, respectively.

Plugging (4.56) into (4.57), one has, for fixed x,

$$\begin{aligned} R_1(x,u) &= E(|x-u|) - \sum_{\nu=1}^{M} \phi_\nu(x) \sum_{i=1}^{N} p_i \phi_\nu(u_i) E(|u_i - u|) + \pi(u) \\ &= E(|x-u|) + \sum_{i=1}^{N} c_i(x) E(|u_i - u|) + \pi(u), \end{aligned}$$

where $\pi(u) \in \mathcal{N}_J$ and $c_i(x) = -\sum_{\nu=1}^{M} p_i \phi_\nu(u_i)\phi_\nu(x)$. From (4.51), it is easy to show that (Problem 4.25)

$$J_m^d\big(\textstyle\sum_{i=1}^{n} c_i E(|x_i - \cdot|), \sum_{j=1}^{p} \tilde{c}_i E(|y_j - \cdot|)\big) = \displaystyle\sum_{i,j} c_i \tilde{c}_j E(|x_i - y_j|), \quad (4.58)$$

for c_i and $\tilde{c}_j$ satisfying $\sum_{i=1}^{n} c_i \phi_\nu(x_i) = \sum_{j=1}^{p} \tilde{c}_j \phi_\nu(y_j) = 0$, $\nu = 1, \ldots, M$, where $J_m^d(f,g)$ denotes the (semi) inner product associated with the square (semi) norm $J_m^d(f)$. It is easy to check that

$$\phi_\nu(x) + \sum_{i=1}^{N} c_i(x)\phi_\nu(u_i) = \phi_\nu(x) - \sum_{\mu=1}^{M} (\phi_\mu, \phi_\nu)_0 \phi_\mu(x) = 0$$

for $\nu = 1, \ldots, M$. Taking $n = p = N+1$, $x_i = y_i = u_i$, $c_i = c_i(x)$, $\tilde{c}_i = c_i(y)$, $i = 1, \ldots, N$, $x_{N+1} = x$, $y_{N+1} = y$, and $c_{N+1} = \tilde{c}_{N+1} = 1$ in (4.58), one has

$$\begin{aligned} &J_m^d(R_1(x,\cdot), R_1(y,\cdot)) \\ &\quad = E(|x-y|) - \sum_{\nu=1}^{M} \phi_\nu(x) \sum_{i=1}^{N} p_i \phi_\nu(u_i) E(|u_i - y|) \\ &\qquad - \sum_{\nu=1}^{M} \phi_\nu(y) \sum_{i=1}^{N} p_i \phi_\nu(u_i) E(|u_i - x|) \\ &\qquad + \sum_{\nu,\mu=1}^{M} \phi_\nu(x)\phi_\mu(y) \sum_{i,j=1}^{N} p_i p_j \phi_\nu(u_i)\phi_\mu(u_j) E(|u_i - u_j|) \\ &\quad = (I - P_{(x)})(I - P_{(y)}) E(|x-y|) = R_1(x,y); \end{aligned} \quad (4.59)$$

see Problem 4.26. It follows from (4.59) that $R_1(x,y)$ is non-negative definite, hence a reproducing kernel (by Theorem 2.3), and that in the corresponding reproducing kernel Hilbert space, $J_m^d(f,g)$ is the inner product. Actually, for all $f \in \mathcal{H} = \{f : J_m^d(f) < \infty\}$, one has

$$J_m^d((I-P)f, R_1(x,\cdot)) = (I-P)f(x),$$

so $R_1(x,y)$ is indeed the reproducing kernel of $\mathcal{H} \ominus \mathcal{N}_J$; further details can be found in Meinguet (1979) and Wahba and Wendelberger (1980).

Write $R_{00}(x,y) = \phi_1(x)\phi_1(y) = 1$ and $R_{01}(x,y) = \sum_{\nu=2}^{M} \phi_\nu(x)\phi_\nu(y)$. The kernel decomposition $R = R_{00} + [R_{01} + R_1]$ defines a one-way ANOVA decomposition on the domain $\mathcal{X} = (-\infty, \infty)^d$ with an averaging operator $Af = \sum_{i=1}^{N} p_i f(u_i)$.

Example 4.12 (Cubic spline)
Consider a cubic spline on the real line with $d = 1$, $m = 2$, and $E(|x-y|) \propto |x-y|^3$. Take $N = 2$, $u_1 = -1$, $u_2 = 1$, $p_1 = p_2 = 0.5$, and $\phi_2 = x$. It is easy to calculate that

$$\begin{aligned} R_1(x,y) \propto |x-y|^3 &- 0.5\{(1-x)|1+y|^3 + (1+x)|1-y|^3\} \\ &- 0.5\{(1-y)|1+x|^3 + (1+y)|1-x|^3\} \\ &+ 2\{(1+x)(1-y) + (1-x)(1+y)\}; \end{aligned} \tag{4.60}$$

see Problem 4.27. □

Whereas the semi-kernel $E(|x-y|)$ is rather convenient to work with, the reproducing kernel $R_1(x,y)$ can be a bit cumbersome to evaluate. With the choices $N = n$, $u_i = x_i$, and $p_i = 1/n$, $i = 1, \ldots, n$, however, efficient algorithms do exist for the calculation of the $n \times n$ matrix Q with the (i,j)th entry $R_1(x_i, x_j)$, and for the calculation of the $n \times 1$ vector $\boldsymbol{\xi}(x)$ with the ith entry $R_1(x_i, x)$. The matrix Q is used in the computation of the fit, and the vector $\boldsymbol{\xi}(x)$ is used in the evaluation of the estimate.

Set $N = n$, $u_i = x_i$, and $p_i = 1/n$. To derive an orthonormal basis ϕ_ν from a set of polynomials ψ_ν that span $\mathcal{N}_J$, one forms the $n \times M$ matrix $\tilde{S}$ with the (i,ν)th entry $\psi_\nu(x_i)$ and calculates a QR-decomposition $\tilde{S} = (F_1, F_2)\binom{R}{O} = F_1 R$. It follows that $\boldsymbol{\phi} = \sqrt{n} R^{-T} \boldsymbol{\psi}$ forms an orthonormal basis in $\mathcal{N}_J$ with the inner product $(f,g)_0 = \sum_{i=1}^{n} f(x_i)g(x_i)/n$ and that F_1 has the (i,ν)th entry $\phi_\nu(x_i)/\sqrt{n}$ (Problem 4.28). From the expression in (4.59), it is easy to see that

$$Q = (I - F_1 F_1^T) K (I - F_1 F_1^T) = F_2 F_2^T K F_2 F_2^T, \tag{4.61}$$

where K is $n \times n$ with the (i,j)th entry $E(|x_i - x_j|)$ (Problem 4.29). To make sure that $\phi_1 = 1$, one needs to set $\psi_1 = 1$ and to exclude the first column of $\tilde{S}$ from pivoting when calculating the QR-decomposition. Similar to (4.61), one has

$$\begin{aligned} \boldsymbol{\xi}(x) &= (I - F_1 F_1^T)(\boldsymbol{\kappa}(x) - K F_1 \boldsymbol{\phi}(x)/\sqrt{n}) \\ &= F_2 F_2^T (\boldsymbol{\kappa}(x) - K F_1 R^{-T} \boldsymbol{\psi}(x)), \end{aligned} \tag{4.62}$$

where $\boldsymbol{\kappa}(x)$ is $n \times 1$ with the ith entry $E(|x_i - x|)$ (Problem 4.30).

4.4.3 Tensor Product Thin-Plate Splines

Using $R_0(x,y)$ of (4.55) and $R_1(x,y)$ of (4.57) in Theorem 2.6, one can construct tensor product splines with thin-plate marginals. Aside from the complication in the evaluation of the reproducing kernels, there is nothing special technically or computationally about thin-plate marginals.

Tensor product splines with thin-plate marginals do offer something conceptually novel, however, albeit technically trivial. The novel feature is the notion of multivariate main effect in an ANOVA decomposition, in a genuine sense. Consider spatial modeling with geography as one of the covariates. Using a $d = 2$ thin-plate marginal on the geography domain, one is able to construct an isotropic geography main effect and interactions involving geography that are rotation invariant in the geography domain. This is often a more natural treatment as compared to breaking the geography into, say, the longitude and the latitude, which would lead to a longitude effect, a latitude effect, plus a longitude-latitude interaction, that may not make much practical sense.

4.4.4 Case Study: Water Acidity in Lakes

We now fill in details concerning the analysis of the EPA lake acidity data discussed in §1.4.1. A subset of the data concerning 112 lakes in the Blue Ridge were read into R as a data frame `LakeAcidity` with components `ph`, `cal`, `lon`, `lat`, and `geog`, where `geog` contains the x-y coordinates (in distance) of the lakes with respect to a local origin. The x-y coordinates in `geog` were converted from the longitude and the latitude as follows. Denote the longitude by ϕ and the latitude by θ, and let ϕ_0 and θ_0 be the middle point of the longitude range and that of the latitude range, respectively. Remember that ϕ and θ are given in degrees. The point (ϕ_0, θ_0) is taken as the local origin, and the x-y coordinates are obtained through

$$
\begin{aligned}
x &= \cos(\pi\theta/180)\sin(\pi(\phi-\phi_0)/180), \\
y &= \sin(\pi(\theta-\theta_0)/180),
\end{aligned}
$$

with the Earth's radius as the unit distance.

A tensor product thin-plate spline can be fitted to the data using `ssanova` in the R package `gss`

```
fit.lake <- ssanova(ph~log(cal)*geog,data=LakeAcidity,
                    type="tp",order=2,method="m")
```

where `type="tp"` specifies thin-plate spline marginals, `order=2` is the default for $m = 2$, and `method="m"` calls for the use of $M(\lambda)$ in smoothing parameter selection; the default `method="v"` was overridden, as it led to apparent undersmoothing. The default choices for entries in (4.54) are $N = n$, $u_i = x_{i\langle\gamma\rangle}$, and $p_i = 1/n$, where $x_{i\langle\gamma\rangle}$, $\gamma = 1, 2$, are the marginal sampling points. Checking the diagnostics

```
sum.lake <- summary(fit.lake,diag=TRUE)
round(sum.lake$kappa,2)
```

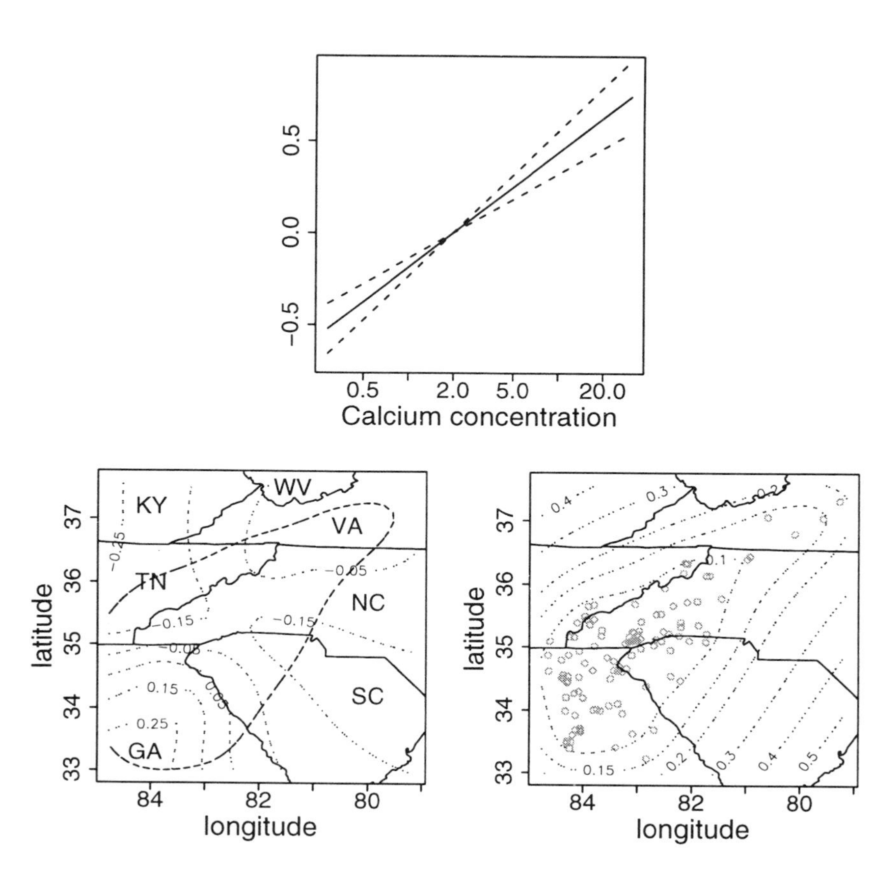

FIGURE 4.5. Water Acidity Fit for Lakes in the Blue Ridge. Top: Calcium effect with 95% Bayesian confidence intervals. Left: Geography effect, with the 0.15 contour of standard errors superimposed as the dashed line. Right: Standard errors of geography effect with the lakes superimposed.

```
#      log(cal)            geog log(cal):geog
#          1.13            1.11            1.15
round(sum.lake$cos,2)
#       log(cal) geog log(cal):geog yhat    y    e
#cos.y      0.65 0.57         -0.06 0.78 1.00 0.67
#cos.e      0.00 0.12          0.00 0.06 0.67 1.00
#norm       2.37 1.50          0.11 3.05 4.10 2.57
```

it is seen that the interaction can be eliminated. An additive model is now fitted to the data, which was plotted in Figure 1.2; the plots are reproduced in Figure 4.5 for convenient reference.

```
fit.lake.add <- update(fit.lake,.~.-log(cal):geog)
```

To obtain the "log(cal)" effect plotted in the top frame, one may use

```
fit.lake.add <- update(fit.lake,.~.-log(cal):geog)
est.cal <- predict(fit.lake.add,se=TRUE,inc="log(cal)",
                   new=model.frame(~log(cal),
                   data=LakeAcidity))
```

where `model.frame(...)` creates the variable named `log(cal)` containing the transformed calcium concentrations. To evaluate the "geog" effect on a grid, the following R commands are used:

```
grid <- seq(-.04,.04,len=31)
new <- model.frame(~geog,list(geog=cbind(rep(grid,31),
                   rep(grid,rep(31,31)))))
est.geog <- predict(fit.lake.add,new,se=TRUE,inc="geog")
```

The fitted values are contoured in the left frame and the standard errors in the right frame using dotted lines, with the x-y grid mapped back to longitude-latitude; the dashed line in the left frame is the 0.15 contour of the standard errors. The R^2 and the decomposition π_β of the "explained" variation in pH can be obtained from the summaries of the fit:

```
sum.lake.add <- summary(fit.lake.add,diag=TRUE)
sum.lake.add$r.squared
# 0.5527218
sum.lake.add$pi
#  log(cal)      geog
# 0.6645443 0.3354557
```

See §3.6 for the definitions of R^2 and π_β.

4.5 Bibliographic Notes

Section 4.1

The idea of partial splines appeared in the literature since the early 1980s in various forms. Extensive discussion on the subject can be found in Wahba (1990, Chapter 6) and Green and Silverman (1994, Chapter 4).

Section 4.2

Fourier series expansion and discrete Fourier transform are among elementary tools in the spectral analysis of time series; see, e.g., Priestley (1981, §4.2, §6.1, §7.6) for comprehensive treatments of related subjects. The spectral decomposition of (4.7) was found in Craven and Wahba (1979), where it was used to analyze the behavior of generalized cross-validation. Some other uses of this decomposition can be found in Gu (1993a) and Stein (1993). The materials of §4.2.3 are largely repackaged arguments found in Craven and Wahba (1979) and Wahba (1985).

Section 4.3

A comprehensive treatment of L-splines from a numerical analytical perspective can be found in Schumaker (1981, Chapters 9–10), upon which a large portion of the technical materials presented here were drawn. The Chebyshev splines of §4.3.2 were found in Kimeldorf and Wahba (1971); see also Wahba (1990, §1.2). Further results on L-splines and their statistical applications can be found in Ramsay and Dalzell (1991), Ansley, Kohn, and Wong (1993), Dalzell and Ramsay (1993), Wang and Brown (1996), and Heckman and Ramsay (2000).

Section 4.4

Standard references on thin-plate splines are Duchon (1977), Meinguet (1979), and Wahba and Wendelberger (1980), upon which much of the materials were drawn. See also Wahba (1990, §§2.4–2.5). Tensor product thin-plate splines were proposed and illustrated by Gu and Wahba (1993b).

4.6 Problems

Section 4.2

4.1 Verify (4.4) for $f \in \mathcal{P}[0,1]$.

4.2 For $f \in \mathcal{P}[0,1]$ and $R_x(y) = R(x,y)$ with $R(x,y)$ as given in (4.5), prove that

$$\left(\int_0^1 f dy\right)\left(\int_0^1 R_x dy\right) + \int_0^1 f^{(m)} R_x^{(m)} dy = f(x).$$

4.3 Compare (4.5) with (2.18) on page 35 to verify that $R(x,y) = 1 + (-1)^{m-1} k_{2m}(x-y)$.

4.4 Let Γ be the Fourier matrix with the (i,j)th entry

$$\frac{1}{\sqrt{n}} \exp\left\{2\pi \mathbf{i} \frac{(i-1)(j-1)}{n}\right\}.$$

(a) Verify that $\Gamma^H \Gamma = \Gamma \Gamma^H = I$.

(b) Verify that (4.7) implies $Q = \Gamma \Lambda \Gamma^H$.

4.5 Verify (4.14) using the orthogonality conditions in (4.13).

4.6 Prove that when (4.15) holds for some $p > 2$ and $B_2 = \lambda^{-2}B(\lambda)|_{\lambda=0} > 0$, then $\lambda^{-2}B(\lambda) - B_2 = o(1)$ for $\lambda = o(1)$.

4.7 Verify (4.16).

4.8 For $c_\nu > 0$ and z_ν and y_ν complex, show that

$$\frac{1}{2}\left|\sum_\nu c_\nu(\bar{z}_\nu y_\nu + z_\nu \bar{y}_\nu)\right| \le \left\{\sum_\nu c_\nu |z_\nu|^2\right\}^{1/2} \left\{\sum_\nu c_\nu |y_\nu|^2\right\}^{1/2},$$

where $\bar{z}$ denotes the conjugate of z.

Section 4.3

4.9 Write $R_x(y) = R_2(x, y)$, where R_2 is given in (4.19). Prove that for $f(x) = \sum_{\mu=2}^{\infty}(a_\mu \cos 2\pi\mu x + b_\mu \sin \pi\mu x)$,

$$\int_0^1 (L_2 f)(y)(L_2 R_x)(y)dy = f(x),$$

where L_2 is given in (4.18).

4.10 Write $R_x(y) = R_{2r}(x, y)$, where R_{2r} is given in (4.22). Prove that for $f(x) = \sum_{\mu=r+1}^{\infty}(a_\mu \cos 2\pi\mu x + b_\mu \sin \pi\mu x)$,

$$\int_0^1 (L_{2r} f)(y)(L_{2r} R_x)(y)dy = f(x),$$

where L_{2r} is given in (4.21).

4.11 Write $R_x(y) = R_3(x, y)$, where R_3 is given in (4.24). Prove that for $f(x) = \sum_{\mu=2}^{\infty}(a_\mu \cos 2\pi\mu x + b_\mu \sin \pi\mu x)$,

$$\int_0^1 (L_3 f)(y)(L_3 R_x)(y)dy = f(x),$$

where L_3 is given in (4.23).

4.12 Write $R_x(y) = R_{2r+1}(x, y)$, where R_{2r+1} is given in (4.27). Prove that for $f(x) = \sum_{\mu=r+1}^{\infty}(a_\mu \cos 2\pi\mu x + b_\mu \sin \pi\mu x)$,

$$\int_0^1 (L_{2r+1} f)(y)(L_{2r+1} R_x)(y)dy = f(x),$$

where L_{2r+1} is given in (4.26).

4.13 Verify (4.30).

4.14 Verify (4.32).

4.15 Consider $R_x(y) = \int_0^1 G(x;u)G(y;u)(h(u))^{-1}du$, with $G(x;u)$ given in (4.31). For L_ν as defined in §4.3.2, verify that $(L_\nu R_x)(0) = 0$, $\nu = 0, \dots, m-1$, and that $(L_m R_x)(y) = G(x;y)/h(y)$.

4.16 In the setting of §4.3.2, set $w_i(x) = 1$, $i = 1, \dots, m$. Verify that $\phi_\nu(x) = x^{\nu-1}/(\nu-1)!$ in (4.29), $\nu = 1, \dots, m$, and that for $u \le x$, $G(x;u) = (x-u)_+^{m-1}/(m-1)!$ in (4.31).

4.17 With $g(\tilde{x}) = f(\phi_2^{-1}(\tilde{x}))$, where ϕ_2^{-1} is the inverse of $\phi_2 = (e^{\theta x}-1)/\theta$, prove that (4.34) reduces to (4.35).

4.18 In the setting of §4.3.2, set $w_1 = 1$ and $w_i = e^{\theta x}$, $i = 2, \dots, m$.

(a) Show that $\phi_\nu(x) = \phi_2^{\nu-1}(x)/(\nu-1)!$ in (4.29), $\nu = 1, \dots, m$, where $\phi_2(x) = (e^{\theta x}-1)/\theta$.

(b) Show that for $u \le x$, $G(x;u) = (\phi_2(x) - \phi_2(u))_+^{m-1}/(m-1)!$ in (4.31).

(c) Given $d\tilde{x}/dx = e^{\theta x}$, show that

$$D_{\tilde{x}}^\nu f = e^{-\nu\theta x}(D_x - (\nu-1)\theta)\cdots D_x f = (L_{\nu(x)} f)(dx/d\tilde{x}),$$

$\nu = 1, \dots, m$, where $D_{\tilde{x}} f = df/d\tilde{x}$, $D_x f = df/dx$, and $L_{\nu(x)}$ is the operator L_ν applied to the variable x.

4.19 In the setting of §4.3.2, set $m = 4$, $w_1 = 1$, $w_2 = w_4 = e^{\theta x}$, and $w_3 = e^{-2\theta x}$.

(a) Show that $\phi_1 = 1$, $\phi_2 = (e^{\theta x}-1)/\theta$, $\phi_3 = (\cosh\theta x - 1)/\theta^2$, and $\phi_4 = (\sinh\theta x - \theta x)/\theta^3$ in (4.29).

(b) Show that for $u \le x$, $G(x,u) = (\sinh\theta(x-u) - \theta(x-u))/\theta^3$ in (4.31).

4.20 Prove equations (4.37) and (4.38) by a change of variable, $\tilde{x} = \phi_2(x) = (e^{\theta x}-1)/\theta$.

4.21 Consider the setting of §4.3.3. For $W(\phi)(0)$ invertible, show that $\sum_{\nu=1}^m f^{(\nu-1)}(0)g^{(\nu-1)}(0)$ forms an inner product in $\text{span}\{\phi_\nu, \nu = 1, \dots, m\}$.

4.22 Verify (4.42).

Section 4.4

4.23 On a d-dimensional real domain, the space of polynomials of up to $(m-1)$ total order is of dimension $M = \binom{d+m-1}{d}$.

(a) Show that the number of polynomials of up to $(m-1)$ total order is the same as the number of ways to choose $m-1$ objects from a set of $d+1$ objects *allowing repeats.*

(b) Show that the number of ways to choose $m-1$ objects from a set of $d+1$ objects *allowing repeats* is the same as the number of ways to choose $m-1$ objects from a set of $(d+1)+(m-1)-1 = d+m-1$ objects *disallowing repeats*, hence is $\binom{d+m-1}{m-1} = \binom{d+m-1}{d}$.

4.24 The functional $J_m^d(f)$ of (4.48) is rotation invariant.

(a) Write $D_i = \partial/\partial x_{\langle i \rangle}$. Show that

$$J_m^d(f) = \int \cdots \int \left\{ \sum_{i_1=1}^{d} \cdots \sum_{i_m=1}^{d} (D_{i_1} \cdots D_{i_m} f)^2 \right\} dx_{\langle 1 \rangle} \cdots dx_{\langle d \rangle}.$$

(b) Let P be a $d \times d$ orthogonal matrix with the (i,j)th entry $p_{i,j}$ and let $y = P^T x$. Note that the Jacobian of the orthogonal transform $y = P^T x$ is 1. Write $\tilde{D}_j = \partial/\partial y_{\langle j \rangle}$. Verify that $\tilde{D}_j = \sum_{i=1}^{d} p_{i,j} D_i$.

(c) Calculating $J_m^d(f)$ with respect to y, the integrand is given by

$$\begin{aligned} \sum_{j_1} \cdots \sum_{j_m} (\tilde{D}_{j_1} \cdots \tilde{D}_{j_m} f)^2 &= \sum_{j_1} \cdots \sum_{j_m} \left\{ \prod_{k=1}^{m} \left(\sum_{i=1}^{d} p_{i,j_k} D_i \right) f \right\}^2 \\ &= \sum_{j_1} \cdots \sum_{j_m} \left\{ \sum_{i_1} \cdots \sum_{i_m} (p_{i_1,j_1} \cdots p_{i_m,j_m})(D_{i_1} \cdots D_{i_m} f) \right\}^2. \end{aligned}$$

Expanding $\{\sum_{i_1} \cdots \sum_{i_m} (p_{i_1,j_1} \cdots p_{i_m,j_m})(D_{i_1} \cdots D_{i_m} f)\}^2$, one gets d^m square terms and $\binom{d^m}{2}$ cross-terms. Summing over $(j_1, \ldots, j_m)$, show that the square terms add up to $\sum_{i_1} \cdots \sum_{i_m} (D_{i_1} \cdots D_{i_m} f)^2$ and the cross-terms all vanish.

4.25 Given (4.51), prove (4.58).

4.26 Verify (4.59).

4.27 Verify (4.60).

4.28 Let ψ_ν, $\nu = 1, \ldots, M$, be a set of polynomials that span $\mathcal{N}_J$ and $\tilde{S}$ an $n \times M$ matrix with the (i, ν)th entry $\psi_\nu(x_i)$. Write $\tilde{S} = F_1 R$ for the QR-decomposition of $\tilde{S}$. Verify that $\boldsymbol{\phi} = \sqrt{n} R^{-T} \boldsymbol{\psi}$ forms an orthonormal basis in $\mathcal{N}_J$ with the inner product $(f, g)_0 = \sum_{i=1}^n f(x_i) g(x_i)/n$ and that F_1 has the (i, ν)th entry $\phi_\nu(x_i)/\sqrt{n}$.

4.29 Verify (4.61).

4.30 Verify (4.62).

5

Regression with Responses from Exponential Families

For responses from exponential family distributions, (1.4) of Example 1.1 defines penalized likelihood regression. Among topics of primary interest are the selection of smoothing parameters, the computation of the estimates, the asymptotic behavior of the estimates, and various data analytical tools.

With a nonquadratic log likelihood, iterations are needed to calculate penalized likelihood regression fit, even for fixed smoothing parameters. Elementary properties concerning the penalized likelihood functional are given in §5.1, followed by discussions in §5.2 of two approaches to smoothing parameter selection. One of the approaches makes use of the scores $U_w(\lambda)$, $V_w(\lambda)$, and $M_w(\lambda)$ of §3.2.4 and the algorithms of §3.4 via iterated reweighted (penalized) least squares, whereas the other implements a version of direct cross-validation. Approximate Bayesian confidence intervals can be calculated through the penalized weighted least squares that approximates the penalized likelihood functional at the converged fit (§5.3). Software implementation is described in §5.4 with simulated examples. Real-data examples are given in §5.5, where it is also shown how the techniques of this chapter can be used to estimate a probability density or to estimate the spectral density of a stationary time series.

The asymptotic convergence of penalized likelihood regression estimates will be discussed in Chapter 8.

5.1 Preliminaries

Consider exponential family distributions with densities of the form

$$f(y|x) = \exp\{(y\eta(x) - b(\eta(x)))/a(\phi) + c(y, \phi)\},$$

where $a > 0$, b, and c are known functions, $\eta(x)$ is the parameter of interest dependent on a covariate x, and ϕ is either known or considered as a nuisance parameter that is independent of x. Observing $Y_i|x_i \sim f(y|x_i)$, $i = 1 \dots, n$, one is to estimate the regression function $\eta(x)$. Parallel to (3.1) on page 60, one has the penalized likelihood functional

$$-\frac{1}{n}\sum_{i=1}^{n}\{Y_i\eta(x_i) - b(\eta(x_i))\} + \frac{\lambda}{2}J(\eta) \tag{5.1}$$

for $\eta \in \mathcal{H} = \oplus_{\beta=0}^{p}\mathcal{H}_\beta$, where $J(f) = J(f,f) = \sum_{\beta=1}^{p}\theta_\beta^{-1}(f,f)_\beta$ and $(f,g)_\beta$ are inner products in $\mathcal{H}_\beta$ with reproducing kernels $R_\beta(x,y)$. The terms $c(Y_i, \phi)$ are independent of $\eta(x)$ and, hence, are dropped from (5.1), and the dispersion parameter $a(\phi)$ is absorbed into λ. The bilinear form $J(f,g)$ is an inner product in $\oplus_{\beta=1}^{p}\mathcal{H}_\beta$ with a reproducing kernel $R_J(x,y) = \sum_{\beta=1}^{p}\theta_\beta R_\beta(x,y)$ and a null space $\mathcal{N}_J = \mathcal{H}_0$. The first term of (5.1) depends on η only through the evaluations $[x_i]\eta = \eta(x_i)$, so the argument of §2.3.2 applies and the minimizer η_λ of (5.1) has an expression

$$\eta(x) = \sum_{\nu=1}^{m} d_\nu\phi_\nu(x) + \sum_{i=1}^{n} c_i R_J(x_i, x) = \boldsymbol{\phi}^T\mathbf{d} + \boldsymbol{\xi}^T\mathbf{c}, \tag{5.2}$$

where $\{\phi_\nu\}_{\nu=1}^{m}$ is a basis of $\mathcal{N}_J = \mathcal{H}_0$, $\boldsymbol{\xi}$ and $\boldsymbol{\phi}$ are vectors of functions, and $\mathbf{c}$ and $\mathbf{d}$ are vectors of coefficients.

Example 5.1 (Gaussian regression)
Consider Gaussian responses with $Y|x \sim N(\eta(x), \sigma^2)$. One has $a(\phi) = \sigma^2$ and $b(\eta) = \eta^2/2$. This reduces to the penalized least squares problem treated in Chapter 3. □

Example 5.2 (Logistic regression)
Consider binary responses with $P(Y = 1|x) = p(x)$ and $P(Y = 0|x) = 1 - p(x)$. The density is

$$f(y|x) = p(x)^y(1 - p(x))^{1-y} = \exp\{y\eta(x) - \log(1 + e^{\eta(x)})\},$$

where $\eta(x) = \log\{p(x)/(1 - p(x))\}$ is the logit function. One has $a(\phi) = 1$ and $b(\eta) = \log(1 + e^\eta)$. This is a special case of penalized likelihood logistic regression with binomial data. □

Example 5.3 (Poisson regression)
Consider Poisson responses with $P(Y = y|x) = \{\lambda(x)\}^y e^{-\lambda(x)}/y!$, $y = 0, 1, \ldots$. The density can be written as

$$f(y|x) = (\lambda(x))^y e^{-\lambda(x)}/y! = \exp\{y\eta(x) - e^{\eta(x)} - \log(y!)\},$$

where $\eta(x) = \log\lambda(x)$ is the log intensity function. One has $a(\phi) = 1$ and $b(\eta) = e^{\eta}$. This defines penalized likelihood Poisson regression. □

By standard exponential family theory, $E[Y|x] = \dot{b}(\eta(x)) = \mu(x)$ and $\mathrm{Var}[Y|x] = \ddot{b}(\eta(x))a(\phi) = v(x)a(\phi)$; see, e.g., McCullagh and Nelder (1989, §2.2.2). The functional $L(f) = -\sum_{i=1}^n \{Y_i f(x_i) - b(f(x_i))\}$ is thus continuous and convex in $f \in \mathcal{H}$. When the matrix S as given in (3.3) on page 60 is of full column rank, one can show that $L(f)$ is strictly convex in $\mathcal{N}_J$, and that (5.1) is strictly convex in $\mathcal{H}$; see Problem 5.1. By Theorem 2.9, the minimizer η_λ of (5.1) uniquely exists when S is of full column rank, which we will assume throughout this chapter.

Fixing the smoothing parameters λ and θ_β hidden in $J(\eta)$, (5.1) is strictly convex in η, of which the minimizer η_λ may be computed via the Newton iteration. Write $\tilde{u}_i = -Y_i + \dot{b}(\tilde{\eta}(x_i)) = -Y_i + \tilde{\mu}(x_i)$ and $\tilde{w}_i = \ddot{b}(\tilde{\eta}(x_i)) = \tilde{v}(x_i)$. The quadratic approximation of $-Y_i\eta(x_i) + b(\eta(x_i))$ at $\tilde{\eta}(x_i)$ is

$$\begin{aligned} & - Y_i\tilde{\eta}(x_i) + b(\tilde{\eta}(x_i)) + \tilde{u}_i\{\eta(x_i) - \tilde{\eta}(x_i)\} + \frac{1}{2}\tilde{w}_i\{\eta(x_i) - \tilde{\eta}(x_i)\}^2 \\ &\qquad = \frac{1}{2}\tilde{w}_i\left\{\eta(x_i) - \tilde{\eta}(x_i) + \frac{\tilde{u}_i}{\tilde{w}_i}\right\}^2 + C_i, \end{aligned}$$

where C_i is independent of $\eta(x_i)$. The Newton iteration updates $\tilde{\eta}$ by the minimizer of the penalized weighted least squares functional

$$\frac{1}{n}\sum_{i=1}^n \tilde{w}_i(\tilde{Y}_i - \eta(x_i))^2 + \lambda J(\eta) \tag{5.3}$$

where $\tilde{Y}_i = \tilde{\eta}(x_i) - \tilde{u}_i/\tilde{w}_i$. Compare (5.3) with (3.9) on page 61.

5.2 Smoothing Parameter Selection

Smoothing parameter selection remains the most important practical issue for penalized likelihood regression. With (5.1) nonquadratic, one needs iterations to compute η_λ, even for fixed smoothing parameters, which adds to the complexity of the problem. Our task here is to devise efficient and effective algorithms to locate good estimates from among the η_λ's with varying smoothing parameters.

The first approach under discussion makes use of the scores $U_w(\lambda)$, $V_w(\lambda)$, and $M_w(\lambda)$ of §3.2.4 through (5.3) in a so-called performance-oriented iteration. The method tracks an appropriate loss in an indirect manner and, hence, may not be the most effective, but the simultaneous updating of (λ, θ_β) and η_λ makes it numerically efficient. To possibly improve on the effectiveness, one may employ the generalized approximate cross-validation of Xiang and Wahba (1996) or its variants, at the cost of less numerical efficiency. The empirical performance of the methods will be illustrated through simple simulation.

As in §3.2, we only make the dependence of various entities on the smoothing parameter λ explicit and suppress their dependence on θ_β in the notation.

5.2.1 Performance-Oriented Iteration

Within an exponential family, the discrepancy between distributions parameterized by (η, ϕ) and (η_λ, ϕ) can be measured by the Kullback-Leibler distance

$$\begin{aligned}\mathrm{KL}(\eta, \eta_\lambda) &= E_\eta[Y(\eta - \eta_\lambda) - (b(\eta) - b(\eta_\lambda))]/a(\phi) \\ &= \{\dot{b}(\eta)(\eta - \eta_\lambda) - (b(\eta) - b(\eta_\lambda))\}/a(\phi),\end{aligned}$$

or its symmetrized version

$$\begin{aligned}\mathrm{SKL}(\eta, \eta_\lambda) &= \mathrm{KL}(\eta, \eta_\lambda) + \mathrm{KL}(\eta_\lambda, \eta) \\ &= (\dot{b}(\eta) - \dot{b}(\eta_\lambda))(\eta - \eta_\lambda)/a(\phi) \\ &= (\mu - \mu_\lambda)(\eta - \eta_\lambda)/a(\phi),\end{aligned}$$

where $\mu = \dot{b}(\eta)$. To measure the performance of $\eta_\lambda(x)$ as an estimate of $\eta(x)$, a natural loss function is given by

$$L(\eta, \eta_\lambda) = \frac{1}{n}\sum_{i=1}^{n}(\mu(x_i) - \mu_\lambda(x_i))(\eta(x_i) - \eta_\lambda(x_i)), \tag{5.4}$$

which is proportional to the average symmetrized Kullback-Leibler distance over the sampling points; (5.4) reduces to (3.13) on page 63 for Gaussian data. The smoothing parameters that minimize $L(\eta, \eta_\lambda)$ represent the ideal choices, given the data, and will be referred to as the optimal smoothing parameters. By the mean value theorem, one has

$$L(\eta, \eta_\lambda) = \frac{1}{n}\sum_{i=1}^{n} w'(x_i)(\eta(x_i) - \eta_\lambda(x_i))^2, \tag{5.5}$$

where $w'(x_i) = \ddot{b}(\eta'(x_i))$ for $\eta'(x_i)$ a convex combination of $\eta(x_i)$ and $\eta_\lambda(x_i)$.

The performance-oriented iteration to be described below operates on (5.3), which has the same numerical structure as (3.9). In fact, (5.3) also has a stochastic structure similar to that of (3.9), as the following lemma asserts.

Lemma 5.1
Suppose $\ddot{b}(\eta(x_i))$ *are bounded away from* 0 *and* $\ddot{b}(\eta'(x_i)) = \ddot{b}(\eta(x_i))(1+o(1))$ *uniformly for* η' *any convex combination of* η *and* $\tilde{\eta}$. *One has*

$$\tilde{Y}_i = \tilde{\eta}(x_i) - \tilde{u}_i/\tilde{w}_i = \eta(x_i) - u_i^o/w_i^o + o_p(1),$$

where $u_i^o = -Y_i + \dot{b}(\eta(x_i))$ *and* $w_i^o = \ddot{b}(\eta(x_i))$.

Proof: We drop the subscripts and write $\tilde{\eta} = \tilde{\eta}(x)$ and $\eta = \eta(x)$. Write

$$\begin{aligned}\delta &= (\tilde{\eta} - \tilde{u}/\tilde{w}) - (\eta - u^o/w^o)\\ &= (\tilde{\eta} - \eta) - (\dot{b}(\tilde{\eta})/\ddot{b}(\tilde{\eta}) - \dot{b}(\eta)/\ddot{b}(\eta)) + Y(1/\ddot{b}(\tilde{\eta}) - 1/\ddot{b}(\eta)).\end{aligned}$$

It is easy to verify that

$$\begin{aligned}E[\delta] &= (\tilde{\eta} - \eta) - (\dot{b}(\tilde{\eta}) - \dot{b}(\eta))/\ddot{b}(\tilde{\eta})\\ &= (\tilde{\eta} - \eta) - (\tilde{\eta} - \eta)(1 + o(1)) = o(\tilde{\eta} - \eta)\end{aligned}$$

and that

$$\mathrm{Var}[\delta] = \{\ddot{b}(\eta)a(\phi)/\ddot{b}^2(\eta)\}o(1) = o(a(\phi)/\ddot{b}(\eta)).$$

The lemma follows. □

Note that $E[u_i^o/w_i^o] = 0$ and $\mathrm{Var}[u_i^o/w_i^o] = a(\phi)/w_i^o$, so (5.3) is almost the same as (3.9), except that u_i^o/w_i^o is not normal and that the weights $\tilde{w}_i$ are not the same as w_i^o. Normality is not needed for Theorem 3.5 of §3.2.4 to hold, but one does need to take care of the "misspecified" weights in (5.3).

Theorem 5.2
Consider the setting of Theorem 3.5. Suppose $\sqrt{w_i}\epsilon_i$ *are independent with mean zero, variances* $v_i\sigma^2$, *and fourth moments that are uniformly bounded. Denote* $R_w(\lambda) = EL_w(\lambda)$ *and* $V = \mathrm{diag}(v_i)$. *If* $nR_w(\lambda) \to \infty$,

$$\{n^{-1}\mathrm{tr}A_w(\lambda)\}^2/n^{-1}trA_w^2(\lambda) \to 0,$$

and $\mathrm{tr}A_w(\lambda) = \mathrm{tr}(VA_w(\lambda))(1 + o(1))$, *then*

$$\begin{aligned}U_w(\lambda) - L_w(\lambda) - n^{-1}\boldsymbol{\epsilon}^T W\boldsymbol{\epsilon} &= o_p(L_w(\lambda)),\\ V_w(\lambda) - L_w(\lambda) - n^{-1}\boldsymbol{\epsilon}^T W\boldsymbol{\epsilon} &= o_p(L_w(\lambda)).\end{aligned}$$

The proof of Theorem 5.2 follows straightforward modifications of the proofs of Theorems 3.1 and 3.3, and is left as an exercise (Problem 5.2).

Theorem 5.2 applies to (5.3) with $w_i = \tilde{w}_i$, $v_i = \tilde{w}_i/w_i^o$, and $\sigma^2 = a(\phi)$. Note that the condition $v_i = 1 + o(1)$ for Lemma 5.1 implies the condition $\mathrm{tr}A_w(\lambda) = \mathrm{tr}(VA_w(\lambda))(1 + o(1))$ for Theorem 5.2.

Denote by $\eta_{\lambda,\tilde{\eta}}$ the minimizer of (5.3) with varying smoothing parameters. By Theorem 5.2, the minimizer of $U_w(\lambda)$ or $V_w(\lambda)$ approximately minimizes $L_w(\lambda) = n^{-1}\sum_{i=1}^n \tilde{w}_i(\eta_{\lambda,\tilde{\eta}}(x_i) - \eta(x_i))^2$, which is a proxy of $L(\eta, \eta_{\lambda,\tilde{\eta}})$; compare with (5.5). The set $\{\eta_{\lambda,\tilde{\eta}}\}$ may not necessarily intersect with the set $\{\eta_\lambda\}$, however.

For $\tilde{\eta} = \eta_{\lambda^o}$ with fixed $(\lambda^o, \theta_\beta^o)$, it is easy to see that $\eta_{\lambda^o,\eta_{\lambda^o}} = \eta_{\lambda^o}$, which is the fixed point of the Newton iteration with the smoothing parameters in (5.1) fixed at $(\lambda^o, \theta_\beta^o)$. Unless $(\lambda^o, \theta_\beta^o)$ minimizes the corresponding $U_w(\lambda)$ or $V_w(\lambda)$ (which are η_{λ^o} dependent), one would not want to use $\eta_{\lambda^o,\eta_{\lambda^o}}$, because it is perceived to be inferior to the $\eta_{\lambda,\eta_{\lambda^o}}$ that minimizes the corresponding $U_w(\lambda)$ or $V_w(\lambda)$. Note that two sets of smoothing parameters come into play here: One set specifies $\tilde{\eta} = \eta_{\lambda^o}$, which, in turn, defines the scores $U_w(\lambda)$ and $V_w(\lambda)$, and the other set indexes $\eta_{\lambda,\tilde{\eta}}$ and is the argument in $U_w(\lambda)$ and $V_w(\lambda)$. The above discussion suggests that one should look for some $\eta_{\lambda^*,\eta_{\lambda^*}} = \eta_{\lambda^*}$ that minimizes the $U_w(\lambda)$ or $V_w(\lambda)$ scores defined by itself, provided such a "self-voting" η_{λ^*} exists. To locate such "self-voting" η_{λ^*}, a performance-oriented iteration procedure was proposed by Gu (1992a), which we discuss next.

In performance-oriented iteration, one iterates on (5.3) with the smoothing parameters updated according to $U_w(\lambda)$ or $V_w(\lambda)$. Instead of moving to a particular Newton update with fixed smoothing parameters, one chooses, from among a family of Newton updates, one that is perceived to be better performing according to $U_w(\lambda)$ or $V_w(\lambda)$. If the smoothing parameters stabilize at, say, $(\lambda^*, \theta_\beta^*)$ and the corresponding Newton iteration converges at η^*, then it is clear that $\eta^* = \eta_{\lambda^*}$ and one has found the solution. Note that the procedure never compares η_λ directly with each other but only tracks $L(\eta, \eta_{\lambda,\tilde{\eta}})$ through $U_w(\lambda)$ or $V_w(\lambda)$ in each iteration. In a neighborhood around η^*, where the corresponding (5.3) is a good approximation of (5.1) for smoothing parameters near $(\lambda^*, \theta_\beta^*)$, η_{λ,η^*}'s are hopefully close approximations of η_λ's, and through indirect comparison, η^*, in turn, is perceived to be better performing among the η_λ's in the neighborhood.

The existence of "self-voting" η_{λ^*} and the convergence of performance-oriented iteration remain open and do not appear to be tractable theoretically. Note that the numerical problem (5.3) as well as the scores $U_w(\lambda)$ and $V_w(\lambda)$ change from iteration to iteration. With proper implementation, performance-oriented iteration is found to converge empirically in most situations, and when it converges, the fixed point of the iteration simply gives the desired "self-voting" η_{λ^*}.

The implementation suggested in Gu (1992a) starts at some $\tilde{\eta} = \eta_\lambda$ with λ large, and it limits the search range for smoothing parameters to a neighborhood of the previous ones during the minimization of $U_w(\lambda)$ or $V_w(\lambda)$ in each iteration. The idea is to start from the numerically more stable end of the trajectory $\{\eta_\lambda\}$ and to stay close to the trajectory, where the final solution will be located. Technical details are to be found in Gu (1992a).

Since $M(\lambda)$ also does a good job in tracking the mean square error loss in penalized least squares regression, as illustrated in simulations (see, e.g., §3.2.5), one may also use $M_w(\lambda)$ to drive the performance-oriented iteration by analogy. Such a procedure does not maximize any likelihood function with respect to the smoothing parameters, however.

5.2.2 Direct Cross-Validation

In order to compare η_λ directly, one needs some computable score that tracks $L(\eta, \eta_\lambda)$ of (5.4). One such score is the generalized approximate cross-validation (GACV) of Xiang and Wahba (1996), to be described below.

Without loss of generality, assume $a(\phi) = 1$. Consider the Kullback-Leibler distance

$$\mathrm{KL}(\eta, \eta_\lambda) = \frac{1}{n}\sum_{i=1}^{n}\{\mu(x_i)(\eta(x_i) - \eta_\lambda(x_i)) - (b(\eta(x_i)) - b(\eta_\lambda(x_i)))\}, \quad (5.6)$$

which is a proxy of $L(\eta, \eta_\lambda)$; roughly, $2\mathrm{KL}(\eta, \eta_\lambda) \approx L(\eta, \eta_\lambda)$. Dropping terms from (5.6) that do not involve η_λ, one gets the relative Kullback-Leibler distance

$$\mathrm{RKL}(\eta, \eta_\lambda) = \frac{1}{n}\sum_{i=1}^{n}\{-\mu(x_i)\eta_\lambda(x_i) + b(\eta_\lambda(x_i))\}. \quad (5.7)$$

Replacing $\mu(x_i)\eta_\lambda(x_i)$ by $Y_i\eta_\lambda^{[i]}(x_i)$, one obtains a cross-validation estimate of $\mathrm{RKL}(\eta, \eta_\lambda)$,

$$V_0(\lambda) = \frac{1}{n}\sum_{i=1}^{n}\{-Y_i\eta_\lambda^{[i]}(x_i) + b(\eta_\lambda(x_i))\}, \quad (5.8)$$

where $\eta_\lambda^{[k]}$ minimizes the "delete-one" version of (5.1),

$$-\frac{1}{n}\sum_{i\neq k}\{Y_i\eta(x_i) - b(\eta(x_i))\} + \frac{\lambda}{2}J(\eta). \quad (5.9)$$

Note that $E[Y_i] = \mu(x_i)$ and that $\eta_\lambda^{[i]}$ is independent of Y_i. Write

$$V_0(\lambda) = -\frac{1}{n}\sum_{i=1}^{n}\{Y_i\eta_\lambda(x_i) - b(\eta_\lambda(x_i))\} + \frac{1}{n}\sum_{i=1}^{n}Y_i(\eta_\lambda(x_i) - \eta_\lambda^{[i]}(x_i)), \quad (5.10)$$

where the first term is readily available, but the second term is impractical to compute. One needs computationally practical approximations of the second term to make use of $V_0(\lambda)$.

Through a series of first-order Taylor expansions, Xiang and Wahba (1996) propose to approximate the second term of (5.10) by

$$\frac{1}{n}\sum_{i=1}^{n}\frac{h_{ii}Y_i(Y_i-\mu_\lambda(x_i))}{1-h_{ii}\tilde{w}_i}, \tag{5.11}$$

where $\tilde{w}_i = \ddot{b}(\eta_\lambda(x_i))$ and h_{ii} is the ith diagonal of a matrix H to be specified below. Recall matrices S and Q from §3.1 and let F_2 be an $n \times (n-m)$ orthogonal matrix satisfying $S^TF_2 = 0$. Write $W = \text{diag}(\tilde{w}_i)$. The matrix H appearing in (5.11) is given by

$$H = (W + n\lambda F_2(F_2^TQF_2)^+F_2^T)^{-1},$$

where $(\cdot)^+$ denotes the Moore-Penrose inverse. Substituting the approximation into (5.10), one gets an approximate cross-validation (ACV) score

$$V_a(\lambda) = -\frac{1}{n}\sum_{i=1}^{n}\{Y_i\eta_\lambda(x_i) - b(\eta_\lambda(x_i))\} + \frac{1}{n}\sum_{i=1}^{n}\frac{h_{ii}Y_i(Y_i-\mu_\lambda(x_i))}{1-h_{ii}\tilde{w}_i}. \tag{5.12}$$

Replacing h_{ii} and $h_{ii}\tilde{w}_i$ in (5.12) by their respective averages $n^{-1}\text{tr}H$ and $1-n^{-1}\text{tr}(HW)$, one obtains the GACV score of Xiang and Wahba (1996),

$$\begin{aligned} V_g(\lambda) = &-\frac{1}{n}\sum_{i=1}^{n}\{Y_i\eta_\lambda(x_i) - b(\eta_\lambda(x_i))\} \\ &+ \frac{\text{tr}H}{n-\text{tr}(HW)}\frac{1}{n}\sum_{i=1}^{n}Y_i(Y_i-\mu_\lambda(x_i)). \end{aligned} \tag{5.13}$$

For n large, Q is often ill-conditioned and the computation of H can be numerically unstable.

As an alternative approach to the approximation of (5.10), Gu and Xiang (2001) substitute $\eta_{\lambda,\eta_\lambda}^{[i]}(x_i)$ for $\eta_\lambda^{[i]}(x_i)$, where $\eta_{\lambda,\eta_\lambda}^{[k]}$ minimizes the "delete-one" version of (5.3),

$$\frac{1}{n}\sum_{i\neq k}\tilde{w}_i(\tilde{Y}_i - \eta(x_i))^2 + \lambda J(\eta), \tag{5.14}$$

for $\tilde{\eta} = \eta_\lambda$. Remember that $\eta_\lambda = \eta_{\lambda,\eta_\lambda}$. Trivial adaptation of Lemma 3.2 of §3.2.2 yields

$$\sqrt{\tilde{w}_i}(\eta_\lambda(x_i) - \eta_{\lambda,\eta_\lambda}^{[i]}(x_i)) = a_{i,i}\sqrt{\tilde{w}_i}(\tilde{Y}_i - \eta_{\lambda,\eta_\lambda}^{[i]}(x_i)),$$

where $a_{i,i}$ is the ith diagonal of the matrix $A_w(\lambda)$; see (3.11) and (3.12) on page 62. It follows that

$$\eta_\lambda(x_i) - \eta_{\lambda,\eta_\lambda}^{[i]}(x_i) = \frac{a_{i,i}}{1-a_{i,i}}(\tilde{Y}_i - \eta_\lambda(x_i)).$$

Recalling that $\tilde{Y}_i = \tilde{\eta}(x_i) - \tilde{u}_i/\tilde{w}_i$ and $-\tilde{u}_i = Y_i - \tilde{\mu}(x_i)$, one has

$$\eta_\lambda(x_i) - \eta^{[i]}_{\lambda,\eta_\lambda}(x_i) = \frac{a_{i,i}}{1-a_{i,i}} \frac{Y_i - \mu_\lambda(x_i)}{\tilde{w}_i}. \tag{5.15}$$

Substituting (5.15) into (5.10), one obtains an alternative ACV score

$$\begin{aligned} V_a^*(\lambda) = &-\frac{1}{n}\sum_{i=1}^n \{Y_i\eta_\lambda(x_i) - b(\eta_\lambda(x_i))\} \\ &+ \frac{1}{n}\sum_{i=1}^n \frac{a_{i,i}}{1-a_{i,i}} \frac{Y_i(Y_i - \mu_\lambda(x_i))}{\tilde{w}_i}. \end{aligned} \tag{5.16}$$

Parallel to (5.13), one may replace $a_{i,i}/\tilde{w}_i$ by $n^{-1}\sum_{i=1}^n a_{i,i}/\tilde{w}_i$ and $1 - a_{i,i}$ by $1 - n^{-1}\mathrm{tr}A_w$ to obtain an alternative GACV score:

$$\begin{aligned} V_g^*(\lambda) = &-\frac{1}{n}\sum_{i=1}^n \{Y_i\eta_\lambda(x_i) - b(\eta_\lambda(x_i))\} \\ &+ \frac{\mathrm{tr}(A_w W^{-1})}{n - \mathrm{tr}A_w} \frac{1}{n}\sum_{i=1}^n Y_i(Y_i - \mu_\lambda(x_i)). \end{aligned} \tag{5.17}$$

It can be shown that when $F_2^T Q F_2$ is nonsingular, $A_w(\lambda) = W^{1/2} H W^{1/2}$; see Problem 5.3. Hence, $V_g(\lambda)$ and $V_g^*(\lambda)$ are virtually the same.

Note that the term $n^{-1}\mathrm{tr}(A_w W^{-1}) = n^{-1}\sum_{i=1}^n a_{i,i}/\tilde{w}_i$ requires $a_{i,i}$ to be calculated individually, which incurs some extra calculation beyond the standard results from the algorithms of §3.4. Replacing

$$\frac{1}{n}\mathrm{tr}(A_w W^{-1}) = \frac{1}{n}\sum_{i=1}^n \frac{a_{i,i}}{\tilde{w}_i}$$

by

$$\frac{1}{n^2}\mathrm{tr}(A_w)\mathrm{tr}(W^{-1}) = \left(\frac{1}{n}\sum_{i=1}^n a_{i,i}\right)\left(\frac{1}{n}\sum_{i=1}^n \frac{1}{\tilde{w}_i}\right),$$

one has a variant of the score $V_g^*(\lambda)$,

$$\begin{aligned} V_g^{**}(\lambda) = &-\frac{1}{n}\sum_{i=1}^n \{Y_i\eta_\lambda(x_i) - b(\eta_\lambda(x_i))\} \\ &+ \frac{\mathrm{tr}A_w}{n - \mathrm{tr}A_w} \frac{\mathrm{tr}(W^{-1})}{n} \frac{1}{n}\sum_{i=1}^n Y_i(Y_i - \mu_\lambda(x_i)), \end{aligned} \tag{5.18}$$

which involves only the trace $\mathrm{tr}(A_w)$. When the sequences $\{a_{i,i}\}$ and $\{\tilde{w}_i^{-1}\}$ are "uncorrelated," $V_g^{**}(\lambda)$ should be a good approximation of $V_g^*(\lambda)$. The terms in (5.17) and (5.18) are numerically stable for all n.

For Gaussian data, $V_g^*(\lambda)$ and $V_g^{**}(\lambda)$ reduce to

$$U^*(\lambda) = \frac{1}{n}\mathbf{Y}^T(I - A(\lambda))^2\mathbf{Y} + \frac{2\mathrm{tr}A(\lambda)}{n}\frac{\mathbf{Y}^T(I - A(\lambda))\mathbf{Y}}{\mathrm{tr}(I - A(\lambda))}. \tag{5.19}$$

Under mild conditions, one can show that

$$U^*(\lambda) - L(\lambda) - n^{-1}\epsilon^T\epsilon = o_p(L(\lambda)).$$

See Problem 5.4.

5.2.3 Empirical Performance

We now present some simulation results concerning the methods discussed above. Among the issues to be addressed are the comparative performance of the various scores and the mechanism that drives the performance-oriented iteration to convergence.

A brief description of the simulation follows. On $x_i = (i - 0.5)/100$, $i = 1, \ldots, 100$, binary data were generated according to a logit function

$$\eta(x) = 3\{10^5x^{11}(1 - x)^6 + 10^3x^3(1 - x)^{10}\} - 2. \tag{5.20}$$

Cubic spline logistic regression estimates were computed on 100 replicates of such data. The minimizer η_λ of (5.1) was calculated for λ on a grid $\log_{10}\lambda = -6(0.1)0$. The scores $V_g^*(\lambda)$ and $V_g^{**}(\lambda)$ were evaluated for each η_λ along with the loss $L(\eta, \eta_\lambda)$ of (5.4). The η_λ selected by these scores on the grid were then identified and the corresponding loss recorded. The relative efficacy of the scores is given by the ratios of the minimum loss on the grid to the losses achieved by the respective scores. The performance-oriented iterations driven by $U_w(\lambda)$ (with $a(\phi) = 1$), $V_w(\lambda)$, and $M_w(\lambda)$ were also conducted on each of the replicates, and the respective losses were obtained and loss ratios calculated.

The five sets of loss ratios are plotted in the top-left frame of Figure 5.1. Although all the methods worked well, it appears that the performance-oriented iteration driven by $V_w(\lambda)$ was less effective as compared to the other methods. Actually, $V_w(\lambda)$ was dominated by $U_w(\lambda)$ on a replicate-by-replicate basis, as the top-center frame of Figure 5.1 shows. Note that the known fact that $a(\phi) = 1$ was used in $U_w(\lambda)$ but not in $V_w(\lambda)$. As expected, $V_g^*(\lambda)$ and $V_g^{**}(\lambda)$ follow each other closely, as seen in the top-right frame. The bottom frames of Figure 5.1 plot the losses of $U_w(\lambda)$, $M_w(\lambda)$, and $V_g^*(\lambda)$ against each other. Clearly, there is no domination among these methods, although V_g^* appears to be the more robust performer.

Although $V_g^*(\lambda)$ and $V_g^{**}(\lambda)$ are more robust than performance-oriented iteration, they are computationally more demanding. In the simulation, the performance-oriented iteration typically converged in 5 to 10 steps, on par with the fixed-λ Newton iteration at a single grid point.

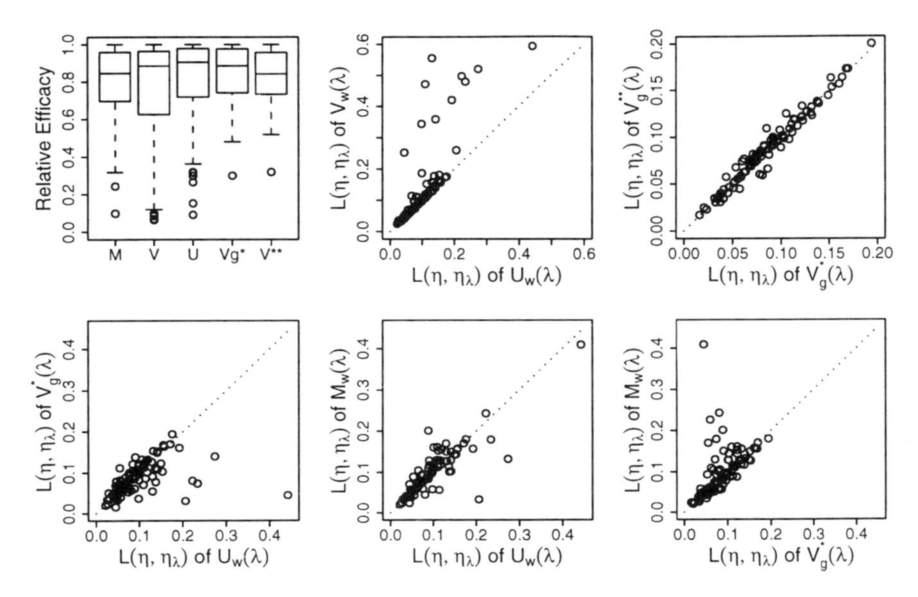

FIGURE 5.1. Effectiveness of $M_w(\lambda)$, $V_w(\lambda)$, $U_w(\lambda)$, $V_g^*(\lambda)$, and $V_g^{**}(\lambda)$.

Let us now explore the mechanism that drives the performance-oriented iteration to convergence. Pick a particular replicate in the above simulation, say the first one, and set $\tilde{\eta} = \eta_{\tilde{\lambda}}$ in (5.3) for $\tilde{\lambda}$ on a grid $\log_{10} \tilde{\lambda} = -6(0.1)0$. The scores $U_w(\lambda)$ (with $a(\phi) = 1$), $V_w(\lambda)$, and $M_w(\lambda)$ were evaluated for λ on a grid $\log_{10} \lambda = -6(0.1)0$. Note that $\tilde{\lambda}$ here indexes $\tilde{\eta} = \eta_{\tilde{\lambda}}$ the minimizer of (5.1) and λ indexes $\eta_{\lambda,\tilde{\eta}}$ the minimizer of (5.3) given $\tilde{\eta}$. This gave 61×61 arrays of $U_w(\lambda)$, $V_w(\lambda)$, and $M_w(\lambda)$. These arrays are contoured in Figure 5.2, where the horizontal axis is λ and the vertical axis is $\tilde{\lambda}$. An $\eta_{\tilde{\lambda}}$ that is not optimal can still be a good approximation of η for the purpose of Lemma 5.1, so for many of the horizontal slices in Figure 5.2, one could expect the minima, marked as a circle or a star in the plots, to provide λ close to optimal for the weighted least squares problem (5.3). The stars in Figure 5.2 indicate the respective "self-voting" λ^*, to which performance-oriented iteration converged. Note that although the iteration in general only visits the slice marked by the solid line on convergence, the scores associated with the intermediate iterates should have behavior similar to the horizontal slices in the plots.

5.3 Approximate Bayesian Confidence Intervals

To obtain interval estimates for penalized likelihood regression with exponential family responses, one may adapt the Bayesian confidence intervals

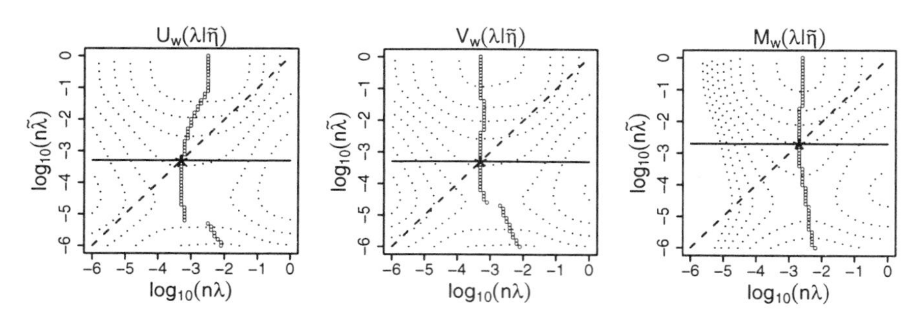

FIGURE 5.2. Contours of $U_w(\lambda|\eta_{\tilde{\lambda}})$, $V_w(\lambda|\eta_{\tilde{\lambda}})$, and $M_w(\lambda|\eta_{\tilde{\lambda}})$. The circles are minima of the horizontal slices with fixed $\tilde{\lambda}$. The star indicates the "self-voting" λ^*. Performance-oriented iteration visits the solid slice on convergence.

of §3.3. Specifically, one calculates the minimizer $\eta_\lambda(x)$ of (5.1), then computes the needed approximate posterior means and variances based on (5.3) at the converged fit $\tilde{\eta} = \eta_\lambda$.

Consider $\eta = \eta_0 + \eta_1$, where η_0 and η_1 have independent mean zero Gaussian process priors with covariances $E[\eta_0(x)\eta_0(y)] = \tau^2 \sum_{\nu=1}^m \phi_\nu(x)\phi_\nu(y)$ and $E[\eta_1(x)\eta_1(y)] = bR_J(x,y)$. Write $\eta_0(x) = \sum_{\nu=1}^m \phi_\nu(x)\beta_\nu$, where $\boldsymbol{\beta} = (\beta_1, \dots, \beta_m)^T \sim N(0, \tau^2 I)$. Write $\boldsymbol{\eta} = (\eta(x_1), \dots, \eta(x_n))^T$ and let $\tau^2 \to \infty$; the likelihood of $(\boldsymbol{\eta}, \boldsymbol{\beta})$ is proportional to

$$\exp\Big\{ -\frac{1}{2b}(\boldsymbol{\eta} - S\boldsymbol{\beta})^T Q^+ (\boldsymbol{\eta} - S\boldsymbol{\beta}) \Big\}, \tag{5.21}$$

where S is $n \times m$ with the (i,ν)th entry $\phi_\nu(x_i)$ and Q^+ is the Moore-Penrose inverse of the $n \times n$ matrix Q with the (i,j)th entry $R_J(x_i, x_j)$; see Problem 5.5. Integrating out $\boldsymbol{\beta}$ from (5.21), the likelihood of $\boldsymbol{\eta}$ is seen to be

$$q(\boldsymbol{\eta}) \propto \exp\Big\{ -\frac{1}{2b}\boldsymbol{\eta}^T (Q^+ - Q^+ S(S^T Q^+ S)^{-1} S^T Q^+)\boldsymbol{\eta} \Big\}; \tag{5.22}$$

see Problem 5.6. The posterior likelihood of $\boldsymbol{\eta}$ given $\mathbf{Y} = (Y_1, \dots, Y_n)^T$ is proportional to the joint likelihood, which is of the form

$$\begin{aligned} p(\mathbf{Y}|\boldsymbol{\eta})q(\boldsymbol{\eta}) \propto \exp\Big\{ &\frac{1}{a(\phi)} \sum_{i=1}^n (Y_i \eta(x_i) - b(\eta(x_i))) \\ &- \frac{1}{2b}\boldsymbol{\eta}^T (Q^+ - Q^+ S(S^T Q^+ S)^{-1} S^T Q^+)\boldsymbol{\eta} \Big\}. \end{aligned} \tag{5.23}$$

The following theorem extends the results of §2.5.

Theorem 5.3

Suppose η_λ minimizes (5.1) with $n\lambda = a(\phi)/b$. For Q nonsingular, the fitted values $\boldsymbol{\eta}^* = (\eta_\lambda(x_1), \dots, \eta_\lambda(x_n))^T$ are the posterior mode of $\boldsymbol{\eta}$ given $\mathbf{Y}$.

Proof: By (5.2), $\boldsymbol{\eta}^* = Q\mathbf{c} + S\mathbf{d}$, where $\mathbf{c} = (c_1, \ldots, c_n)^T$, $\mathbf{d} = (d_1, \ldots, d_m)^T$ minimize

$$-\frac{1}{n}\sum_{i=1}^{n}\{Y_i(\boldsymbol{\xi}_i^T\mathbf{c} + \boldsymbol{\phi}_i^T\mathbf{d}) - b(\boldsymbol{\xi}_i^T\mathbf{c} + \boldsymbol{\phi}_i^T\mathbf{d})\} + \frac{\lambda}{2}\mathbf{c}^T Q\mathbf{c}, \tag{5.24}$$

with $\boldsymbol{\xi}_i = (R_J(x_1, x_i), \ldots, R_J(x_n, x_i))^T$ and $\boldsymbol{\phi}_i = (\phi_1(x_i), \ldots, \phi_m(x_i))^T$. Taking derivatives of (5.24) with respect to $\mathbf{c}$ and $\mathbf{d}$ and setting them to zero, one has

$$\begin{aligned} Q\mathbf{u} + n\lambda Q\mathbf{c} &= 0, \\ S^T\mathbf{u} &= 0, \end{aligned} \tag{5.25}$$

where $\mathbf{u} = (u_1, \ldots, u_n)^T$ with $u_i = -Y_i + \dot{b}(\eta_\lambda(x_i))$. For Q nonsingular, $Q^+ = Q^{-1}$. Taking derivatives of $-a(\phi)\log p(\mathbf{Y}|\boldsymbol{\eta})q(\boldsymbol{\eta})$ as given in (5.23) with respect to $\boldsymbol{\eta}$, and plugging in $\boldsymbol{\eta}^* = Q\mathbf{c} + S\mathbf{d}$ with $\mathbf{c}$ and $\mathbf{d}$ satisfying (5.25), one has

$$\begin{aligned} &\mathbf{u} + n\lambda(Q^{-1} - Q^{-1}S(S^TQ^{-1}S)^{-1}S^TQ^{-1})(Q\mathbf{c} + S\mathbf{d}) \\ &\qquad = \mathbf{u} + n\lambda(\mathbf{c} - Q^{-1}S(S^TQ^{-1}S)^{-1}S^T\mathbf{c}) = 0. \end{aligned}$$

The theorem follows. □

Replacing the exponent of $p(\mathbf{Y}|\boldsymbol{\eta})$ by its quadratic approximation at $\boldsymbol{\eta}^*$, one gets a Gaussian likelihood with observations $\tilde{Y}_i$ and variances $a(\phi)/\tilde{w}_i$, where $\tilde{Y}_i$ and $\tilde{w}_i$ are as specified in (5.3), all evaluated at $\tilde{\eta} = \eta_\lambda$. With such a Gaussian approximation of the sampling likelihood $p(\mathbf{Y}|\boldsymbol{\eta})$, the results of §3.3 yield approximate posterior means and variances for $\eta(x)$ and its components, which can be used to construct approximate Bayesian confidence intervals.

On the sampling points, for Q nonsingular, such an approximate posterior analysis of $\boldsymbol{\eta}$ is simply Laplace's method applied to the posterior distribution of $\boldsymbol{\eta}$, as ascertained by Theorem 5.3; see, e.g., Tierney and Kadane (1986) and Leonard et al. (1989) for discussions on Laplace's method. The statement, however, is generally not true even for a subset of $\boldsymbol{\eta}$, as the corresponding subset of $\boldsymbol{\eta}^*$ are, in general, not the exact mode of the respective likelihood. It appears that the exact Bayesian calculation can be sensitive to parameter specification. This also serves to explain why one would need Q to be nonsingular for Theorem 5.3 to hold.

5.4 Software: R Package gss

It is straightforward to implement the performance-oriented iteration using RKPACK facilities. Given a starting $\tilde{\eta}$, the iteration can be driven by $U_w(\lambda)$, $V_w(\lambda)$, or $M_w(\lambda)$ through RKPACK routines `dsidr` or `dmudr`; see

§3.5.1. The `gssanova` suite in the R package `gss` provides a front end to penalized likelihood regression via performance-oriented iteration. Implemented are distribution families binomial, Poisson, gamma, inverse Gaussian, and negative binomial. The syntax mimics that of the `glm` suite.

For each of the families, only one link is provided, one that is free of constraint. This is not much of a restriction, however, as splines are flexible.

5.4.1 Binomial Family

The binomial distribution Binomial(m, p) has a density

$$\binom{m}{y} p^y (1-p)^{m-y}$$

and a log likelihood

$$y\eta - m\log(1+e^\eta), \tag{5.26}$$

where $\eta = \log\{p/(1-p)\}$ is the logit. The binary data of Example 5.2 is a special case with $m = 1$. To iterate on (5.3), it is easy to calculate $\tilde{u}_i = -Y_i + m_i\, p(x_i)$ and $\tilde{w}_i = m_i\, p(x_i)(1-p(x_i))$; see Problem 5.7. The syntax of `gssanova` for the binomial family is similar to that of `glm`, as shown in the following example.

Example 5.4 (Logistic regression)
The following R commands generate some synthetic data on a grid and fit a cubic spline logistic regression model:

```
set.seed(5732)
test <- function(x)
        {.3*(1e6*(x^11*(1-x)^6)+1e4*(x^3*(1-x)^10))-2}
x <- (0:100)/100
p <- 1-1/(1+exp(test(x)))
y <- rbinom(x,3,p)
logit.fit <- gssanova(cbind(y,3-y)~x,family="binomial")
```

Note that `type="cubic"` is omitted in the call, as it is the default. Equivalently, one may use a one-column response Y_i/m_i and enter $m_i = 3$ as weights:

```
logit.fit1 <- gssanova(y/3~x,"binomial",
                       weights=rep(3,101))
```

To evaluate the fit on the grid, use

```
est <- predict(logit.fit,data.frame(x=x),se=TRUE)
```

The fit is plotted in the left frame of Figure 5.3, with the data and the test function superimposed:

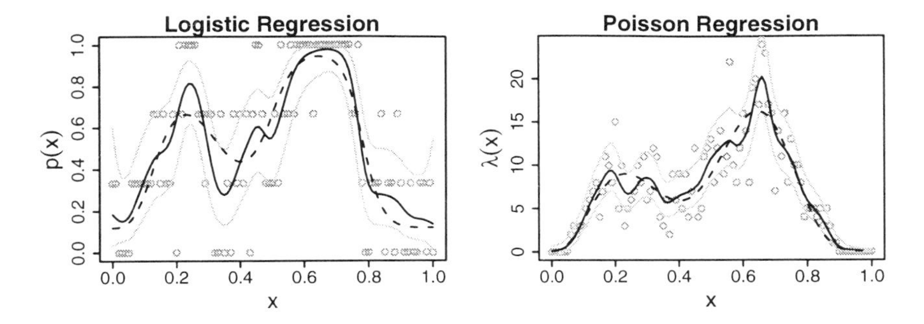

FIGURE 5.3. Cubic Spline Logistic and Poisson Regression. The test functions are indicated by dashed lines, the fits are indicated by solid lines, and the 95% Bayesian confidence intervals are indicated by faded solid lines. The data are superimposed as circles.

```
plot(x,y/3,ylab="p"); lines(x,p,lty=2)
lines(x,1-1/(1+exp(est$fit)))
lines(x,1-1/(1+exp(est$fit+1.96*est$se)),col=5)
lines(x,1-1/(1+exp(est$fit-1.96*est$se)),col=5)
```

Note that the predictions are on the logit scale. The working residuals and deviance residuals are also available:

```
resid(logit.fit)
resid(logit.fit,type="dev")
```

By default, the performance-oriented iteration is driven by the score $U_w(\lambda)$ (`method="u"`) with the dispersion $a(\phi)$ set to 1 (`varht=1`) for the binomial family. □

5.4.2 Poisson Family

For the Poisson regression of Example 5.3 with a log likelihood

$$y \log \lambda - \lambda, \tag{5.27}$$

$\tilde{u}_i = -Y_i + \lambda(x_i)$ and $\tilde{w}_i = \lambda(x_i)$; see Problem 5.8. The use of `gssanova` is illustrated in the following example.

Example 5.5 (Poisson regression)
Similar to the commands used in Example 5.4, the following sequence fits a cubic spline Poisson regression model to a synthetic data set and plots the fit in the right frame of Figure 5.3, with the data and the test function superimposed:

```
# generate the data
set.seed(5732)
test <- function(x)
        {(1e6*(x^11*(1-x)^6)+1e4*(x^3*(1-x)^10))+.1}
x <- (0:100)/100
lam <- test(x)
y <- rpois(x,lam)
# calculate the fit
pois.fit <- gssanova(y~x,family="poisson")
# predict
est <- predict(pois.fit,data.frame(x=x),se=TRUE)
# plot the fit
plot(x,y); lines(x,lam,lty=2)
lines(x,exp(est$fit))
lines(x,exp(est$fit+1.96*est$se),col=5)
lines(x,exp(est$fit-1.96*est$se),col=5)
```

By default, the performance-oriented iteration is driven by the score $U_w(\lambda)$ with $a(\phi) = 1$ for the Poisson family. □

5.4.3 Gamma Family

The gamma distribution $\text{Gamma}(\alpha, \beta)$ has a density

$$\frac{1}{\beta^\alpha \Gamma(\alpha)} y^{\alpha-1} e^{-y/\beta}, \quad \alpha, \beta, y > 0,$$

where α is the shape parameter and β is the scale parameter. When $\alpha = 1$, the gamma distribution reduces to the exponential distribution. Reparameterizing by (α, μ), where $\mu = \alpha\beta = E[Y]$, and dropping terms that do not involve μ, one has a log likelihood

$$\Big\{ -\frac{y}{\mu} - \log \mu \Big\} \alpha, \tag{5.28}$$

with α^{-1} being the dispersion parameter; see Problem 5.9. To avoid the constraint associated with the canonical parameter $-\mu^{-1}$, we choose to work with the log link $\eta = \log \mu$. To iterate on (5.3), one has $\tilde{u}_i = -Y_i/\mu(x_i) + 1$ and $\tilde{w}_i = Y_i/\mu(x_i)$; see Problem 5.10. Note that $E[u] = 0$, $\text{Var}[u] = \alpha^{-1}$, and $E[w] = \alpha^{-1}$.

Example 5.6 (Gamma regression)
Similar to earlier examples, the following sequence fits a cubic spline gamma regression model to a synthetic data set and plots the fit in the left frame of Figure 5.4:

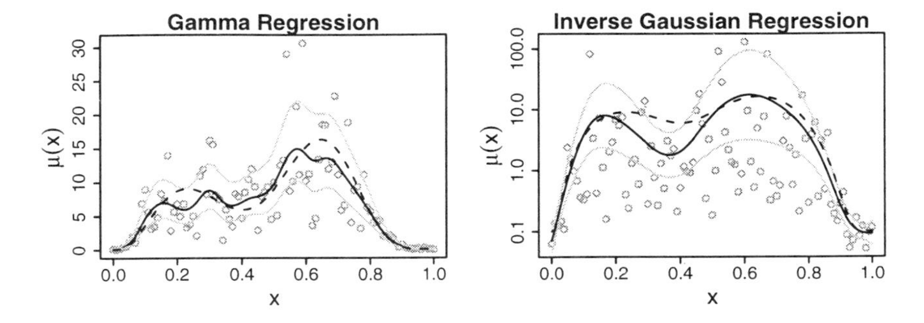

FIGURE 5.4. Cubic Spline Gamma and Inverse Gaussian Regression. The test functions are indicated by dashed lines, the fits are indicated by solid lines, and the 95% Bayesian confidence intervals are indicated by faded solid lines. The data are superimposed as circles.

```
# generate the data
set.seed(5732)
test <- function(x)
        {(1e6*(x^11*(1-x)^6)+1e4*(x^3*(1-x)^10))+.1}
x <- (0:100)/100
mu <- test(x)
y <- rgamma(x,4,mu/4)
# calculate the fit
gamma.fit <- gssanova(y~x,family="Gamma")
# predict
est <- predict(gamma.fit,data.frame(x=x),se=TRUE)
# plot the fit
plot(x,y); lines(x,mu,lty=2)
lines(x,exp(est$fit))
lines(x,exp(est$fit+1.96*est$se),col=5)
lines(x,exp(est$fit-1.96*est$se),col=5)
```

By default, the performance-oriented iteration is driven by the score $V_w(\lambda)$ (`method="v"`) for the gamma family, with the dispersion parameter estimated by the weighted version of σ_v^2 on convergence; see (3.26) on page 67 for σ_v^2. □

5.4.4 Inverse Gaussian Family

The inverse Gaussian distribution $\mathrm{IG}(\mu, \sigma^2)$ has a density

$$\frac{1}{\sqrt{2\pi\sigma^2}} y^{-3/2} e^{-(y-\mu)^2/2\sigma^2\mu^2 y}, \quad \mu, \sigma^2, y > 0,$$

where $E[Y] = \mu$ and $\text{Var}[Y] = \sigma^2\mu^3$. Dropping terms that do not involve μ, one has a log likelihood

$$\Big\{ -\frac{y}{2\mu^2} + \frac{1}{\mu} \Big\} \frac{1}{\sigma^2}, \tag{5.29}$$

with σ^2 the dispersion parameter; see Problem 5.11. Working with the log link $\eta = \log\mu$, one has $\tilde{u}_i = -Y_i/\mu^2(x_i) + 1/\mu(x_i)$ and $\tilde{w}_i = 2Y_i/\mu^2(x_i) - 1/\mu(x_i)$. Note that $E[u] = 0$, $\text{Var}[u] = \sigma^2/\mu$, and $E[w] = 1/\mu$. Since $\tilde{w}_i$ can be negative, we use its expected value $\tilde{w}_i^* = 1/\mu(x_i)$ in (5.3).

Example 5.7 (Inverse Gaussian regression)
Similar to earlier examples, the following sequence fits a cubic spline inverse Gaussian regression model to a synthetic data set and plots the fit in the right frame of Figure 5.4:

```
# generate the data
set.seed(5732)
test <- function(x)
        {(1e6*(x^11*(1-x)^6)+1e4*(x^3*(1-x)^10))+.1}
x <- (0:100)/100
mu <- test(x)
y <- rinvgauss(x,mu,1)
# calculate the fit
ig.fit <- gssanova(y~x,family="inverse.gaussian")
# predict
est <- predict(ig.fit,data.frame(x=x),se=TRUE)
# plot the fit
plot(x,y,log="y"); lines(x,mu,lty=2)
lines(x,exp(est$fit))
lines(x,exp(est$fit+1.96*est$se),col=5)
lines(x,exp(est$fit-1.96*est$se),col=5)
```

To date, R does not have the distribution facilities for inverse Gaussian. The `rinvgauss` function was found at

`http://www.maths.uq.edu.au/~gks/s/invgauss.html`

By default, the performance-oriented iteration is driven by the score $V_w(\lambda)$ (`method="v"`) for the inverse Gaussian family, with the dispersion parameter estimated by the weighted version of σ_v^2 on convergence. □

5.4.5 Negative Binomial Family

The negative binomial distribution has a density

$$\frac{\Gamma(\alpha + y)}{y!\,\Gamma(\alpha)} p^\alpha (1-p)^y, \quad \alpha > 0, \; p \in (0,1), \; y = 0, 1, \ldots . \tag{5.30}$$

For α an integer, the distribution describes the number of failures before the αth success in a sequence of Bernoulli trials with a success probability p. The distribution also describes the behavior of composite Poisson data with $Y \sim \text{Poisson}(\lambda)$ and $\lambda \sim \text{Gamma}(\alpha, (1-p)/p)$; see Problem 5.13. It can be shown that $E[Y] = \alpha(1-p)/p$ and $\text{Var}[Y] = \alpha(1-p)/p^2$. Taking $\eta = \log\{p/(1-p)\}$ and dropping terms that do not involve η, one has a log likelihood

$$-(\alpha + y)\log(1 + e^{\eta}) + \alpha\eta; \tag{5.31}$$

see Problem 5.14. It follows that $\tilde{u}_i = (\alpha_i + Y_i)p(x_i) - \alpha_i$ and $\tilde{w}_i = (\alpha_i + Y_i)p(x_i)(1-p(x_i))$; see Problem 5.15. Note that $E[u] = 0$, $\text{Var}[u] = \alpha(1-p)$, and $E[w] = \alpha(1-p)$.

The implementation of the negative binomial family has two options implicitly invoked through the composition of the responses. The first option assumes known α_i's, which are to be provided in the second column of the response data. The second option assume common but unknown α, and for every iterate $\tilde{\eta}$, an estimate of α that maximizes

$$\textstyle\sum_{i=1}^{n}\{\log\Gamma(\alpha + Y_i) - \log\Gamma(\alpha) + \alpha\log p_i\}$$

is calculated and used in $\tilde{u}_i$ and $\tilde{w}_i$ for the updating of $\tilde{Y}_i$. The following example illustrates the use of gssanova for the negative binomial family.

Example 5.8 (Logistic regression with negative binomial data)
First, generate the data:

```
set.seed(5732)
test <- function(x)
        {1e+06*(x^11*(1-x)^6)+1e4*(x^3*(1-x)^10)+.1}
x <- (0:100)/100
mu <- test(x)
p <- 1/(mu/2+1)
y <- rnbinom(x,2,p)
```

Now, fit the model with known and unknown α. For unknown α, the iteration moves along slowly in tiny steps, so a large number of iterations are needed:

```
nb.fit1 <- gssanova(cbind(y,2)~x,family="nbinomial")
nb.fit2 <- gssanova(y~x,family="nbinomial",maxiter=70)
```

The fits can now be evaluated and plotted as in Figure 5.5:

```
# known alpha
est1 <- predict(nb.fit1,data.frame(x=x),se=TRUE)
plot(x,y); lines(x,2*(1-p)/p,lty=2)
lines(x,2/exp(est1$fit))
lines(x,2/exp(est1$fit+est1$se),col=5)
```

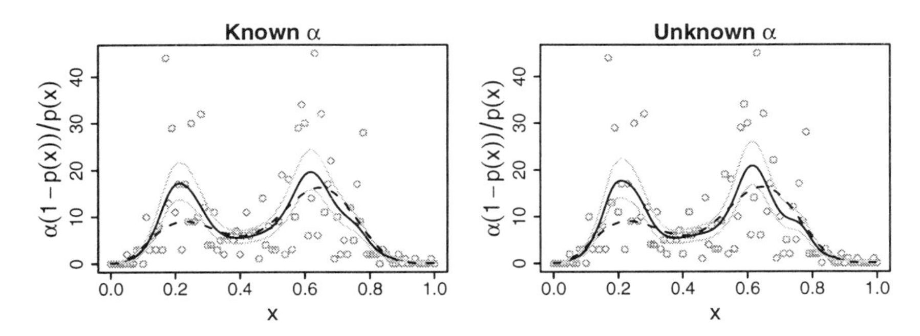

FIGURE 5.5. Cubic Spline Logistic Regression with Negative Binomial Data. The test function is indicated by dashed lines, the fits are indicated by solid lines, and the 95% Bayesian confidence intervals are indicated by faded solid lines. The data are superimposed as circles.

```
lines(x,2/exp(est1$fit-est1$se),col=5)
# unknown alpha
est2 <- predict(nb.fit2,data.frame(x=x),se=TRUE)
alpha <- nb.fit2$alpha
plot(x,y); lines(x,2*(1-p)/p,lty=2)
lines(x,alpha/exp(est2$fit))
lines(x,alpha/exp(est2$fit+est2$se),col=5)
lines(x,alpha/exp(est2$fit-est2$se),col=5)
```

By default, the performance-oriented iteration is driven by the score $U_w(\lambda)$ with $a(\phi) = 1$ for the negative binomial family. □

5.5 Case Studies

We now apply the techniques developed in this chapter to analyze a few real data sets. It will be seen that Poisson regression can be used to estimate a probability density and that gamma regression can be used to estimate the spectral density of a stationary time series.

5.5.1 Eruption Time of Old Faithful

Listed in Härdle (1991) are the duration and the waiting time to the next eruption gathered from 272 consecutive eruptions of the Old Faithful geyser in Yellowstone National Park. The data are available in R as a data frame `faithful` with components `eruptions` and `waiting`, both in minutes. In this study, we use the first component to estimate a continuous "mass spectrum" of the eruption duration.

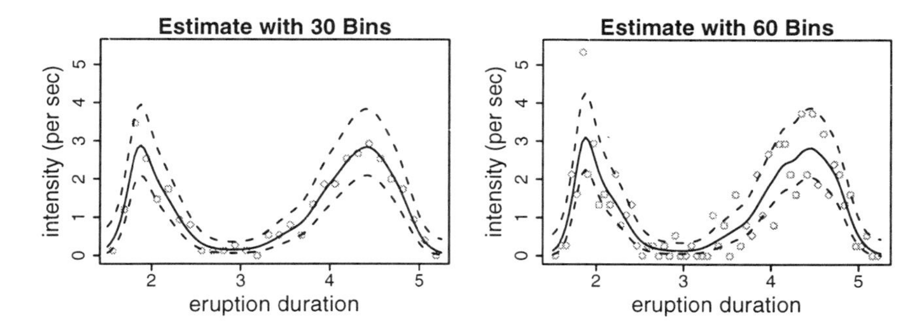

FIGURE 5.6. Mass Spectrum of Eruption Duration of Old Faithful. The estimated Poisson intensity is indicated by solid lines and the 95% Bayesian confidence intervals are indicated by dashed lines. The data are superimposed as circles.

The range of the eruption times is $[1.6, 5.1]$. Rounding the data to a histogram of 30 bins on $[1.5, 5.25]$, each of length 0.125, one has x_i as the middle points of the bins and Y_i as the frequencies of the bins:

```
jk <- hist(faithful$erup,breaks=seq(1.5,5.25,length=31),
           freq=TRUE,plot=FALSE)
x <- jk$mids
y <- jk$counts
```

The continuous "mass spectrum" can be estimated through a Poisson regression, which is plotted in the left frame of Figure 5.6:

```
faithful.fit <- gssanova(y~x,family="poisson",
                         offset=rep(log(60/8),30))
xx <- seq(1.5,5.25,length=101)
est <- predict(faithful.fit,
               data.frame(x=xx,offset=0),se=TRUE)
# plot the fit
plot(x,y*8/60)
lines(xx,exp(est$fit))
lines(xx,exp(est$fit+1.96*est$se),lty=2)
lines(xx,exp(est$fit-1.96*est$se),lty=2)
```

The `offset` term scales the estimate to the unit of per-second intensity; note that Y_i are counts per 1/8 minute. For the evenly binned data given here, the offset is not necessary, as one can always rescale the fit afterward, but if the data are given in heterogeneous units, the offset provides a convenient device to align them to a common scale; see Problem 5.16.

Repeating the process with a histogram of 60 bins on $[1.5, 5.25]$, one gets the estimate in the right frame of Figure 5.6.

Scaling the Poisson intensity to integrate to 1 on the domain $[1.5, 5.25]$, one gets a probability density of eruption duration; see §6.2. The Bayesian confidence intervals lose their meaning for a density, however. An analysis of the data using density estimation techniques will be shown in §6.5.2.

5.5.2 Spectrum of Yearly Sunspots

The yearly number of sunspots from 1700 to 1988 can be found in Tong (1990, page 471). Our task here is to estimate the frequency spectrum of the series.

For a stationary time series X_t, $t = 0, \pm 1, \pm 2, \ldots$, with covariance function $\gamma_k = \text{Cov}(X_t, X_{t+k})$, the spectral density is defined by

$$f(\omega) = \frac{1}{\gamma_0} \sum_{k=-\infty}^{\infty} \gamma_k e^{-\mathbf{i}2\pi k\omega}, \quad \omega \in (-0.5, 0.5),$$

where $\mathbf{i} = \sqrt{-1}$, which satisfies

$$\gamma_k = \gamma_0 \int_{-0.5}^{0.5} f(\omega) e^{\mathbf{i}2\pi k\omega} d\omega.$$

See, e.g., Priestley (1981, §4.8.3) and Brockwell and Davis (1991, §4.3), where the frequency is parameterized by $\tilde{\omega} = 2\pi\omega \in (-\pi, \pi)$. The spectral density is an even function, so one only needs to estimate $f(\omega)$ on $(0, 0.5)$. Observing x_t, $t = 1, \ldots, T$, one may calculate the discrete Fourier transform (cf. §4.2.2)

$$\tilde{x}_\nu = \frac{1}{\sqrt{T}} \sum_{t=1}^{T} x_t e^{-\mathbf{i}2\pi t\nu/T}, \quad \nu = 0, 1, \ldots, T-1, \tag{5.32}$$

which yields the periodogram $I(\omega_\nu) = |\tilde{x}_\nu|^2$ on the so-called Fourier frequencies $\omega_\nu = \nu/T$. Note that $I(\omega_\nu) = I(\omega_{T-\nu})$. For T large, it can be shown that $I(\omega_\nu)$, $\omega_\nu \in (0, 0.5)$, are asymptotically independent exponential random variables with means $E[I(\omega_\nu)] \propto f(\omega_\nu)$; see, e.g., Priestley (1981, page 425) and Brockwell and Davis (1991, Theorem 10.3.2). The estimation of the spectrum can thus be obtained from a gamma regression with $x_\nu = \omega_\nu$ and $Y_\nu = I(\omega_\nu)$.

The observed series are available as an R object `sunspot.year` from package `ts`. The following commands load the data, calculate the periodogram, and set up x_ν and Y_ν for gamma regression:

```
data(sunspot,package=ts)
n <- length(sunspot.year)
ind <- 1:(ceiling(n/2)-1)
y <- (abs(fft(sunspot.year))^2/n)[-1][ind]
x <- ind/n
```

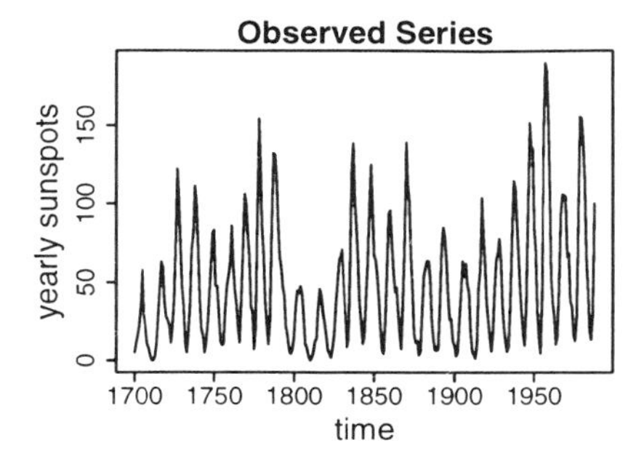

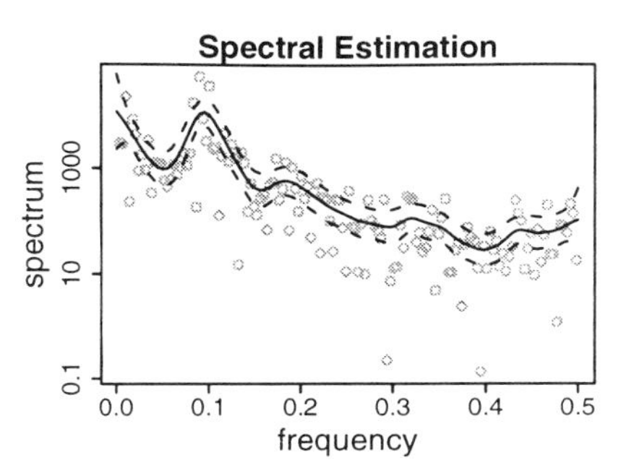

FIGURE 5.7. Spectrum of Yearly Sunspots. Left: Observed series. Right: Spectral estimate with 95% Bayesian confidence intervals; the periodogram is superimposed as circles.

The `fft` function calculates an unscaled discrete Fourier transform [i.e., the transform given in (5.32) but without $1/\sqrt{T}$]. A gamma regression model can now be fitted to the periodogram and plotted as in the right frame of Figure 5.7.

```
sunspot.fit <- gssanova(y~x,family="Gamma",
                        method="u",varht=1)
xx <- seq(0,.5,length=101)
est <- predict(sunspot.fit,data.frame(x=xx),se=TRUE)
# plot the fit
plot(x,y,log="y")
lines(xx,exp(est$fit))
lines(xx,exp(est$fit+1.96*est$se),lty=2)
lines(xx,exp(est$fit-1.96*est$se),lty=2)
```

Note that the dispersion parameter is known to be 1 for the exponential distribution and that $U_w(\lambda)$ can be used when the dispersion is known. Scaling the estimate to integrate to 0.5 on $(0, 0.5)$, one gets the spectral density. The Bayesian confidence intervals lose their meaning for a spectral density, however.

5.5.3 Progression of Diabetic Retinopathy

The Wisconsin Epidemiological Study of Diabetic Retinopathy (WESDR) is an on-going epidemiological study of a cohort of patients receiving their medical care in an 11-county area in southern Wisconsin, who were first examined in 1980-1982, then again in 1984-1986, 1990-1992, and 1994-1996. A subset derived from the WESDR data is distributed in GRKPACK (Wang 1997), to be found at

```
ftp://ftp.stat.wisc.edu/pub/wahba/software
```

which consists of the baseline measures of duration of diabetes in years, percent of glycosylated hemoglobin, body mass index, and a binary indicator of retinopathy progression at the first follow-up, of 669 patients. There were 278 positive cases among the 669 patients.

The data were read into R as a data frame `wesdr` with components `dur`, `gly`, `bmi`, and `ret`. We first fit a tensor product linear spline with all interactions included and inspect the diagnostics; the diagnostics are based on the weighted least squares at the fit:

```
wesdr.fit0 <- gssanova(ret~dur*bmi*gly,data=wesdr,
                       family="binomial",type="linear")
sum.fit0 <- summary(wesdr.fit0,diag=TRUE)
round(sum.fit0$pi,2)
#   dur  bmi  gly dur:bmi dur:gly bmi:gly dur:bmi:gly
#  0.14 0.06 0.80    0.00    0.00    0.00        0.00
round(sum.fit0$cos,2)
#         dur  bmi  gly dur:bmi dur:gly bmi:gly
# cos.y  0.21 0.13 0.36    0.22    0.35    0.29
# cos.e  0.07 0.07 0.05    0.10    0.07    0.13
# norm   3.56 2.99 9.28    0.00    0.00    0.00
#          dur:bmi:gly    yhat       y       e
# cos.y           0.38    0.45    1.00    0.93
# cos.e           0.15    0.09    0.93    1.00
# norm            0.00   10.08   27.41   24.60
```

The calibrations of the cosines for Gaussian data, as discussed in §3.6, may not be appropriate in this context, but the norms clearly indicate that all the interactions are effectively eliminated during the performance-oriented iteration driven by $U_w(\lambda)$. The deviance of the fit is 736.60.

One can now fit a cubic spline additive model and evaluate the components on the sampling points:

```
wesdr.fit <- gssanova(ret~dur+bmi+gly,data=wesdr,
                      family="binomial")
sum.fit <- summary(wesdr.fit,diag=TRUE)
round(sum.fit$kappa,2)
#  dur  bmi  gly
# 1.01 1.04 1.03
round(sum.fit$pi,2)
#  dur  bmi  gly
# 0.14 0.07 0.79
round(sum.fit$cos,2)
#         dur  bmi  gly  yhat     y     e
# cos.y  0.18 0.10 0.32  0.39  1.00  0.93
# cos.e  0.03 0.02 0.00  0.02  0.93  1.00
# norm   3.66 3.75 9.79 10.50 28.01 25.76
```

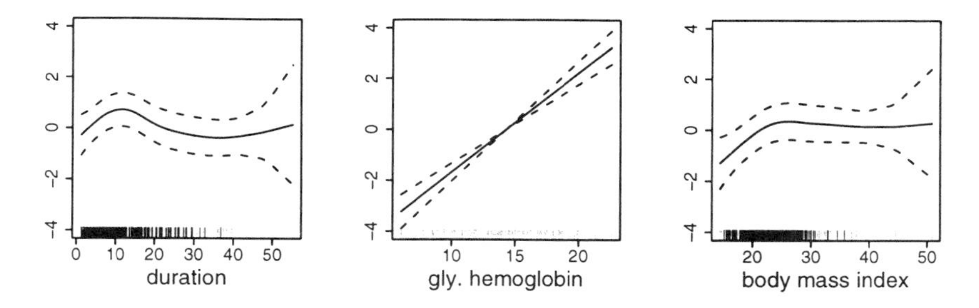

FIGURE 5.8. Factors Affecting Diabetic Retinopathy Progression. Left: Effect of duration of diabetes. Center: Effect of percent of glycosylated hemoglobin. Right: Effect of body mass index. The logit components are indicated by solid lines and the 95% Bayesian confidence intervals are indicated by dashed lines. The rugs on the bottom mark the sampling points.

```
est.dur <- predict(wesdr.fit,wesdr,se=TRUE,inc="dur")
est.bmi <- predict(wesdr.fit,wesdr,se=TRUE,inc="bmi")
est.gly <- predict(wesdr.fit,wesdr,se=TRUE,inc="gly")
```

The fitted logit components are plotted in Figure 5.8 with Bayesian confidence intervals. The effect of glycosylated hemoglobin was linear and was dominant. The rugs on the bottom of the plots mark the marginal sampling points, and it is comforting to see that the standard errors are larger in sparse areas. The deviance of the fit is 741.16.

For comparison, a linear logistic regression yields a deviance of 780.98, with the duration effect insignificant (p-value 0.45).

5.6 Bibliographic Notes

Section 5.1

Penalized likelihood regression was formulated and studied by O'Sullivan, Yandell, and Raynor (1986); see also Silverman (1978) and Green and Yandell (1985). Fits with multiple penalty terms were found in Gu (1990) and Wahba, Wang, Gu, Klein, and Klein (1995), among others.

A standard reference on linear parametric regression with exponential family responses, better known as generalized linear models, is McCullagh and Nelder (1989), where extensive discussions can be found on the properties of exponential families and on the use of iterated weighted least squares in the fitting of generalized linear models.

Section 5.2

A suggestion in the early literature was to compute the minimizer η_λ of (5.1) for fixed λ, evaluate $V_w(\lambda|\tilde{\eta})$ at $\tilde{\eta} = \eta_\lambda$, and compare such $V_w(\lambda)$ values on a grid of λ. This amounts to comparing the scores on the dashed slice in Figure 5.2. Since $V_w(\lambda|\tilde{\eta})$ with different $\tilde{\eta}$ are not comparable, this approach is ineffective, as was shown in Gu (1992a).

Performance-oriented iteration was used implicitly by Gu (1990), but the mechanism and the related issues were not understood until Gu (1992a). The direct cross-validation through $V_0(\lambda)$ of (5.8) was proposed by Cox and Chang (1990). Xiang and Wahba (1996) derived the more effective and computable GACV score $V_g(\lambda)$. Gu and Xiang (2001) derived the numerically stable, readily computable $V_g^*(\lambda)$ and $V_g^{**}(\lambda)$ and proved the equivalence of $V_g(\lambda)$ and $V_g^*(\lambda)$.

Section 5.3

The adaptation of Bayesian confidence intervals for non-Gaussian regression was proposed and illustrated by Gu (1992c). Examples of component-wise intervals were shown in Wahba, Wang, Gu, Klein, and Klein (1995).

Section 5.4

The `gssanova` suite in the R package `gss` was released to the public in 1999. GRKPACK, a collection of RATFOR routines implementing the performance-oriented iteration, was put together earlier by Wang (1997).

Extensive discussion of binomial, Poisson, and gamma distributions can be found in McCullagh and Nelder (1989). Facts concerning the inverse Gaussian distribution can be found in Chhikara and Folks (1989). Generalized linear model for the negative binomial family is discussed in Venables and Ripley (1999, §7.4).

Section 5.5

Various versions of the Old Faithful eruption data have been used in the literature to showcase regression and density estimation techniques; see, e.g., Azzalini and Bowman (1990), Härdle (1991), and Scott (1992), among others. A nice discussion of density estimation through Poisson regression can be found in Lindsey (1997, Chapter 3).

The sunspot data or subsets thereof are among the most popular examples being used in textbooks and research articles on time series analysis. Spectral estimation through gamma regression with unit dispersion was studied by Pawitan and O'Sullivan (1994); see also Cogburn and Davis (1974) and Wahba (1980).

Detailed descriptions of the WESDR data can be found in, e.g., Klein, Klein, Moss, Davis, and DeMets (1988, 1989), among others. The analysis presented here differs slightly from the one found in Wahba, Wang, Gu, Klein, and Klein (1995).

5.7 Problems

Section 5.1

5.1 Consider the functional $L(f) = -\sum_{i=1}^{n}\{Y_i f(x_i) - b(f(x_i))\}$ in a reproducing kernel Hilbert space $\mathcal{H}$ with a square seminorm $J(f)$.

(a) Prove that $L(f)$ is continuous, convex, and Fréchet differentiable.

(b) Let $\{\phi_\nu, \nu = 1, \ldots, m\}$ be a basis of $\mathcal{N}_J = \{f : J(f) = 0\}$ and S be $n \times m$ with the (i, ν)th entry $\phi_\nu(x_i)$. Prove that if S is of full column rank, then $L(f)$ is strictly convex in $\mathcal{N}_J$.

(c) Prove that if S is of full column rank, then $L(f) + \lambda J(f)$ is strictly convex in $\mathcal{H}$.

Section 5.2

5.2 Prove Theorem 5.2.

5.3 Rewrite (3.12) on page 62 as

$$A_w(\lambda) = I - n\lambda W^{-1/2}(M^{-1} - M^{-1}S(S^T M^{-1}S)^{-1}S^T M^{-1})W^{-1/2},$$

where $M = Q + n\lambda W^{-1}$. Let $S = FR^* = (F_1, F_2)\binom{R}{O} = F_1 R$ be the QR-decomposition of S with F orthogonal and R upper-triangular.

(a) Show that

$$\begin{aligned} &M^{-1} - M^{-1}S(S^T M^{-1}S)^{-1}S^T M^{-1} \\ &\qquad = F_2(F_2^T Q F_2 + n\lambda F_2^T W^{-1} F_2)^{-1} F_2^T. \end{aligned}$$

(b) Let $H = (W + n\lambda F_2(F_2^T Q F_2)^+ F_2^T)^{-1}$. For $F_2^T Q F_2$ nonsingular, verify that $A_w(\lambda)(W^{1/2} H W^{1/2})^{-1} = I$.

5.4 Consider $U^*(\lambda)$ as given in (5.19) for penalized least squares regression.

(a) Show that $V_g^*(\lambda)$ of (5.17) reduces to $U^*(\lambda)$.

(b) Assume $n^{-1}\boldsymbol{\eta}^T(I - A(\lambda))\boldsymbol{\eta} = o(1)$, where $\boldsymbol{\eta}^T = (\eta(x_1), \ldots, \eta(x_n))$. If, in addition, Condition 3.2.1 of §3.2.1 also holds, that $nR(\lambda) \to \infty$, show that $U^*(\lambda) - L(\lambda) - n^{-1}\boldsymbol{\epsilon}^T\boldsymbol{\epsilon} = o_p(L(\lambda))$.

Section 5.3

5.5 Prove (5.21).

5.6 Prove (5.22).

Section 5.4

5.7 Show that $\tilde{u}_i = -Y_i + m_i\, p(x_i)$ and $\tilde{w}_i = m_i\, p(x_i)(1 - p(x_i))$ for the binomial log likelihood (5.26) with $\eta = \log\{p/(1-p)\}$.

5.8 Show that $\tilde{u}_i = -Y_i + \lambda(x_i)$ and $\tilde{w}_i = \lambda(x_i)$ for the Poisson log likelihood (5.27) with $\eta = \log \lambda$.

5.9 Derive the log likelihood (5.28) for the gamma family.

5.10 Show that $\tilde{u}_i = -Y_i/\mu(x_1) + 1$ and $\tilde{w}_i = Y_i/\mu(x_1)$ for the gamma log likelihood (5.28) with $\eta = \log \mu$.

5.11 Derive the log likelihood (5.29) for the inverse Gaussian family.

5.12 Show that $\tilde{u}_i = -Y_i/\mu^2(x_i) + 1/\mu(x_i)$ and $\tilde{w}_i = 2Y_i/\mu^2(x_i) - 1/\mu(x_i)$ for the inverse Gaussian log likelihood (5.29) with $\eta = \log \mu$.

5.13 Derive the probability density (5.30) for composite Poisson data with $Y \sim \text{Poisson}(\lambda)$ and $\lambda \sim \text{Gamma}(\alpha, (1-p)/p)$.

5.14 Derive the log likelihood (5.31) for the negative binomial family with $\eta = \log\{p/(1-p)\}$.

5.15 Show that $\tilde{u}_i = (\alpha_i + Y_i)p(x_i) - \alpha_i$ and $\tilde{w}_i = (\alpha_i + Y_i)p(x_i)(1 - p(x_i))$ for the negative binomial log likelihood (5.31) with $\eta = \log\{p/(1-p)\}$.

Section 5.5

5.16 Round the `faithful` data into an uneven histogram using the break points `seq(1.5,5.25,length=61)[-(1:20)*3]`. Estimate the per-second intensity using the uneven histogram and compare the estimate with the ones plotted in Figure 5.6.

6

Probability Density Estimation

For observational data, (1.5) of Example 1.2 defines penalized likelihood density estimation. Of interest are the selection of smoothing parameters, the computation of the estimates, and the asymptotic behavior of the estimates. Variants of (1.5) are also called for to accommodate data subject to selection bias and data from conditional distributions.

The precise formulation, the existence and uniqueness, and the computability of penalized likelihood density estimates are discussed in §6.1, and it is noted in §6.2 that the technique can be used to estimate inhomogeneous Poisson processes. The selection of smoothing parameters are discussed in §6.3, where a simple cross-validation score and its modifications are derived and illustrated. The computation of the estimates is briefly discussed in §6.4, and the techniques are applied to analyze a few real data sets in §6.5. Discussed in §6.7 is the estimation of the conditional density $f(y|x)$, which, with x and y on generic domains, can be specialized to treat all sorts of regression problems. Recipes for dealing with various forms of selection bias are covered in §6.6 and §6.8.

Missing from the discussion is user-friendly software, which is still under development. The computability of the estimates relies on the asymptotic convergence rates, which will be discussed in Chapter 8.

6.1 Preliminaries

Let X_i, $i = 1, \ldots, n$, be independent and identically distributed (i.i.d.) random samples from a probability density $f(x)$ on a bounded domain $\mathcal{X}$. One is to estimate $f(x)$ from the observations X_i. When some parametric form of $f(x)$ is assumed, say $f \in P_\theta = \{f(x;\theta) : \theta \in \Theta\}$, where $f(x;\theta)$ is known up to a finite-dimensional parameter θ, density estimation reduces to parameter estimation, and the maximum likelihood method is the standard technique, which possesses many favorable properties. When a parametric form is not available, however, a naive maximum likelihood density estimate without any nonintrinsic constraint (see the following paragraph for the intrinsic constraints) is a sum of delta function spikes at the sample points, which, apparently, is not an appealing estimate when the domain $\mathcal{X}$ is continuous. In between the two extremes, one may use the penalized likelihood estimate.

Two intrinsic constraints that a probability density must satisfy are the positivity constraint that $f \geq 0$ and the unity constraint that $\int_\mathcal{X} f dx = 1$. Assuming $f > 0$ on $\mathcal{X}$, one can make a logistic density transform $f = e^\eta / \int_\mathcal{X} e^\eta dx$ and estimate η instead, which is free of the positivity and unity constraints. To make the transform one-to-one, one may impose a side condition on η, say $A\eta = 0$, where A is an averaging operator on $\mathcal{X}$; see §1.3.1 for a discussion of averaging operators. The estimate of η can then be obtained by minimizing the penalized (negative) log likelihood functional,

$$-\frac{1}{n}\sum_{i=1}^{n} \eta(X_i) + \log \int_\mathcal{X} e^\eta dx + \frac{\lambda}{2} J(\eta), \tag{6.1}$$

in a reproducing kernel Hilbert space $\mathcal{H}$, in which the roughness penalty $J(\eta)$ is a square (semi) norm. The members of $\mathcal{H}$ have to comply with a side condition mentioned above to make the first term of (6.1) strictly convex. It is easy to construct such an $\mathcal{H}$ by dropping the constant term in a (one-way) ANOVA decomposition.

Let $L(f) = -n^{-1}\sum_{i=1}^{n} f(X_i) + \log \int_\mathcal{X} e^f dx$ be the negative log likelihood. When the maximum likelihood estimate exists in the null space $\mathcal{N}_J = \{f : Af = 0, J(f) = 0\}$, the following lemmas establish the existence and uniqueness of the minimizer of (6.1) via Theorem 2.9.

Lemma 6.1
$L(f)$ is strictly convex for $f \in \mathcal{H} \subseteq \{f : Af = 0\}$.

Proof: By Hölder's inequality, for $\alpha, \beta > 0$, $\alpha + \beta = 1$, and $f, g \in \mathcal{H}$,

$$\log \int_\mathcal{X} e^{\alpha f + \beta g} dx \leq \alpha \log \int_\mathcal{X} e^f dx + \beta \log \int_\mathcal{X} e^g dx,$$

where the equality holds if and only if $e^f \propto e^g$, which amounts to $f = g$ with $Af = Ag = 0$. □

Lemma 6.2
If $e^{|f|}$ are Riemann integrable on $\mathcal{X}$ for all $f \in \mathcal{H}$, then $L(f)$ is continuous in $\mathcal{H}$. Furthermore, $L(f + \alpha g)$, $\forall f, g \in \mathcal{H}$, is infinitely differentiable as a function of α real.

Proof: The claims follow from the Riemann sum approximations of related integrals and the continuity of evaluation. □

A simple example follows.

Example 6.1 (Cubic spline)
Let $\mathcal{X} = [0, 1]$ and $J(\eta) = \int_0^1 \ddot{\eta}^2 dx$. The null space of $J(\eta)$ without side condition is $\text{span}\{1, x\}$. One has the choice of at least two different formulations.

The first formulation employs the construction of §2.3.1. Take $Af = f(0)$. One has

$$\mathcal{H} = \{f : f(0) = 0, \int_0^1 \ddot{f}^2 dx < \infty\} = \mathcal{N}_J \oplus \mathcal{H}_J,$$

where $\mathcal{N}_J = \text{span}\{x\}$ and

$$\mathcal{H}_J = \{f : f(0) = \dot{f}(0) = 0, \int_0^1 \ddot{f}^2 dx < \infty\},$$

with $R_J(x, y) = \int_0^1 (x - u)_+ (y - u)_+ du$.

The second formulation employs the construction of §2.3.3. Take $Af = \int_0^1 f dx$. One has

$$\mathcal{H} = \{f : \int_0^1 f dx = 0, \int_0^1 \ddot{f}^2 dx < \infty\} = \mathcal{N}_J \oplus \mathcal{H}_J,$$

where $\mathcal{N}_J = \text{span}\{x - .5\}$ and

$$\mathcal{H}_J = \{f : \int_0^1 f dx = \int_0^1 \dot{f} dx = 0, \int_0^1 \ddot{f}^2 dx < \infty\},$$

with $R_J(x, y) = k_2(x)k_2(y) - k_4(x - y)$; see (2.27) on page 37 for $k_2(x)$ and $k_4(x)$. □

With the same data and the same penalty, one would naturally expect that the two formulations of Example 6.1 would yield the same density estimate. It is indeed the case, as assured by the following proposition.

Proposition 6.3
Let $\mathcal{H} \subseteq \{f : J(f) < \infty\}$ and suppose that $J(f)$ annihilates constant. For any two different averaging operators A_1 and A_2, if η_1 minimizes (6.1) in $\mathcal{H}_1 = \mathcal{H} \cap \{A_1 f = 0\}$ and η_2 minimizes (6.1) in $\mathcal{H}_2 = \mathcal{H} \cap \{A_2 f = 0\}$, then $e^{\eta_1} / \int_{\mathcal{X}} e^{\eta_1} dx = e^{\eta_2} / \int_{\mathcal{X}} e^{\eta_2} dx$.

Proof: For any $f \in \mathcal{H}_1$, it is easy to check that $Pf = f - A_2 f \in \mathcal{H}_2$, $L(Pf) = L(f)$, and $J(Pf) = J(f)$. Similarly, for any $g \in \mathcal{H}_2$, $Qg = g - A_1 g \in \mathcal{H}_1$, $L(Qg) = L(g)$, and $J(Qg) = J(g)$. Now, for $f \in \mathcal{H}_1$,

$Q(Pf) = Pf - A_1(Pf) = (f - A_2 f) - A_1(f - A_2 f) = f$, so there is an isomorphism between $\mathcal{H}_1$ and $\mathcal{H}_2$. Clearly, $e^f / \int_{\mathcal{X}} e^f dx = e^{Pf} / \int_{\mathcal{X}} e^{Pf} dx$. The proposition follows. □

Example 6.2 (Tensor product spline)
Consider the domain $\mathcal{X} = [0,1]^3$. Multiple-term models can be constructed using the tensor product splines of §2.4, with an ANOVA decomposition

$$f = f_\emptyset + f_1 + f_2 + f_3 + f_{1,2} + f_{1,3} + f_{2,3} + f_{1,2,3},$$

where terms other than the constant $f_\emptyset$ satisfy certain side conditions. The constant shall be dropped for density estimation to maintain a one-to-one logistic density transform. The remaining seven components can all be included or excluded separately, resulting in 2^7 possible models of different complexities. The additive model implies the independence of the three coordinates, and it is easily seen to be equivalent to solutions of three separate problems on individual axes. Less trivial probability structures may also be built in via selective inclusion of the ANOVA terms. For example, the conditional independence of $x_{\langle 1 \rangle}$ and $x_{\langle 2 \rangle}$ given $x_{\langle 3 \rangle}$ may be incorporated by excluding $f_{1,2}$ and $f_{1,2,3}$ from the model.

The above discussion is simply a partial repeat of §1.3.3, where more discussions can be found. □

In addition to the evaluations $[x_i]\eta = \eta(x_i)$, the first term of (6.1) depends on η also through the integral $\int_{\mathcal{X}} e^\eta dx$. This breaks the argument of §2.3.2, so the solution expression (3.2) on page 60 no longer holds for the minimizer η_λ of (6.1) in the space $\mathcal{H} = \{f : Af = 0, J(f) < \infty\}$. Actually, η_λ is, in general, not computable. Nevertheless, one may substitute (3.2) into (6.1) anyway and calculate the minimizer η_λ^* in the (data-adaptive) finite-dimensional space

$$\mathcal{H}^* = \mathcal{N}_J \oplus \text{span}\{R_J(X_i, \cdot), i = 1, \dots, n\}. \tag{6.2}$$

It will be shown in §8.2 that η_λ^* and η_λ share the same asymptotic convergence rates, so there is no loss of asymptotic efficiency in the substitution. When the maximum likelihood estimate exists in the null space $\mathcal{N}_J$, the existence and uniqueness of η_λ^* follow from Lemmas 6.1 and 6.2.

Proposition 6.3 does not apply to η_λ^*, however. For the two different formulations in Example 6.1, $\mathcal{H}^*$ are different. The situation here is different from that in regression, where the condition $S^T\mathbf{c} = 0$ satisfied by the minimizing coefficients (cf. (3.4) and (3.10) on pages 61 and 62) effectively restricts the different linear spaces $\text{span}\{R_J(X_i, \cdot), i = 1, \dots, n\}$ with different R_J's to a common subspace. The different $\mathcal{H}^*$, however, only differ in m dimensions, m being the dimension of $\mathcal{N}_J$ including the constant, and the asymptotic convergence results of §8.2 hold regardless which R_J is used. Hence, the nonuniqueness due to different choices of $\mathcal{H}^*$ poses no practical problem.

In the rest of the chapter, we shall focus on η_λ^* but drop the star from the notation. Plugging the expression

$$\eta(x) = \sum_{\nu=1}^{m} d_\nu \phi_\nu(x) + \sum_{i=1}^{n} c_i R_J(X_i, x) = \boldsymbol{\phi}^T \mathbf{d} + \boldsymbol{\xi}^T \mathbf{c} \tag{6.3}$$

into (6.1), the calculation of η_λ reduces to the minimization of

$$A_\lambda(\mathbf{c}, \mathbf{d}) = -\frac{1}{n}\mathbf{1}^T(Q\mathbf{c} + S\mathbf{d}) + \log \int_{\mathcal{X}} \exp(\boldsymbol{\xi}^T\mathbf{c} + \boldsymbol{\phi}^T\mathbf{d})dx + \frac{\lambda}{2}\mathbf{c}^T Q \mathbf{c} \tag{6.4}$$

with respect to $\mathbf{c}$ and $\mathbf{d}$, where Q is $n \times n$ with the (i,j)th entry $\xi_i(X_j) = R_J(X_i, X_j)$ and S is $n \times m$ with the (i,ν)th entry $\phi_\nu(X_i)$.

Write $\mu_f(g) = \int g e^f dx / \int e^f dx$, $V_f(g,h) = \mu_f(gh) - \mu_f(g)\mu_f(h)$, and $V_f(g) = V_f(g,g)$. Taking derivatives at $\tilde{\eta} = \boldsymbol{\xi}^T\tilde{\mathbf{c}} + \boldsymbol{\phi}^T\tilde{\mathbf{d}} \in \mathcal{H}^*$, one has

$$\begin{aligned}
\frac{\partial A_\lambda}{\partial \mathbf{c}} &= -Q\mathbf{1}/n + \mu_{\tilde{\eta}}(\boldsymbol{\xi}) + \lambda Q\tilde{\mathbf{c}} = -Q\mathbf{1}/n + \mu_\xi + \lambda Q \tilde{\mathbf{c}}, \\
\frac{\partial A_\lambda}{\partial \mathbf{d}} &= -S^T\mathbf{1}/n + \mu_{\tilde{\eta}}(\boldsymbol{\phi}) = -S^T\mathbf{1}/n + \mu_\phi, \\
\frac{\partial^2 A_\lambda}{\partial \mathbf{c}\partial \mathbf{c}^T} &= V_{\tilde{\eta}}(\boldsymbol{\xi}, \boldsymbol{\xi}^T) + \lambda Q = V_{\xi,\xi} + \lambda Q, \\
\frac{\partial^2 A_\lambda}{\partial \mathbf{d}\partial \mathbf{d}^T} &= V_{\tilde{\eta}}(\boldsymbol{\phi}, \boldsymbol{\phi}^T) = V_{\phi,\phi}, \\
\frac{\partial^2 A_\lambda}{\partial \mathbf{c}\partial \mathbf{d}^T} &= V_{\tilde{\eta}}(\boldsymbol{\xi}, \boldsymbol{\phi}^T) = V_{\xi,\phi};
\end{aligned} \tag{6.5}$$

see Problem 6.1. The Newton updating equation is thus

$$\begin{pmatrix} V_{\xi,\xi} + \lambda Q & V_{\xi,\phi} \\ V_{\phi,\xi} & V_{\phi,\phi} \end{pmatrix} \begin{pmatrix} \mathbf{c} - \tilde{\mathbf{c}} \\ \mathbf{d} - \tilde{\mathbf{d}} \end{pmatrix} = \begin{pmatrix} Q\mathbf{1}/n - \mu_\xi - \lambda Q\tilde{\mathbf{c}} \\ S^T\mathbf{1}/n - \mu_\phi \end{pmatrix}. \tag{6.6}$$

After rearranging terms, (6.6) becomes

$$\begin{pmatrix} V_{\xi,\xi} + \lambda Q & V_{\xi,\phi} \\ V_{\phi,\xi} & V_{\phi,\phi} \end{pmatrix} \begin{pmatrix} \mathbf{c} \\ \mathbf{d} \end{pmatrix} = \begin{pmatrix} Q\mathbf{1}/n - \mu_\xi + V_{\xi,\eta} \\ S^T\mathbf{1}/n - \mu_\phi + V_{\phi,\eta} \end{pmatrix}, \tag{6.7}$$

where $V_{\xi,\eta} = V_{\tilde{\eta}}(\xi, \tilde{\eta})$ and $V_{\phi,\eta} = V_{\tilde{\eta}}(\phi, \tilde{\eta})$; see Problem 6.2. Fixing the smoothing parameter λ, and θ_β hidden in Q for multiple-term models, one may iterate on (6.7) to calculate η_λ.

For prebinned data with replicate counts k_i at X_i, (6.4) becomes

$$-\frac{1}{N}\mathbf{k}^T(Q\mathbf{c} + S\mathbf{d}) + \log \int_{\mathcal{X}} \exp(\boldsymbol{\xi}^T\mathbf{c} + \boldsymbol{\phi}^T\mathbf{d})dx + \frac{\lambda}{2}\mathbf{c}^T Q\mathbf{c}, \tag{6.8}$$

where $\mathbf{k} = (k_1, \dots, k_n)^T$ and $N = \sum_{i=1}^n k_i$, and (6.7) changes to

$$\begin{pmatrix} V_{\xi,\xi} + \lambda Q & V_{\xi,\phi} \\ V_{\phi,\xi} & V_{\phi,\phi} \end{pmatrix} \begin{pmatrix} \mathbf{c} \\ \mathbf{d} \end{pmatrix} = \begin{pmatrix} Q\mathbf{k}/N - \mu_\xi + V_{\xi,\eta} \\ S^T\mathbf{k}/N - \mu_\phi + V_{\phi,\eta} \end{pmatrix}. \tag{6.9}$$

6.2 Poisson Intensity

Consider a Poisson counting process on $\mathcal{X}$ with an intensity function $\lambda(x)$, where $\lambda(x)$ is not to be confused with the smoothing parameter λ. Observing N occurrences X_i, $i = 1, \dots, N$, from the process, the joint likelihood of N and X_i can be shown to be

$$\left\{\prod_{i=1}^{N} \lambda(X_i)\right\} \exp\left\{-\int_{\mathcal{X}} \lambda(x)dx\right\} = \left\{\prod_{i=1}^{N} \lambda_0(X_i)\right\}(\Lambda^N e^{-\Lambda}),$$

where $\Lambda = \int_{\mathcal{X}} \lambda(x)dx$ is the overall intensity of the process on $\mathcal{X}$ and $\lambda_0(x) = \lambda(x)/\Lambda$ is the occurrence density; see, e.g., Snyder (1975, §2.3). N is statistically sufficient for Λ and has a Poisson distribution with intensity Λ, and $X_i|N$ are conditionally independent with a probability density $\lambda_0(x)$. A penalized likelihood estimate of the Poisson intensity can be defined as the minimizer of

$$-\sum_{i=1}^{N} \log \lambda_0(X_i) - N \log \Lambda + \Lambda + J(\log \lambda_0(x) + \log \Lambda), \tag{6.10}$$

for $\log \lambda(x) \in \tilde{\mathcal{H}} \supset \{1\}$, where $\tilde{\mathcal{H}}$ is a general reproducing kernel Hilbert space and the smoothing parameter is absorbed into the roughness penalty $J(f)$ to avoid confusion with the intensity $\lambda(x)$. Decompose $\tilde{\mathcal{H}} = \{1\} \oplus \mathcal{H}$, where $\mathcal{H}$ satisfies a side condition, and write $\log \lambda(x) = C + \eta$, where C is a constant and $\eta \in \mathcal{H}$. Since $\log \lambda_0 = \eta - \log \int_{\mathcal{X}} e^{\eta} dx$ and $\log \Lambda = C + \log \int_{\mathcal{X}} e^{\eta} dx$, (6.10) can be written as

$$\begin{aligned}\left[-\sum_{i=1}^{N} \eta(X_i) + N \log \int_{\mathcal{X}} e^{\eta} dx + J(C + \eta)\right]& \\ + \left[-N\left(C + \log \int_{\mathcal{X}} e^{\eta} dx\right) + \exp\left(C + \log \int_{\mathcal{X}} e^{\eta} dx\right)\right]&;\end{aligned} \tag{6.11}$$

see Problem 6.3. Naturally, $J(f)$ should annihilate constant since smoothing should only apply to the occurrence density, so $J(C + \eta) = J(\eta)$. The minimization of (6.11) can then be achieved in two steps: first to minimize the sum in the first pair of square brackets in (6.11) with respect to $\eta \in \mathcal{H}$ to estimate the occurrence density $\lambda_0(x)$ and, second, to minimize the sum in the second pair of square brackets with respect to C to estimate the overall intensity Λ. The former is simply a penalized likelihood density estimation through (6.1) based on X_i, $i = 1, \dots, N$, and the latter is a Poisson density estimation based on a single observation N.

When $J(f)$ annihilates constant, the two-step estimation in (6.11) may be manipulated to enforce an arbitrary positive value on Λ by modifying the second part. Specifically, replacing $-N \log \Lambda + \Lambda$ by $-N \log \Lambda + N\Lambda$ in (6.11), one effectively enforces $\Lambda = 1$. Dividing the functional thus modified by N, one has

$$-\frac{1}{N}\sum_{i=1}^{N} \tilde{\eta}(X_i) + \int_{\mathcal{X}} e^{\tilde{\eta}} dx + \tilde{J}(\tilde{\eta}), \tag{6.12}$$

where $\tilde{\eta} = \log \lambda(x)$ and $\tilde{J}(f) = J(f)/N$. Obviously, the minimizer $\tilde{\eta}^*$ of (6.12) satisfies $\int_{\mathcal{X}} e^{\tilde{\eta}^*} dx = 1$; see Problem 6.4. This device was proposed by Silverman (1982) to enforce the unity constraint without imposing any side condition on the log density. Were a probability density defined to integrate to 2, one would use $\int_{\mathcal{X}} e^{\eta} dx/2$ in (6.12) instead of $\int_{\mathcal{X}} e^{\eta} dx$ to enforce the "unity" constraint $\int_{\mathcal{X}} e^{\tilde{\eta}^*} dx = 2$.

6.3 Smoothing Parameter Selection

As with regression, smoothing parameter selection holds the key to any practical success of penalized likelihood density estimation. Similar to the situation with non-Gaussian regression in Chapter 5, the convex but non-quadratic functional (6.1) has to be solved iteratively, even with λ and θ_β fixed. Needed are effective and efficient methods to locate good estimates from among the η_λ's with varying smoothing parameters.

Parallel to the methods developed in §5.2, one has the performance-oriented iteration and the direct cross-validation. The former works fine with a single smoothing parameter, but is numerically less efficient with multiple ones. In comparison, the later is as effective yet much easier to implement. In this section, we derive a cross-validation score and various modifications thereof, and we illustrate and compare their effectiveness through simple simulations.

As in §3.2 and §5.2, we only make the dependence of various entities on the smoothing parameter λ explicit, suppressing their dependence on θ_β in the notation.

6.3.1 Kullback-Leibler Loss and Cross-Validation

To measure the proximity of the estimate $f_\lambda = e^{\eta_\lambda} / \int_{\mathcal{X}} e^{\eta_\lambda} dx$ to the true density $f = e^{\eta} / \int_{\mathcal{X}} e^{\eta} dx$, consider the Kullback-Leibler distance

$$\mathrm{KL}(\eta, \eta_\lambda) = E_f[\log(f/f_\lambda)] = \mu_\eta(\eta - \eta_\lambda) - \log \int_{\mathcal{X}} e^{\eta} dx + \log \int_{\mathcal{X}} e^{\eta_\lambda} dx,$$

where $\mu_f(g) = \int g e^f dx / \int e^f dx$, as defined in §6.1. The smoothing parameters that minimize $\mathrm{KL}(\eta, \eta_\lambda)$ are considered the optimal ones. Ignoring terms that do not involve η_λ, one has the relative Kullback-Leibler distance,

$$\text{RKL}(\eta, \eta_\lambda) = \log \int_{\mathcal{X}} e^{\eta_\lambda} dx - \mu_\eta(\eta_\lambda).$$

The first term is readily computable, but the second term, $\mu_\eta(\eta_\lambda)$, involves the unknown density and will have to be estimated.

A naive estimate of $\mu_\eta(\eta_\lambda)$ is the sample mean $n^{-1}\sum_{i=1}^n \eta_\lambda(X_i)$, but the resulting estimate of the relative Kullback-Leibler distance would simply be the minus log likelihood, clearly favoring $\lambda = 0$. The naive sample mean is biased because the samples X_i contribute to the estimate η_λ. Standard cross-validation suggests an estimate $\tilde{\mu}_\eta(\eta_\lambda) = n^{-1}\sum_{i=1}^n \eta_\lambda^{[i]}(X_i)$, where $\eta_\lambda^{[i]}$ minimizes the delete-one version of (6.1),

$$-\frac{1}{n-1}\sum_{j\neq i} \eta(X_j) + \log \int_{\mathcal{X}} e^{\eta} dx + \frac{\lambda}{2} J(\eta). \tag{6.13}$$

Note that X_i does not contribute to $\eta_\lambda^{[i]}$, although $\eta_\lambda^{[i]}$ is not quite the same as η_λ. The delete-one estimates $\eta_\lambda^{[i]}$ are not analytically available, however; so it is impractical to compute $\tilde{\mu}_\eta(\eta_\lambda)$ directly.

We now discuss a certain analytically tractable approximation of $\eta_\lambda^{[i]}$. Consider the quadratic approximation of (6.1) at η_λ. For $f, g \in \mathcal{H}$ and α real, define $L_{f,g}(\alpha) = \log \int_{\mathcal{X}} e^{f+\alpha g} dx$ as a function of α. It is easy to show that $\dot{L}_{f,g}(0) = \mu_f(g)$ (hence $L(f) = \log \int_{\mathcal{X}} e^f dx$ is Fréchet differentiable) and that $\ddot{L}_{f,g}(0) = V_f(g)$; see Problem 6.5. Setting $f = \tilde{\eta}$, $g = \eta - \tilde{\eta}$, and $\alpha = 1$, one has the Taylor expansion

$$\log \int_{\mathcal{X}} e^{\eta} dx = L_{\tilde{\eta},\eta-\tilde{\eta}}(1) \approx L_{\tilde{\eta},\eta-\tilde{\eta}}(0) + \mu_{\tilde{\eta}}(\eta - \tilde{\eta}) + \frac{1}{2} V_{\tilde{\eta}}(\eta - \tilde{\eta}). \tag{6.14}$$

Substituting the right-hand side of (6.14) for the term $\log \int_{\mathcal{X}} e^{\eta} dx$ in (6.1) and dropping terms that do not involve η, one obtains the quadratic approximation of (6.1) at $\tilde{\eta}$:

$$-\frac{1}{n}\sum_{i=1}^n \eta(X_i) + \mu_{\tilde{\eta}}(\eta) - V_{\tilde{\eta}}(\tilde{\eta}, \eta) + \frac{1}{2} V_{\tilde{\eta}}(\eta) + \frac{\lambda}{2} J(\eta). \tag{6.15}$$

Plugging (6.3) into (6.15) and solving for $\mathbf{c}$ and $\mathbf{d}$, one obtains (6.7); see Problem 6.6.

The delete-one version of (6.15),

$$-\frac{1}{n-1}\sum_{j\neq i} \eta(X_j) + \mu_{\tilde{\eta}}(\eta) - V_{\tilde{\eta}}(\tilde{\eta}, \eta) + \frac{1}{2} V_{\tilde{\eta}}(\eta) + \frac{\lambda}{2} J(\eta), \tag{6.16}$$

only involves changes in the first term. Set $\tilde{\eta} = \eta_\lambda$ and write $\breve{\boldsymbol{\xi}} = (\boldsymbol{\xi}^T, \boldsymbol{\phi}^T)^T$ and $\breve{\mathbf{c}} = (\mathbf{c}^T, \mathbf{d}^T)^T$. Rewrite (6.7) as $H\breve{\mathbf{c}} = \breve{Q}\mathbf{1}/n + \mathbf{g}$, where $H = V_{\tilde{\eta}}(\breve{\boldsymbol{\xi}}, \breve{\boldsymbol{\xi}}^T) +$

$\text{diag}(\lambda Q, O)$, $\breve{Q} = (\breve{\boldsymbol{\xi}}(X_1), \ldots, \breve{\boldsymbol{\xi}}(X_n))$, and $\mathbf{g} = V_{\tilde{\eta}}(\breve{\boldsymbol{\xi}}, \tilde{\eta}) - \mu_{\tilde{\eta}}(\breve{\boldsymbol{\xi}})$. The minimizer $\eta_{\lambda,\tilde{\eta}}^{[i]}$ of (6.16) has the coefficient

$$\breve{\mathbf{c}}^{[i]} = H^{-1}\left(\frac{\breve{Q}\mathbf{1} - \breve{\boldsymbol{\xi}}(X_i)}{n-1} + \mathbf{g}\right) = \breve{\mathbf{c}} + \frac{H^{-1}\breve{Q}\mathbf{1}}{n(n-1)} - \frac{H^{-1}\breve{\boldsymbol{\xi}}(X_i)}{n-1},$$

so

$$\eta_{\lambda,\tilde{\eta}}^{[i]}(X_i) = \breve{\boldsymbol{\xi}}(X_i)^T \breve{\mathbf{c}}^{[i]} = \breve{\boldsymbol{\xi}}(X_i)^T \breve{\mathbf{c}} - \frac{1}{n-1}\breve{\boldsymbol{\xi}}(X_i)^T H^{-1}(\breve{\boldsymbol{\xi}}(X_i) - \breve{Q}\mathbf{1}/n). \tag{6.17}$$

Noting that $\breve{Q}\mathbf{1}/n = n^{-1}\sum_{i=1}^n \breve{\boldsymbol{\xi}}(X_i)$, this leads to a cross-validation estimate of $\mu_\eta(\eta_\lambda)$,

$$\hat{\mu}_\eta(\eta_\lambda) = \frac{1}{n}\sum_{i=1}^n \eta_{\lambda,\tilde{\eta}}^{[i]}(X_i) = \frac{1}{n}\sum_{i=1}^n \eta_\lambda(X_i) - \frac{\text{tr}(P_{\mathbf{1}}^{\perp}\breve{Q}^T H^{-1}\breve{Q}P_{\mathbf{1}}^{\perp})}{n(n-1)}, \tag{6.18}$$

where $P_{\mathbf{1}}^{\perp} = I - \mathbf{1}\mathbf{1}^T/n$, and the corresponding estimate of the relative Kullback-Leibler distance,

$$-\frac{1}{n}\sum_{i=1}^n \eta_\lambda(X_i) + \log\int_{\mathcal{X}} e^{\eta_\lambda}dx + \frac{\text{tr}(P_{\mathbf{1}}^{\perp}\breve{Q}^T H^{-1}\breve{Q}P_{\mathbf{1}}^{\perp})}{n(n-1)}. \tag{6.19}$$

Note that $\eta_{\lambda,\tilde{\eta}}^{[i]}$ is simply the one-step Newton update from η_λ for the minimization of (6.13).

For prebinned data, the delete-one operation should be done on the individual observations instead of the bins, yielding

$$\begin{aligned}\hat{\mu}_\eta(\eta_\lambda) &= \frac{1}{N}\sum_{i=1}^n k_i \eta_{\lambda,\tilde{\eta}}^{[i]}(X_i)\\ &= \frac{1}{N}\sum_{i=1}^n k_i\eta_\lambda(X_i) - \frac{\text{tr}(P_{\tilde{\mathbf{k}}}^{\perp}\tilde{K}\breve{Q}^T H^{-1}\breve{Q}\tilde{K}P_{\tilde{\mathbf{k}}}^{\perp})}{N(N-1)},\end{aligned} \tag{6.20}$$

where $P_{\tilde{\mathbf{k}}}^{\perp} = I - \tilde{\mathbf{k}}\tilde{\mathbf{k}}^T/N$ with $\tilde{\mathbf{k}} = (\sqrt{k_1}, \ldots, \sqrt{k_n})^T$, $\tilde{K} = \text{diag}(\sqrt{k_i})$, and $\eta_{\lambda,\tilde{\eta}}^{[i]}$ minimizes

$$-\frac{1}{N-1}\left\{\sum_{j=1}^n k_j\eta(X_j) - \eta(X_i)\right\} + \mu_{\tilde{\eta}}(\eta) - V_{\tilde{\eta}}(\tilde{\eta},\eta) + \frac{1}{2}V_{\tilde{\eta}}(\eta) + \frac{\lambda}{2}J(\eta);$$

see Problem 6.7.

6.3.2 Modifications of Cross-Validation Score

Fixing $\tilde{\eta} = \eta_{\tilde{\lambda}}$, the minimizer $\eta_{\lambda,\tilde{\eta}}$ of (6.15) defines a family of estimates with a varying λ, with $\eta_{\tilde{\lambda},\tilde{\eta}} = \eta_{\tilde{\lambda}}$. The cross-validation score (6.19) as a function of λ for fixed $\tilde{\eta}$, to be denoted by $V(\lambda|\tilde{\eta})$, compares the performances of the $\eta_{\lambda,\tilde{\eta}}$'s directly, and in case $\tilde{\lambda}$ does not minimize $V(\lambda|\tilde{\eta})$, $\eta_{\tilde{\lambda}} = \eta_{\tilde{\lambda},\tilde{\eta}}$ is likely to underperform the $\eta_{\lambda,\tilde{\eta}}$ associated with the minimizing λ, which is nothing but the one-step Newton update of η_λ from $\eta_{\tilde{\lambda}}$. This is virtually the self-voting argument presented in §5.2.1, which leads to the selection of $\tilde{\eta}$ that actually minimizes $V(\lambda|\eta_{\tilde{\lambda}})$.

Consider a single smoothing parameter λ and write

$$\eta_{\lambda,\tilde{\eta}}(x) = \breve{\xi}(x)^T H^{-1}(\breve{Q}\mathbf{1}/n + V_{\tilde{\eta}}(\breve{\boldsymbol{\xi}},\tilde{\eta}) - \mu_{\tilde{\eta}}(\breve{\boldsymbol{\xi}})).$$

Fixing $\tilde{\eta}$, λ appears in (6.19) only through $H = V + \text{diag}(\lambda Q, O)$, where $V = V_{\tilde{\eta}}(\breve{\boldsymbol{\xi}}, \breve{\boldsymbol{\xi}}^T)$. It is easy to show that (Problem 6.8)

$$\left.\frac{dH^{-1}}{d\log\lambda}\right|_{\lambda=\tilde{\lambda}} = H^{-1} - H^{-1}VH^{-1},$$

and it follows that

$$\left.\frac{dV(\lambda|\tilde{\eta})}{d\log\lambda}\right|_{\lambda=\tilde{\lambda}} = -(\mu_{\tilde{\eta}}(\breve{\boldsymbol{\xi}}) - \breve{Q}\mathbf{1}/n)^T H^{-1}(\mu_{\tilde{\eta}}(\breve{\boldsymbol{\xi}}) - \breve{Q}\mathbf{1}/n)^T + \frac{\text{tr}(P_{\mathbf{1}}^{\perp}\breve{Q}^T(H^{-1} - H^{-1}VH^{-1})\breve{Q}P_{\mathbf{1}}^{\perp})}{n(n-1)}; \tag{6.21}$$

see Problem 6.9. The root of (6.21) as a function of $\tilde{\lambda}$ is self-voting.

This derivative approach is not practical for multiple smoothing parameters, however, as (6.21) only gives the directional derivative with fixed θ_β ratios; the root of (6.21) is not necessarily self-voting with multiple smoothing parameters. Multivariate root-finding is also a more challenging numerical problem, although the issue is of secondary importance here.

As will be shown in §6.3.3, the practical performances of the cross-validation score (6.19) and the derivative self-voting criterion are generally adequate, but there is ample room for improvement. As is typical with cross-validation techniques, the methods may severely undersmooth up to about 10% of the replicates in simulation studies. To circumvent the problem, a simple modification of the cross-validation score proves to be remarkably effective. The cross-validation score is seen to have a rather simple structure, where the minus log likelihood monotonically decreases as λ decreases whereas the trace term moves in the opposite direction. To force smoother estimates, one may simply attach a constant α larger than 1 to the trace term,

$$-\frac{1}{n}\sum_{i=1}^{n}\eta_\lambda(X_i) + \log\int_{\mathcal{X}} e^{\eta_\lambda}dx + \alpha\frac{\text{tr}(P_{\mathbf{1}}^{\perp}\breve{Q}^T H^{-1}\breve{Q}P_{\mathbf{1}}^{\perp})}{n(n-1)}. \tag{6.22}$$

Simulation studies suggest that an α around 1.4 would be most effective, curbing undersmoothing on "bad" replicates while sacrificing minimal performance degradation on "good" ones.

6.3.3 Empirical Performance

We now illustrate the effectiveness of cross-validation through some simple simulation. Samples of size $n = 100$ were drawn from

$$f(x) = \left\{ \frac{1}{3} e^{-50(x-0.3)^2} + \frac{2}{3} e^{-50(x-0.7)^2} \right\} I_{[0<x<1]},$$

which is a mixture of $N(0.3, 0.01)$ and $N(0.7, 0.01)$ truncated to $[0, 1]$. Using the second formulation of cubic spline on $\mathcal{X} = [0, 1]$, as discussed in Example 6.1, estimates η_λ were calculated on the grid $\log_{10} \lambda = (-7)(0.1)(-3)$ for 100 replicates. Recorded for each of the estimates were the Kullback-Leibler distance $\text{KL}(\eta, \eta_\lambda)$, the cross-validation score (6.19), the trace term $\text{tr}(P_1^\perp \breve{Q}^T H^{-1} \breve{Q} P_1^\perp)$, and the derivative (6.21). The trace term was needed to evaluate the modified cross-validation scores.

Plotted in the left and middle frames of Figure 6.1 are $\text{KL}(\eta, \eta_\lambda)$ of the cross-validated λ's with $\alpha = 1$ (CV) and $\alpha = 1.4$ (modified CV) versus the minimum $\text{KL}(\eta, \eta_\lambda)$ on the grid. The relative efficacy of the methods, defined as the ratio of the horizontal axis to the vertical axis, are shown in the right frame in box plots along with those of $\alpha = 1.2$ and 1.6 and that of the self-voting λ.

In the left frame of Figure 6.2, $\text{KL}(\eta, \eta_\lambda)$ of the indirect cross-validation (the self-voting λ) is plotted against that of the direct cross-validation of (6.19), showing that the two are largely equivalent. The middle frame compares the modified score ($\alpha = 1.4$) with the unmodified one ($\alpha = 1$), suggesting that the modified score may gain significantly over the unmodified one on some replicates but only lose minimally on some others. The right frame plots the $\log_{10} \lambda$ of the modified score versus that of the unmodified one, where it is evident that the gains by the modified score were achieved through correcting the occasional severe undersmoothing by the unmodified score.

Plotted in Figure 6.3 are the fits for two replicates corresponding to the plus and the star in the middle frame of Figure 6.2. Fits returned by the modified ($\alpha = 1.4$) and the unmodified cross-validation are represented by the solid and long dashed curves, respectively, and the best possible fits are shown by the short dashed curves. The data are superimposed as the finely binned histograms and the test density is indicated by dotted lines. The solid and short dashed lines are visually indistinguishable from each other in the left frame. The data in the right frame appear to be a "bad" sample, based on which even the best possible estimate is no good.

6.4 Computation

Fixing smoothing parameters, the computation involves the Newton iteration via (6.6) and the evaluation of the cross-validation scores given by

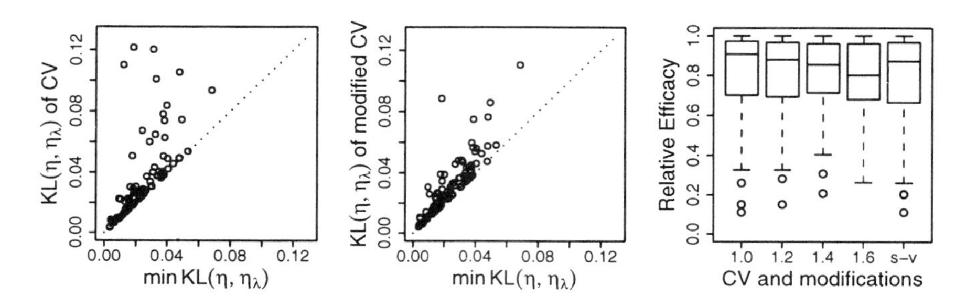

FIGURE 6.1. Performance of Cross-Validation and Modifications Thereof. Left: Loss achieved by unmodified cross-validation of (6.19). Center: Loss achieved by modified cross-validation of (6.22) with $\alpha = 1.4$. Right: Relative efficacy of (6.22) with $\alpha = 1, 1.2, 1.4, 1.6$ and of the self-voting λ.

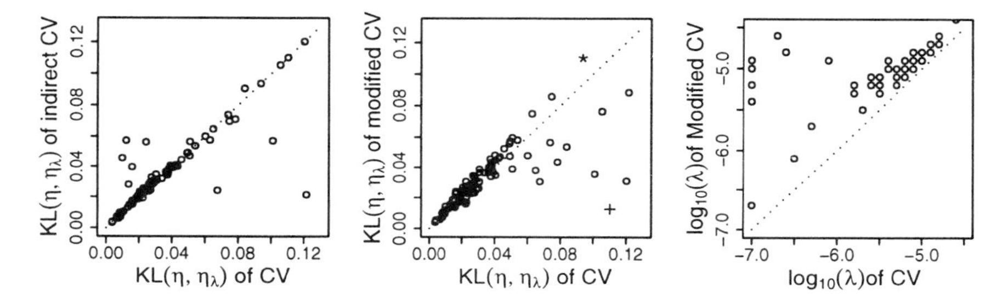

FIGURE 6.2. Comparison of Cross-Validation Scores. Left: Loss of the indirect cross-validation versus that of the direct one. Center: Loss of the modified cross-validation with $\alpha = 1.4$ versus that of the unmodified. Right: $\log_{10} \lambda$ of the modified cross-validation with $\alpha = 1.4$ versus that of the unmodified.

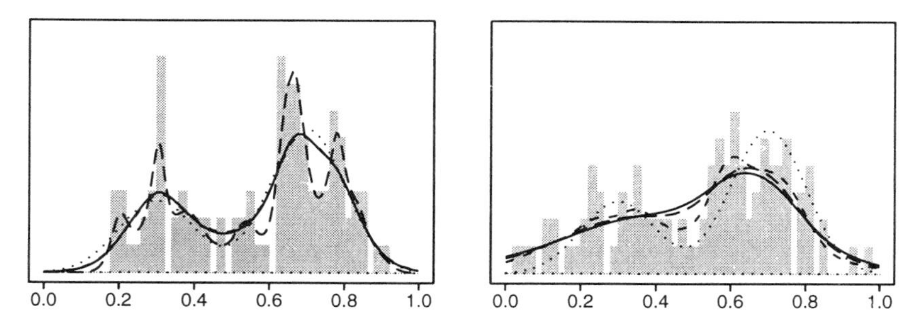

FIGURE 6.3. Selected Fits from Simulation. Left: The plus in Figure 6.2. Right: The star in Figure 6.2. The solid lines are the fits with modified cross-validation ($\alpha = 1.4$) and the long dashed lines are those with unmodified cross-validation. The short dashed lines are the best possible fits for the data, and the dotted lines plot the test density. The data are superimposed as finely binned histograms.

(6.22). To select smoothing parameters by cross-validation, appropriate optimization algorithms will be employed to minimize (6.22) with respect to the smoothing parameters, for some $\alpha \geq 1$.

To perform the Newton iteration, one calculates the Cholesky decomposition of $H = V_{\tilde{\eta}}(\breve{\boldsymbol{\xi}}, \breve{\boldsymbol{\xi}}^T) + \text{diag}(\lambda Q, O)$, $H = G^T G$, where G is upper-triangular, and then uses a back substitution, followed by a forward substitution, to calculate the update. Standard safeguard procedures such as step-halving might be called upon to ensure decreasing minus log likelihood in each step, and the iteration usually takes 5 to 10 steps to converge given reasonable starting values. The Cholesky decomposition takes $n^3/3 + O(n^2)$ flops, and both the back and forward substitutions take n^2 flops, so each step of the Newton iteration takes $n^3/3 + O(n^2)$ flops. The above flop count does not include the operations for the formation of $V_{\tilde{\eta}}(\breve{\boldsymbol{\xi}}, \breve{\boldsymbol{\xi}}^T)$, which tally to $qn^2/2 + O(qn) + O(n^2)$ flops, where q is the size of the quadrature.

On convergence of the Newton iteration, the Cholesky decomposition $H = G^T G$ has already been computed. Back substitution yields $G^{-T}\breve{Q}$ in n^3 flops, from which $\text{tr}(P_{\mathbf{1}}^{\perp}\breve{Q}^T H^{-1}\breve{Q}P_{\mathbf{1}}^{\perp})$ is available in $O(n^2)$ flops.

For the minimization of (6.19) or modifications thereof, standard optimization tools such as those developed in Dennis and Schnabel (1996) can be used, which employ a certain quasi-Newton approach with numerical derivatives.

Care must be taken with numerically singular H, which may arise when $\xi_i(x) = R_J(X_i, x)$ are linearly dependent. Rearranging (6.7) as

$$\begin{pmatrix} V_{\phi,\phi} & V_{\phi,\xi} \\ V_{\xi,\phi} & V_{\xi,\xi} + \lambda Q \end{pmatrix} \begin{pmatrix} \mathbf{d} \\ \mathbf{c} \end{pmatrix} = \begin{pmatrix} S^T\mathbf{1}/n - \mu_\phi + V_{\phi,\eta} \\ Q\mathbf{1}/n - \mu_\xi + V_{\xi,\eta} \end{pmatrix} \tag{6.23}$$

and writing the Cholesky decomposition of the rearranged H as

$$\begin{pmatrix} V_{\phi,\phi} & V_{\phi,\xi} \\ V_{\xi,\phi} & V_{\xi,\xi} + \lambda Q \end{pmatrix} = \begin{pmatrix} G_1^T & O \\ G_2^T & G_3^T \end{pmatrix} \begin{pmatrix} G_1 & G_2 \\ O & G_3 \end{pmatrix},$$

one has $V_{\phi,\phi} = G_1^T G_1$, $V_{\phi,\xi} = G_1^T G_2$, and $(V_{\xi,\xi} - V_{\xi,\phi}V_{\phi,\phi}^{-1}V_{\phi,\xi}) + \lambda Q = G_3^T G_3$. For H singular, with a possible permutation of the data index $i = 1, \ldots, n$, called pivoting in LINPACK (Dongarra et al. 1979), G_3 can be written as

$$G_3 = \begin{pmatrix} C_1 & C_2 \\ O & O \end{pmatrix} = \begin{pmatrix} C \\ O \end{pmatrix},$$

where C is of full row rank and $G_3^T G_3 = C^T C$. Define

$$\tilde{G}_3 = \begin{pmatrix} C_1 & C_2 \\ O & \delta I \end{pmatrix}, \quad \tilde{G} = \begin{pmatrix} G_1 & G_2 \\ O & \tilde{G}_3 \end{pmatrix},$$

for some $\delta > 0$, and partition $\tilde{G}_3^{-1} = (K, L)$. It is easy to see that $CK = I$ and $CL = O$. This leads to $L^T G_3^T G_3 L = O$, and since $V_{\xi,\xi} - V_{\xi,\phi}V_{\phi,\phi}^{-1}V_{\phi,\xi}$

is non-negative definite, $L^TQL = O$. Note that $J(f)$ is a norm in the space span$\{\xi_1, \dots \xi_n\}$ and $J(\boldsymbol{\xi}^T \mathbf{l}) = \mathbf{l}^T Q \mathbf{l}$, $L^TQL = O$ implies $L^T\boldsymbol{\xi} = \mathbf{0}$, and, consequently, $L^TV_{\xi,\xi} = O$, $L^TV_{\xi,\phi} = O$, $L^TV_{\xi,\eta} = \mathbf{0}$, and $L^T\mu_\xi = \mathbf{0}$. Premultiply (6.23) by $\tilde{G}^{-T}$ and write $\tilde{\mathbf{c}} = \tilde{G}\left(\begin{smallmatrix}\mathbf{d}\\ \mathbf{c}\end{smallmatrix}\right)$; straightforward algebra yields

$$\begin{pmatrix} I & O & O \\ O & I & O \\ O & O & O \end{pmatrix} \begin{pmatrix} \tilde{\mathbf{c}}_1 \\ \tilde{\mathbf{c}}_2 \\ \tilde{\mathbf{c}}_3 \end{pmatrix} = \begin{pmatrix} * \\ * \\ \mathbf{0} \end{pmatrix}, \tag{6.24}$$

where $(\tilde{\mathbf{c}}_1^T, \tilde{\mathbf{c}}_2^T, \tilde{\mathbf{c}}_3^T)^T$ form the partitioned $\tilde{\mathbf{c}}$; see Problem 6.10. This is a solvable linear system, but the solution is not unique, where $\tilde{\mathbf{c}}_3$ can be arbitrary. Replacing the lower-right block O by I in the matrix on the left-hand side, one effectively fixes $\tilde{\mathbf{c}}_3 = 0$, and (6.24) thus modified is equivalent to

$$\begin{pmatrix} G_1^T & O \\ G_2^T & \tilde{G}_3^T \end{pmatrix} \begin{pmatrix} G_1 & G_2 \\ O & \tilde{G}_3 \end{pmatrix} \begin{pmatrix} \mathbf{d} \\ \mathbf{c} \end{pmatrix} = \begin{pmatrix} S^T\mathbf{1}/n - \mu_\phi + V_{\phi,\eta} \\ Q\mathbf{1}/n - \mu_\xi + V_{\xi,\eta} \end{pmatrix}.$$

In actual computation, one performs the Cholesky decomposition of H with pivoting, replaces the trailing O by δI with an appropriate δ, and proceeds as if H were nonsingular.

6.5 Case Studies

We now apply the techniques developed so far to analyze a few real-life data sets. It will be seen that the specification of the domain $\mathcal{X}$ carries a rather heavy weight in the estimation process.

6.5.1 Buffalo Snowfall

The annual snowfall accumulations in Buffalo from 1910 to 1973 are listed in Scott (1985). The data range from 25.0 to 126.4. To see how the domain $\mathcal{X}$ affects the estimate, three fits were calculated using $\mathcal{X} = [0, 150]$, $[10, 140]$, and $[20, 130]$. For each of the specifications, the domain $\mathcal{X}$ was mapped onto $[0, 1]$ and the cubic spline of Example 6.1 (the second formulation) was used in the estimation. Unmodified cross-validation of (6.19) was used to select the smoothing parameter, which yielded $\log_{10}\lambda = -5.1934$, -3.8856, and -3.4668 respectively for the three domain specifications. Modified cross-validation with $\alpha = 1.4$ yielded the smoother $\log_{10}\lambda = -3.9646$, -3.6819, and -3.3287, but as the data appear to be a "good replicate" not bothered by undersmoothing, we chose to plot fits based on the unmodified score. The three fits are plotted in Figure 6.4, along with the data as the finely binned histogram.

It is clear from the plot that as the domain $\mathcal{X}$ extends farther into the no-data area, the cross-validation tries harder to take away the mass assigned

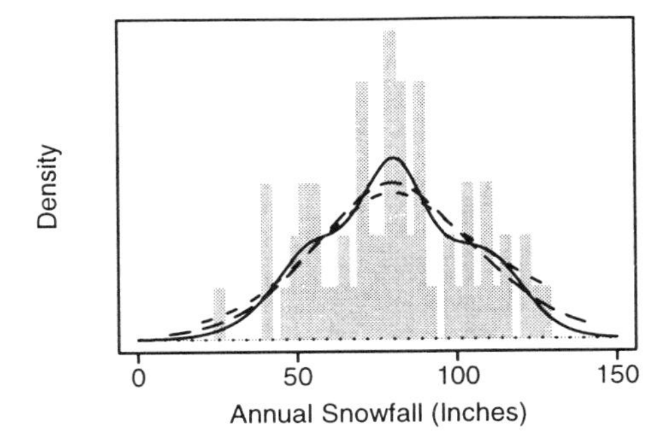

FIGURE 6.4. Distribution of Buffalo Annual Snowfall. The solid, long dashed, and short dashed lines are the cross-validated fits with the three domain specifications indicated by their running lengths. The data are superimposed as a finely binned histogram.

to the empty space by the smoothness of the estimate, resulting in less smoothing. From the fits presented in Figure 6.4, it is fairly safe to conclude that a unimodal density fits the data reasonably well. Support from the data for the shoulders appearing in the $\mathcal{X} = [0, 150]$ fit, however, appears marginal at best.

6.5.2 Eruption Time of Old Faithful

We now estimate the density of Old Faithful eruption duration on $\mathcal{X} = [1.5, 5.25]$ using the data discussed in §5.5.1. Three fits were calculated: one using the original data, another using the histogram with 30 bins, and the third using the histogram with 60 bins. Modified cross-validation with $\alpha = 1.4$ was used to select the smoothing parameter for all three fits. Plotted in the left frame of Figure 6.5 are the fit based on the exact data represented by the solid line, the fit based on the 30-bin histogram represented by the dashed line, and the fit scaled from the Poisson regression fit of §5.5.1 based on the 30-bin histogram represented by the dotted line; the histogram is superimposed. The right frame duplicates the procedure but with the 30-bin histogram replaced by the 60-bin histogram.

6.5.3 AIDS Incubation

Details are in order concerning the AIDS incubation study discussed in §1.4.2. Assume pretruncation independence and condition on the truncation mechanism. The density of (X, Y) is given by

$$f(x, y) = \frac{e^{\eta_x(x)+\eta_y(y)}}{\int_{\mathcal{T}} e^{\eta_x(x)+\eta_y(y)}},$$

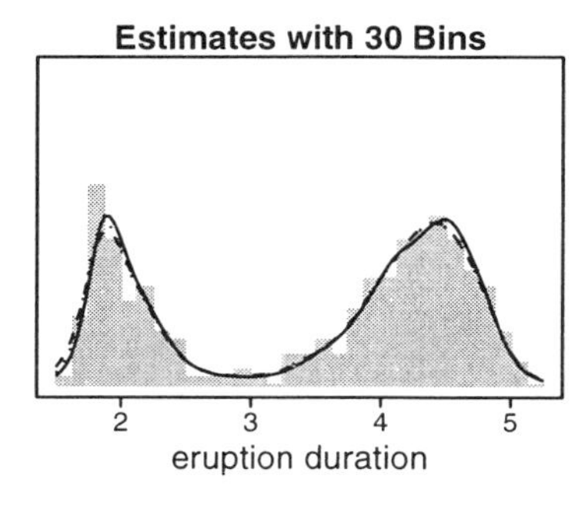

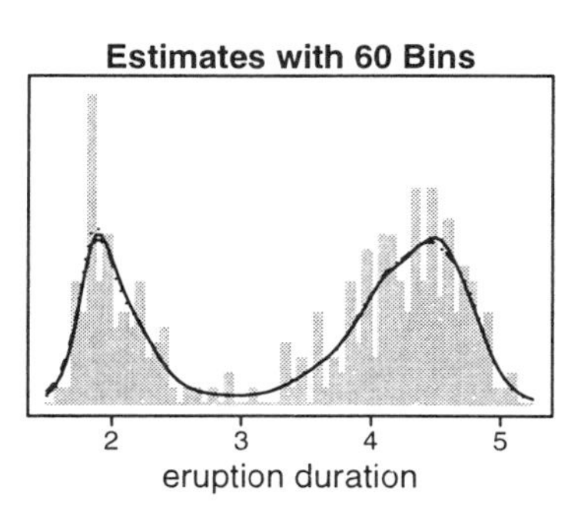

FIGURE 6.5. Density of Eruption Duration of Old Faithful. The fit based on the original data is indicated by solid lines, the fits based on the histograms are indicated by dashed lines, and the fits based on Poisson regression are indicated by dotted lines. The histograms are superimposed on the probability scale.

where $\mathcal{T} = \{x < y\}$. Mapping $[0, 100]$ onto $[0, 1]$ and using the second formulation of Example 6.1 on the marginal domains, cubic spline additive models were fitted to the three subsets as well as the whole data set. Unmodified cross-validation was leading to interpolation, so the modified score (6.22) with $\alpha = 1.4$ was used to select the smoothing parameters. Plotted in Figure 6.6 are the four fits in contours, with the marginal densities $f(x) = e^{\eta_x} / \int_0^{100} e^{\eta_x} dx$ and $f(y) = e^{\eta_y} / \int_0^{100} e^{\eta_y} dy$ plotted in the empty space on their respective axes; the data are also superimposed.

Based on only 38 observations, the fit for the youth group is not to be taken too seriously. Due to the lack of information from the samples, $f(x)$ at the upper end and $f(y)$ at the lower end cannot be estimated accurately, and, indeed, the marginal estimates plotted near the lower-right corner demonstrate less consistency among different data sets. An interesting observation is the bump in $f(y)$ in the fit for the elderly, which appears to suggest that at the vicinity of January 1984 (30 months before July 1986), a batch of contaminated blood might have been distributed in the population from which the elderly data were collected.

6.6 Biased Sampling and Random Truncation

Independent and identically distributed samples may not always be available or may not be all that are available concerning the density $f(x)$. Biased sampling and random truncation are two sources from which non-i.i.d. samples may result.

A simple general formulation provides a unified framework for treating such data, and (6.1) can be easily modified to combine information from heterogeneous samples. The computation and smoothing parameter selection require only trivial modifications to the algorithms designed for (6.1).

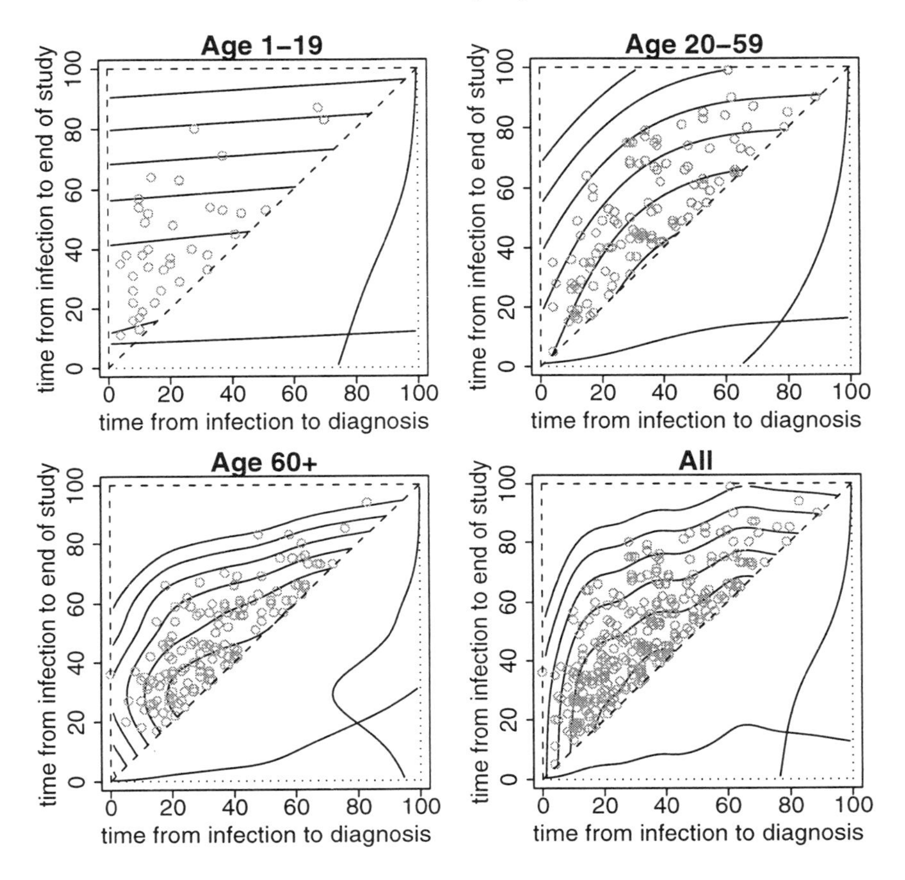

FIGURE 6.6. AIDS Incubation and HIV Infection. Contours are estimated density on the observable region surrounded by the dashed lines. Circles mark the observations. Curves over the dotted lines in the empty space are the estimated marginal densities.

A simple simulation and a case study will be presented to illustrate the technique.

6.6.1 Biased and Truncated Samples

Consider independent observations X_i on $\mathcal{X}$ sampled from densities proportional to $w_i(x)f(x)$, where $w_i(x) \geq 0$ are *known* biasing functions and $f(x)$ is to be estimated. Note that the data are actually the pairs (w_i, X_i). Let $\mathcal{T}$ be an index set and $w(t,x)$ a known function on $\mathcal{T} \times \mathcal{X}$ such that the set $\{w(t,\cdot), t \in \mathcal{T}\}$ includes all possible biasing functions and $w(t,\cdot) \neq w(t',\cdot)$ when $t \neq t'$. The "observed" biasing function w_i can then be written as $w(t_i,\cdot)$ for some $t_i \in \mathcal{T}$, and the data are now

(t_i, X_i). Assume $0 < \int_{\mathcal{X}} w(t,x)f(x)dx < \infty$, $\forall t \in \mathcal{T}$, so that the densities $w(t,x)f(x)/\int_{\mathcal{X}} w(t,x)f(x)dx$ are well defined. Take t_i as observations from a probability density $m(t)$ on $\mathcal{T}$. The data (t_i, X_i) can then be treated as from a two-stage sampling.

Example 6.3 (Ordinary samples)
Let $\mathcal{T} = \{1\}$ be a singleton and $w(1,x) = 1$. X_i are i.i.d. samples from $f(x)$. □

Example 6.4 (Length-biased samples)
Let $\mathcal{T} = \{1\}$ be a singleton, $\mathcal{X} = [0,1]$, and $w(1,x) = x$. X_i are i.i.d. length-biased samples from the probability density $xf(x)/\int_0^1 xf(x)dx$. □

Example 6.5 (Ordinary and length-biased samples)
Let $\mathcal{T} = \{1,2\}$, $\mathcal{X} = [0,1]$, $w(1,x) = 1$, and $w(2,x) = x$. $X_i|(t_i = 1)$ are ordinary samples from $f(x)$ and $X_i|(t_i = 2)$ are length-biased samples from $xf(x)/\int_0^1 xf(x)dx$. Examples 6.3 and 6.4 are special cases with $m(1) = 1$ and $m(1) = 0$, respectively. □

Example 6.6 (Finite-strata biased samples)
Let $\mathcal{T} = \{1, \ldots, s\}$ and $\mathcal{X} = \cup_{t:m(t)>0}\{x : w(t,x) > 0\}$, where $w(t,x) \geq 0$ but otherwise arbitrary. $X_i|t_i$ are from the densities

$$\frac{w(t_i,x)f(x)}{\int_{\mathcal{X}} w(t_i,x)f(x)dx}.$$

Example 6.5 is a special case with $s = 2$. □

Example 6.7 (Truncated samples)
Paired data (t, X) are generated from a joint density $g(t)f(x)$ on $\mathcal{T} \times \mathcal{X}$, but only those that fall on an observable region $A \subset \mathcal{T} \times \mathcal{X}$ are recorded and those that fall on A^c are lost. Of interest is the estimation of $f(x)$. It follows that $w(t,x) = I_{[(t,x)\in A]}$ and $m(t) \propto g(t)\int_{\mathcal{X}} I_{[(t,x)\in A]}f(x)dx$. Note that t and X are interchangeable and that the truncation scheme is virtually arbitrary in this setting.

For a specific case, consider $\mathcal{T} = \mathcal{X} = [0,1]$ and $A = \{t < x\}$. One has $w(t,x) = I_{[t<x]}$ and $m(t) \propto g(t)\int_t^1 f(x)dx$. □

6.6.2 Penalized Likelihood Estimation

Write $f(x) = e^{\eta(x)}/\int_{\mathcal{X}} e^{\eta(x)}dx$; the sampling likelihood of $X|t$ is seen to be

$$\frac{w(t,x)f(x)}{\int_{\mathcal{X}} w(t,x)f(x)dx} = \frac{w(t,x)e^{\eta(x)}}{\int_{\mathcal{X}} w(t,x)e^{\eta(x)}dx},$$

which leads to the penalized log likelihood functional

$$-\frac{1}{n}\sum_{i=1}^{n}\left\{\eta(X_i)-\log\int_{\mathcal{X}}w(t_i,x)e^{\eta(x)}dx\right\}+\frac{\lambda}{2}J(\eta). \tag{6.25}$$

For a singleton $\mathcal{T}$ such as the case with the length-biased samples of Example 6.4, (6.25) virtually reduces to (6.1) but with $\int_{\mathcal{X}} e^{\eta(x)}dx$ replaced by $\int_{\mathcal{X}} e^{\eta(x)}w(x)dx$, a substitution of the integration measure.

Removing dx from the notation and writing the integral as $\int_{\mathcal{X}} e^{\eta}$, (6.1) covers more ground than it first appears. Note that a probability density $f=e^{\eta}/\int_{\mathcal{X}} e^{\eta}$ is the Radon-Nikodym derivative of the probability measure with respect to a base measure, the integration measure that defines $\int_{\mathcal{X}} e^{\eta}$. By the chain rule of the Radon-Nikodym derivative, biased samples from $w(x)f(x)$ with respect to the uniform integration measure are simply ordinary samples from $f(x)$ with respect to the "biased" integration measure $\nu_w(A)=\int_A w(x)dx$. With such a change in notation, one no longer needs the domain $\mathcal{X}$ to be bounded but only the integral $\int_{\mathcal{X}} e^{\eta}$ over the domain to be finite, so that the uniform distribution (with respect to the integration measure) is properly defined.

The minimizer of (6.25) in $\mathcal{H}=\{f: J(f)<\infty\}$ is generally not computable, but one may calculate that in

$$\mathcal{H}^*=\mathcal{N}_J\oplus\text{span}\{R_J(X_i,\cdot),i=1,\ldots,n\},$$

which shares the same convergence rates; see §8.2. Define

$$\mu_f(g|t)=\frac{\int_{\mathcal{X}} g(x)w(t,x)e^{f(x)}}{\int_{\mathcal{X}} w(t,x)e^{f(x)}}$$

and write $v_f(g,h|t)=\mu_f(gh|t)-\mu_f(g|t)\mu_f(h|t)$. Modify the definitions of $\mu_f(g)$ and $V_f(g,h)$ as

$$\mu_f(g)=\frac{1}{n}\sum_{i=1}^{n}\mu_f(g|t_i), \qquad V_f(g,h)=\frac{1}{n}\sum_{i=1}^{n}v_f(g,h|t_i). \tag{6.26}$$

The Newton updating formula (6.7) on page 181 holds verbatim for the minimization of (6.25) in $\mathcal{H}^*$, with the entries defined by the modified $\mu_f(g)$ and $V_f(g,h)$ (Problem 6.11).

Taking into account the sampling mechanism, the Kullback-Leibler distance of $e^{\eta_\lambda}/\int_{\mathcal{X}} e^{\eta_\lambda}$ from $e^{\eta}/\int_{\mathcal{X}} e^{\eta}$ should be modified as

$$\begin{aligned}\text{KL}(\eta,\eta_\lambda)=\int_{\mathcal{T}} m(t)\Big\{&\mu_\eta(\eta-\eta_\lambda|t)\\ &-\log\int_{\mathcal{X}} w(t,x)e^{\eta(x)}+\log\int_{\mathcal{X}} w(t,x)e^{\eta_\lambda(x)}\Big\},\end{aligned} \tag{6.27}$$

with the relative Kullback-Leibler distance

$$\mathrm{RKL}(\eta, \eta_\lambda) = \int_{\mathcal{T}} m(t) \log \int_{\mathcal{X}} w(t,x) e^{\eta_\lambda(x)} - \int_{\mathcal{T}} m(t) \mu_\eta(\eta_\lambda | t). \tag{6.28}$$

The first term of (6.28) can be estimated by $n^{-1} \sum_{i=1}^{n} \log \int_{\mathcal{X}} w(t_i, x) e^{\eta_\lambda(x)}$. For the second term, $E[\eta_\lambda(X)]$, where X follows the marginal distribution under the sampling mechanism,

$$X \sim \int_{\mathcal{T}} m(t) \frac{w(t,x) e^{\eta(x)}}{\int_{\mathcal{X}} w(t,x) e^{\eta(x)}},$$

one may use the cross-validation estimate given by (6.18) on page 185, with the entries in the relevant matrices defined by the modified $\mu_f(g)$ and $V_f(g,h)$. The counterparts of (6.19) and (6.22) are easy to work out (Problem 6.12), and the computation following these lines can be accomplished via trivial modifications of the algorithms developed for (6.1).

6.6.3 Empirical Performance

We now study the empirical performance of the techniques outlined above through simple simulation. Samples (t_i, X_i) of size $n = 100$ were generated according to Example 6.7 with $A = \{t < x\}$, $g(t) = I_{[0<t<1]}$ uniform, and

$$f(x) = \left\{ \frac{1}{3} e^{-50(x-0.3)^2} + \frac{2}{3} e^{-50(x-0.7)^2} \right\} I_{[0<x<1]}.$$

Note that the X_i thus generated are length-biased (Problem 6.13). Cubic splines of Example 6.1 (the second formulation) were fitted to 100 replicates, with two different biasing functions $w_i(x) = x$ and $w_i(x) = I_{[x>t_i]}$, for λ on the grid $\log_{10} \lambda = (-7)(0.1)(-3)$. Recorded for each of the estimates were the Kullback-Leibler distance $\mathrm{KL}(\eta, \eta_\lambda)$ of (6.27) and the cross-validation scores. Note that the definitions of $\mathrm{KL}(\eta, \eta_\lambda)$ and the cross-validation scores depend on the biasing function used. The counterpart of Figure 6.1 for $w_i(x) = x$ is given in Figure 6.7 and that for $w_i(x) = I_{[x>t_i]}$ is given in Figure 6.8. With the biasing function $w_i(x) = x$, one incorporates knowledge of $g(t)$ but discards t_i, whereas with the biasing function $w_i(x) = I_{[x>t_i]}$, one relies solely on the observed t_i.

The fits for two replicates corresponding to the plus and the star in Figures 6.7 and 6.8 are plotted in Figure 6.9, where the solid lines are the fits from the modified ($\alpha = 1.4$) cross-validation and the dashed lines are the fits from the unmodified one. The data are superimposed as the finely binned histograms and the test density is indicated by dotted lines. The plus is not visible in the middle frames of Figures 6.7 and 6.8, as it falls into the crowds. The undersmoothing in the right frame of Figure 6.9 is likely caused by the empty space around 0.45. Note that the plotted estimates are for $f(x)$, but the histograms are from a density proportional to $xf(x)$.

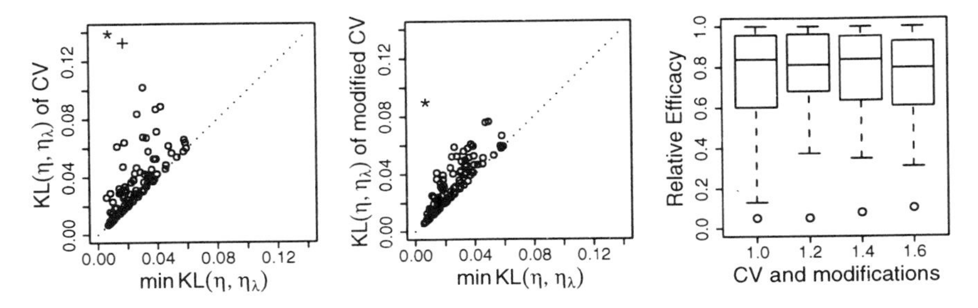

FIGURE 6.7. Performance of Cross-Validation and Modifications Thereof for Length-Biased Data. Left: Loss achieved by unmodified cross-validation. Center: Loss achieved by modified cross-validation with $\alpha = 1.4$. Right: Relative efficacy of cross-validation with $\alpha = 1, 1.2, 1.4, 1.6$.

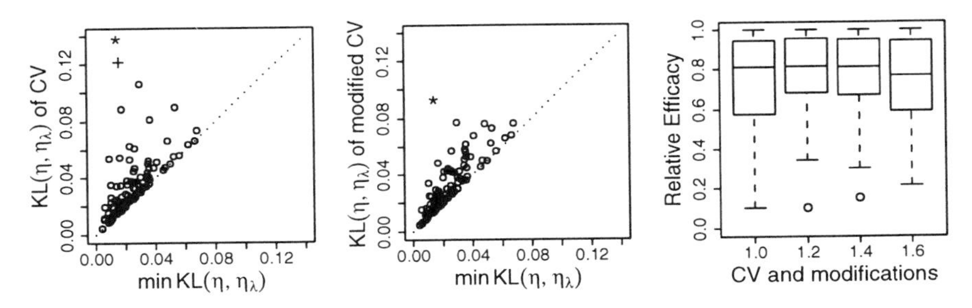

FIGURE 6.8. Performance of Cross-Validation and Modifications Thereof for Truncated Data. Left: Loss achieved by unmodified cross-validation. Center: Loss achieved by modified cross-validation with $\alpha = 1.4$. Right: Relative efficacy of cross-validation with $\alpha = 1, 1.2, 1.4, 1.6$.

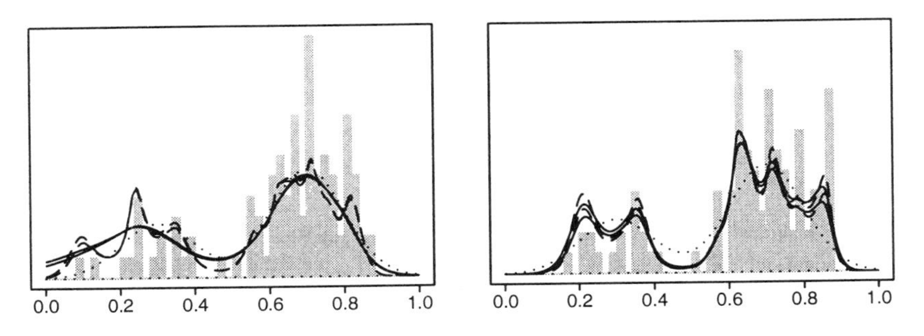

FIGURE 6.9. Selected Fits from Simulation with Length-Biased and Truncated Data. Left: The plus in Figures 6.7 and 6.8. Right: The star in Figures 6.7 and 6.8. The solid lines are the fits with modified cross-validation ($\alpha = 1.4$) and the dashed lines are those with unmodified cross-validation. The dotted lines plot the test density. The data are superimposed as a finely binned histogram.

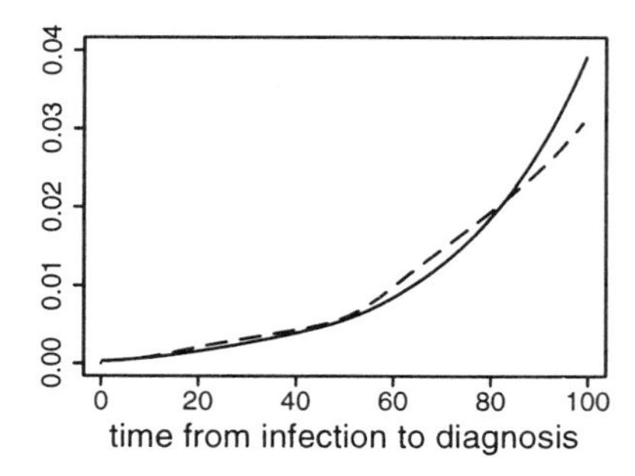

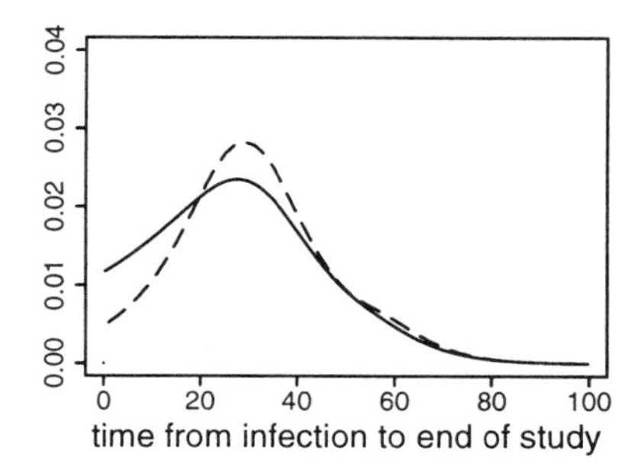

FIGURE 6.10. AIDS Incubation and HIV Infection for the Elderly. Left: Incubation density $f(x)$ of X. Right: Infection density $f(y)$ of Y. The solid lines are the fits with modified cross-validation ($\alpha = 1.4$) through (6.25). The dashed lines are taken from Figure 6.6, lower-left frame.

6.6.4 Case Study: AIDS Incubation

We now apply the techniques developed in this section to the CDC blood transfusion data and compare the estimates of $f(x)$ and $f(y)$ with the ones obtained from the joint density in §6.5.3. The cubic splines were used, with the biasing function $w_i(x) = I_{[x<y_i]}$ for $f(x)$ and the biasing function $w_i(y) = I_{[x_i<y]}$ for $f(y)$. Plotted in Figure 6.10 are the cross-validated ($\alpha = 1.4$) estimates for the elderly group (age 60+) indicated by solid lines, with the marginals of the joint density estimate from Figure 6.6 superimposed as dashed lines. Remember that information from data is scarce on the upper end of $f(x)$ and on the lower end of $f(y)$.

6.7 Conditional Densities

On a product domain $\mathcal{X} \times \mathcal{Y}$, the primary interest is often the estimation of the conditional density $f(y|x)$. Such a problem is typically known as regression, but unlike the formulations of Chapters 3 and 5, no parametric assumption is made here on a generic $\mathcal{Y}$ axis, and the function to be estimated is "bivariate" in (x, y) instead of "univariate" only in x.

A logistic conditional density transform can be made one-to-one through side conditions on the $\mathcal{Y}$ axis, with which the penalized likelihood estimation is straightforward. The computation and smoothing parameter selection follow trivial modifications of the algorithms designed for (6.1). The analysis of a real data set is presented, where an adaptation of the partial spline technique of §4.1 is explored. With $\mathcal{Y}$ discrete, logistic regression is formulated for multinomial responses, which reduces to the formulation of Example 5.2 for $\mathcal{Y}$ binary.

6.7.1 Penalized Likelihood Estimation

Consider independent observations (X_i, Y_i) on a product domain $\mathcal{X} \times \mathcal{Y}$ from a density $f(x,y) = f(x)f(y|x)$. Of interest is the estimation of the conditional density $f(y|x) = f(x,y)/\int_{\mathcal{Y}} f(x,y)$ of Y given X. Since the marginal density $f(x)$ of X is only a nuisance parameter, the sampling of X_i can actually be arbitrary, random or deterministic, so long as $Y|X \sim f(y|x)$. For notational convenience, however, $f(x)$ will still be used to denote the "limiting distribution" of X_i's, even when they are deterministic.

The logistic conditional density transform, $f(y|x) = e^{\eta(x,y)}/\int_{\mathcal{Y}} e^{\eta(x,y)}$, can be employed to enforce the positivity and unity constraints. To make the transform one-to-one, $\eta(x,y)$ has to satisfy certain side conditions, say $A_y\eta(x,y) = 0$, $\forall x \in \mathcal{X}$, where the averaging operator A_y on the domain $\mathcal{Y}$ can, in principal, depend on x. A simple approach to achieving a one-to-one logistic conditional density transform is through term elimination in an ANOVA decomposition, as discussed in §1.3.2: For $\eta(x,y) = \eta_\emptyset + \eta_x + \eta_y + \eta_{x,y}$ with averaging operators A_x and A_y,

$$f(y|x) = \frac{e^{\eta_\emptyset+\eta_x+\eta_y+\eta_{x,y}}}{\int_{\mathcal{Y}} e^{\eta_\emptyset+\eta_x+\eta_y+\eta_{x,y}}} = \frac{e^{\eta_y+\eta_{x,y}}}{\int_{\mathcal{Y}} e^{\eta_y+\eta_{x,y}}},$$

where $A_y(\eta_y + \eta_{x,y}) = 0$, $\forall x \in \mathcal{X}$; the side condition here is independent of x. Eliminating $\eta_\emptyset + \eta_x$ from $\eta(x,y)$, one may estimate $f(y|x) = e^{\eta(x,y)}/\int_{\mathcal{Y}} e^{\eta(x,y)}$ by minimizing

$$-\frac{1}{n}\sum_{i=1}^{n}\left\{\eta(X_i,Y_i) - \log\int_{\mathcal{Y}} e^{\eta(X_i,y)}\right\} + \frac{\lambda}{2}J(\eta) \tag{6.29}$$

in an appropriately assembled tensor product reproducing kernel Hilbert space.

Example 6.8 (Tensor product cubic spline)
Consider $\mathcal{X} = \mathcal{Y} = [0,1]$. Use the construction of Example 2.5, with (x,y) replacing $(x_{\langle 1\rangle}, x_{\langle 2\rangle})$ in the notation. Eliminating $\eta_\emptyset$ and η_x, one has the space

$$\mathcal{H} = \mathcal{H}_{00\langle x\rangle} \otimes (\mathcal{H}_{01\langle y\rangle} \oplus \mathcal{H}_{1\langle y\rangle}) \oplus (\mathcal{H}_{01\langle x\rangle} \oplus \mathcal{H}_{1\langle x\rangle}) \otimes (\mathcal{H}_{01\langle y\rangle} \oplus \mathcal{H}_{1\langle y\rangle}).$$

In the notation of Example 2.8, one may set

$$J(f,g) = \theta_{00,1}^{-1}(f,g)_{00,1} + \theta_{01,1}^{-1}(f,g)_{01,1} + \theta_{1,01}^{-1}(f,g)_{1,01} + \theta_{1,1}^{-1}(f,g)_{1,1},$$

which has the null space $\mathcal{N}_J = (\mathcal{H}_{00\langle x\rangle} \otimes \mathcal{H}_{01\langle y\rangle}) \oplus (\mathcal{H}_{01\langle x\rangle} \otimes \mathcal{H}_{01\langle y\rangle})$ spanned by $\phi_1 = y - 0.5$ and $\phi_2 = (x-0.5)(y-0.5)$, and the reproducing kernel

$$R_J = \theta_{00,1}R_{00,1} + \theta_{01,1}R_{01,1} + \theta_{1,01}R_{1,01} + \theta_{1,1}R_{1,1}.$$

Clearly, one has $\int_0^1 \eta(x,y)dy = 0$, $\forall x \in [0,1]$, for $\eta \in \mathcal{H}$. □

The minimizer of (6.29) in $\mathcal{H} = \{f : J(f) < \infty\}$ is generally not computable, but one may calculate that in

$$\mathcal{H}^* = \mathcal{N}_J \oplus \text{span}\{R_J((X_i, Y_i), \cdot), i = 1, \ldots, n\},$$

which shares the same asymptotic convergence rates; see §8.2. Now, define $\mu_f(g|x) = \int_{\mathcal{Y}} g e^f / \int_{\mathcal{Y}} e^f$ and $v_f(g, h|x) = \mu_f(gh|x) - \mu_f(g|x)\mu_f(h|x)$. The Newton updating formula (6.7) on page 181 again holds verbatim for the minimization of (6.29) in $\mathcal{H}^*$, with $\mu_f(g)$ and $V_f(g, h)$ modified as follows,

$$\mu_f(g) = \frac{1}{n}\sum_{i=1}^{n} \mu_f(g|X_i), \qquad V_f(g, h) = \frac{1}{n}\sum_{i=1}^{n} v_f(g, h|X_i); \tag{6.30}$$

see Problem 6.15.

Weighted by the sampling proportion $f(x)$, the aggregated Kullback-Leibler distance of $f_\lambda(y|x) = e^{\eta_\lambda} / \int_{\mathcal{Y}} e^{\eta_\lambda}$ from $f(y|x) = e^{\eta} / \int_{\mathcal{Y}} e^{\eta}$ is now

$$\text{KL}(\eta, \eta_\lambda) = \int_{\mathcal{X}} f(x)\Big\{\mu_\eta(\eta - \eta_\lambda|x) - \log \int_{\mathcal{Y}} e^{\eta} + \log \int_{\mathcal{Y}} e^{\eta_\lambda}\Big\}, \tag{6.31}$$

with the relative Kullback-Leibler distance

$$\text{RKL}(\eta, \eta_\lambda) = \int_{\mathcal{X}} f(x) \log \int_{\mathcal{Y}} e^{\eta_\lambda} - \int_{\mathcal{X}} f(x)\mu_\eta(\eta_\lambda|x). \tag{6.32}$$

The first term can be estimated by $n^{-1}\sum_{i=1}^{n} \log \int_{\mathcal{Y}} e^{\eta_\lambda(X_i, y)}$. The second term $E[\eta_\lambda(X, Y)]$, where $(X, Y) \sim f(x)f(y|x) = f(x, y)$, can be estimated by $n^{-1}\sum_{i=1}^{n} \eta_{\lambda,\tilde{\eta}}^{[i]}(X_i, Y_i)$, which is given by (6.18) on page 185 with the entries in the relevant matrices defined by the modified $\mu_f(g)$ and $V_f(g, h)$. The derivation of cross-validation scores and the structure of the numerical problem parallel those in §6.6.2.

6.7.2 Case Study: Penny Thickness

The thickness in mils of a sample of 90 U.S. Lincoln pennies is listed in Scott (1992, Appendix B.4). Two pennies from each year between 1945 and 1989 were measured. After mapping $\mathcal{X} \times \mathcal{Y} = [1944.5, 1989.5] \times [49, 61]$ onto $[0, 1]^2$, the construction of Example 6.8 was used in (6.29) for the estimation of the conditional density of the penny thickness given the year, and the modified cross-validation with $\alpha = 1.4$ was used to select the smoothing parameters. The fit is plotted in the left frame of Figure 6.11, where the solid line marks the conditional medians, the dashed lines the conditional quartiles, and the horizontal dotted lines the conditional 5th and 95th percentiles. The data are superimposed as circles, with the x coordinate jittered to unmask a few overlaps. The estimate is under the assumption of smoothness of the log conditional density on both axes, despite the apparent

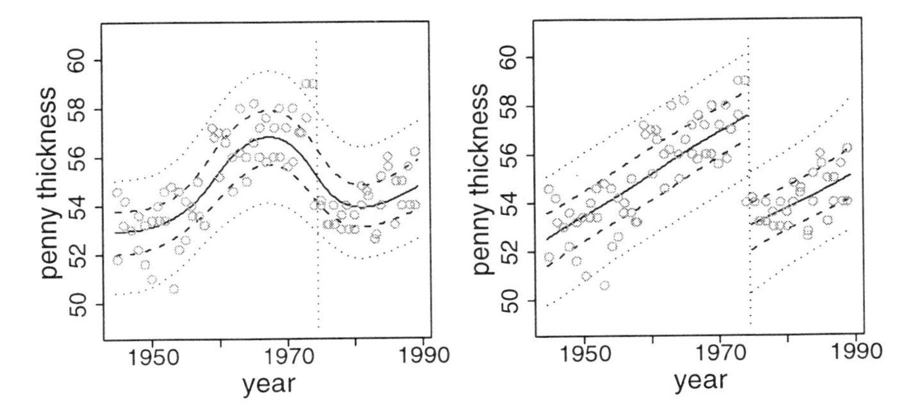

FIGURE 6.11. Thickness of U.S. Lincoln Pennies. Left: Continuous fit. Right: Fit with built-in break. The solid lines are conditional medians, the dashed lines quartiles, and the dotted lines 5th and 95th percentiles. The circles are the data with the year jittered. The vertical dotted lines mark the position of the break.

abrupt downward shift of penny thickness from 1974 to 1975. A vertical dotted line is superimposed to mark the break.

Adapting the partial spline technique discussed in §4.1, it is possible to build the break into the fit. Consider the tensor sum of four reproducing kernel Hilbert spaces on $[0,1]$ with the reproducing kernels $R_{00l} = I_{[x_1 \in L]} I_{[x_2 \in L]}$, $R_{00u} = I_{[x_1 \in U]} I_{[x_2 \in U]}$, $R_{01} = k_1(x_1)k_1(x_2)$, and

$$R_1 = k_2(x_1)k_2(x_2) - k_4(x_1 - x_2),$$

where $L = [0, 2/3]$, $U = (2/3, 1]$, and $k_1(x)$, $k_2(x)$, and $k_4(x)$ are as given in §2.3.3. Using the tensor product space

$$\mathcal{H} = (\mathcal{H}_{00l\langle x \rangle} \oplus \mathcal{H}_{00u\langle x \rangle} \oplus \mathcal{H}_{01\langle x \rangle} \oplus \mathcal{H}_{1\langle x \rangle}) \otimes (\mathcal{H}_{01\langle y \rangle} \oplus \mathcal{H}_{1\langle y \rangle})$$

in (6.29), one allows a break at $x = 2/3$, which is 1974.5 when $[0,1]$ is mapped back to $[1944.5, 1989.5]$. An ANOVA decomposition is no longer available on $\mathcal{X}$, but one does not really need one. Of the eight subspaces of $\mathcal{H}$, five involve $\mathcal{H}_{1\langle x \rangle}$ or $\mathcal{H}_{1\langle y \rangle}$ and have to be penalized. The other three form the null space

$$\mathcal{N}_J = \text{span}\{I_{[x \in L]}(y - 0.5), I_{[x \in U]}(y - 0.5), (x - 0.5)(y - 0.5)\}.$$

The modified cross-validation with $\alpha = 1.4$ was used to select the smoothing parameters, and the fit is plotted in the right frame of Figure 6.11. Apart from the downward shift, the pennies appeared to get thicker at a steady pace as time progressed.

6.7.3 Logistic Regression

The domain $\mathcal{Y}$ in (6.29) is generic. In particular, it can be discrete. Consider $\mathcal{Y} = \{1, \ldots, K\}$ and define $\int_{\mathcal{Y}} f = \sum_{y=1}^{K} f(y)$. As discussed in §2.2, the reproducing kernel $R_{1\langle y \rangle}(y_1, y_2) = I_{[y_1=y_2]} - (1/K)$ generates the contrast space $\mathcal{H}_{1\langle y \rangle}$ in a one-way ANOVA decomposition, satisfying $\int_{\mathcal{Y}} f = 0$, $\forall f \in \mathcal{H}_{1\langle y \rangle}$.

Let $\mathcal{H}_{0\langle x \rangle} \oplus \mathcal{H}_{1\langle x \rangle}$ be a tensor sum reproducing kernel Hilbert space on $\mathcal{X}$, where $\mathcal{H}_{0\langle x \rangle}$ has a basis $\{\phi_\nu\}_{\nu=1}^{m}$ including the constant and $\mathcal{H}_{1\langle x \rangle}$ has the square norm $J_{\langle x \rangle}(f)$ and the reproducing kernel $R_{1\langle x \rangle}(x_1, x_2)$. The tensor product space $\mathcal{H} = (\mathcal{H}_{0\langle x \rangle} \oplus \mathcal{H}_{1\langle x \rangle}) \otimes \mathcal{H}_{1\langle y \rangle}$ contains the ANOVA terms $\eta_y + \eta_{x,y}$ for use in (6.29). The space $\mathcal{H}_{0\langle x \rangle} \otimes \mathcal{H}_{1\langle y \rangle}$ is $m(K-1)$-dimensional with a basis

$$\{\phi_\nu(x) R_{1\langle y \rangle}(\mu, y),\ \nu = 1, \ldots, m,\ \mu = 1, \ldots, K-1\},$$

and the space $\mathcal{H}_{1\langle x \rangle} \otimes \mathcal{H}_{1\langle y \rangle}$ has the square norm

$$J(\eta) = \frac{1}{K} \sum_{y=1}^{K} J_{\langle x \rangle}((I - A_y)\eta(x, y)) = \frac{1}{K} \sum_{y=1}^{K} J_{\langle x \rangle}(\eta(x, y))$$

with the reproducing kernel

$$R_J(z_1, z_2) = R_{1\langle x \rangle}(x_1, x_2) R_{1\langle y \rangle}(y_1, y_2),$$

where $z = (x, y) \in \mathcal{X} \times \mathcal{Y}$ and $A_y f = \int_{\mathcal{Y}} f/K$; see Problem 6.14. The conditional probability $P(y|x)$ is given by

$$P(y|x) = \frac{e^{\eta(x,y)}}{\sum_{u=1}^{K} e^{\eta(x,u)}},$$

the multinomial logistic transform.

For $K = 2$, one has $\eta(x, 1) = -\eta(x, 2)$, $\forall \eta \in \mathcal{H}$. It follows that

$$\begin{aligned} \eta(X, Y) - \log \int_{\mathcal{Y}} e^{\eta(X,y)} \\ &= (I_{[Y=1]} - I_{[Y=2]})\eta(X, 1) - \log(e^{\eta(X,1)} + e^{-\eta(X,1)}) \\ &= I_{[Y=1]}(2\eta(X, 1)) - \log(1 + e^{2\eta(X,1)}). \end{aligned}$$

Since $J(\eta) = J_{\langle x \rangle}(\eta(x, 1))$, (6.29) becomes

$$-\frac{1}{n} \sum_{i=1}^{n} \{I_{[Y_i=1]}(2\eta(X_i, 1)) - \log(1 + e^{2\eta(X_i,1)})\} + \frac{\lambda}{8} J_{\langle x \rangle}(2\eta(x, 1)). \quad (6.33)$$

Noting that

$$P(1|x) = \frac{e^{\eta(x,1)}}{e^{\eta(x,1)} + e^{-\eta(x,1)}} = \frac{e^{2\eta(x,1)}}{e^{2\eta(x,1)} + 1}$$

and that $\lambda > 0$ is arbitrary, (6.33) is simply the logistic regression of Example 5.2, with $I_{[Y=1]}$ being the Y in Example 5.2 and $2\eta(x,1)$ being the $\eta(x)$ there. Hence, (6.29) with $\mathcal{Y} = \{1,\ldots,K\}$ naturally extends penalized likelihood logistic regression from binomial responses to multinomial responses.

We now take a look at the cross-validation scores for logistic regression with $K = 2$. Since

$$\log \int_{\mathcal{Y}} e^{\eta_\lambda(X_i,y)} = \log(e^{2\eta_\lambda(X_i,1)} + 1) - \eta_\lambda(X_i,1)$$

and

$$\eta_{\lambda,\tilde{\eta}}^{[i]}(X_i,Y_i) = (2I_{[Y_i=1]} - 1)\eta_{\lambda,\tilde{\eta}}^{[i]}(X_i,1),$$

the relative Kullback-Leibler distance of (6.32) is estimated by

$$\frac{1}{n}\sum_{i=1}^{n}\{\log(e^{2\eta_\lambda(X_i,1)} + 1) - I_{[Y_i=1]}(2\eta_\lambda(X_i,1))\} \\ + \frac{1}{n}\sum_{i=1}^{n}(2I_{[Y_i=1]} - 1)(\eta_\lambda(X_i,1) - \eta_{\lambda,\tilde{\eta}}^{[i]}(X_i,1)), \quad (6.34)$$

which, in the notation of §5.2.2, can be written as

$$-\frac{1}{n}\sum_{i=1}^{n}\{Y_i\eta_\lambda(x_i) - b(\eta_\lambda(x_i))\} + \frac{\alpha}{n}\sum_{i=1}^{n}\frac{a_{i,i}}{1-a_{i,i}}\frac{(Y_i-.5)(Y_i-\mu_\lambda(x_i))}{\tilde{w}_i} \quad (6.35)$$

for $\alpha = 1$; see Problem 6.16. Compare (6.35) with (5.16) on page 157. From (6.18), it is easy to see that the second term of (6.34) is the trace term, so the modified cross-validation scores are given by (6.35) for $\alpha > 1$.

Setting $\int_{\mathcal{X}} f(x)h(x) = n^{-1}\sum_{i=1}^{n} h(x_i)$ in (6.31) for $K = 2$, it can be shown that $\mathrm{KL}(\eta,\eta_\lambda)$ is given by (5.6) on page 155 in the notation of §5.2.2; see Problem 6.17. In the simulation of §5.2.3, $\mathrm{KL}(\eta,\eta_\lambda)$ and (6.35) with $\alpha = 1, 1.2, 1.4, 1.6$ were also calculated for the estimates on the same replicates. Plotted in the left frame of Figure 6.12 are the ratios of the minimum $\mathrm{KL}(\eta,\eta_\lambda)$ loss to the losses achieved by the cross-validation scores with $\alpha = 1, 1.2, 1.4, 1.6$ as well as by $U_w(\lambda)$ and $V_g^*(\lambda)$ of §5.2. The losses achieved by (6.35) with $\alpha = 1$ and by $U_w(\lambda)$ of §5.2 are plotted in the middle frame, suggesting the practical equivalence of the two methods. The right frame compares the modified cross-validation score for $\alpha = 1.4$ with $V_g^*(\lambda)$ of §5.2, showing the superiority of the latter. Note that $V_g^*(\lambda)$ of §5.2 was derived for the estimation of a "univariate" function $\eta(x)$, with parametric assumptions on the $\mathcal{Y}$ axis, but (6.35) was specialized from a score designed for the estimation of a "bivariate" function $\eta(x,y)$, with no parametric assumption on $\mathcal{Y}$.

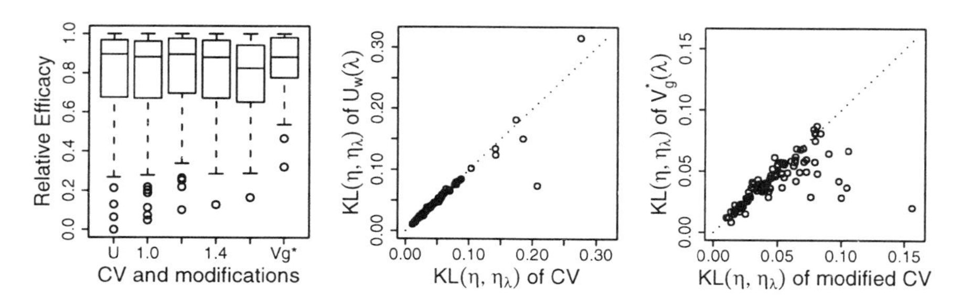

FIGURE 6.12. Performance of Cross-Validation and Modifications Thereof for Binomial Logistic Regression. Left: Relative efficacy of cross-validation with $\alpha = 1, 1.2, 1.4, 1.6$ along with that of $U_w(\lambda)$ and $V_g^*(\lambda)$ of §5.2. Center: Comparison of (6.35) for $\alpha = 1$ with $U_w(\lambda)$ of §5.2. Right: Comparison of (6.35) for $\alpha = 1.4$ with $V_g^*(\lambda)$ of §5.2.

6.8 Response-Based Sampling

In studies of rare events, data are often subject to a form of selection bias known as choice-based sampling in econometrics or case-control sampling in biostatistics. Samples largely from $f(x|y)$ have to be used to estimate $f(y|x)$ or part of it.

Because of the selection bias, $f(y|x)$ is estimable only through the joint density $f(x, y)$. The joint density is not always estimable, but when it is, the estimation through penalized likelihood method is straightforward. The odds ratio is available through either $f(x|y)$ or $f(y|x)$, so is always estimable.

6.8.1 Response-Based Samples

Consider a probability density $f(x, y)$ on a product domain $\mathcal{X} \times \mathcal{Y}$, where $\mathcal{Y} = \{1, \ldots, K\}$ is discrete. Let $\mathcal{Y}_j \subseteq \mathcal{Y}$, $j = 1, \ldots, s$, be s strata; $\cup_{j=1}^s \mathcal{Y}_j = \mathcal{Y}$. A stratum $\mathcal{Y}_j$ is selected with probability π_j, $\sum_{i=1}^s \pi_j = 1$, and given the stratum, observations are taken from $f(x, y)$ but restricted to the stratum $\mathcal{X} \times \mathcal{Y}_j$. Such data are known as choice-based samples in econometrics or case-control samples in biostatistics. Of interest is the estimation of the conditional density $f(y|x)$. Since the strata are defined by restricted y values, the sampling scheme is called response-based sampling.

Example 6.9 (Separate sampling)
With $s = K$ and $\mathcal{Y}_j = \{y = j\}$, one gets a separate sample for case-control studies (Anderson 1972). □

Example 6.10 (Enriched choice-based sampling)
With $s = K+1$, $\mathcal{Y}_j = \{y = j\}$, $j = 1, \ldots, K$, and $\mathcal{Y}_{K+1} = \mathcal{Y}$, one obtains an enriched choice-based sample (Cosslett 1981). □

With response-based sampling, the data are largely from the the "wrong" conditional distribution $f(x|y)$. Such sampling strategy is necessary when the categories of interest are rare in the population, in which case an informative random sample from $f(x,y)$ or $f(y|x)$ can be astronomical.

From $f(x,y) = e^{\eta_x+\eta_y+\eta_{x,y}} / \int_{\mathcal{X}} \int_{\mathcal{Y}} e^{\eta_x+\eta_y+\eta_{x,y}}$, one has

$$f(y|x) = \frac{e^{\eta_y+\eta_{x,y}}}{\int_{\mathcal{Y}} e^{\eta_y+\eta_{x,y}}}, \qquad f(x|y) = \frac{e^{\eta_x+\eta_{x,y}}}{\int_{\mathcal{X}} e^{\eta_x+\eta_{x,y}}}.$$

Separate sampling does not warrant the estimation of $f(y|x)$ unless an independent estimate of the marginal density $f(y) \propto e^{\eta_y} \int_{\mathcal{X}} e^{\eta_x+\eta_{x,y}}$ is available, whereas an enriched sample does carry information about $f(x,y)$ and, hence, about $f(y|x)$. Note that the empirical π_j cannot be used to estimate the marginal density $f(y)$ due to the very selection bias in the sampling scheme. It is easy to verify, however, that an odds ratio

$$\frac{f(y_1|x_1)/f(y_2|x_1)}{f(y_1|x_2)/f(y_2|x_2)} = \frac{f(y_1|x_1)f(y_2|x_2)}{f(y_1|x_2)f(y_2|x_1)}, \tag{6.36}$$

depends only on the interaction $\eta_{x,y}$, and, hence, is always estimable; see Problem 6.18.

In the case that none of the partitions $\{1, \ldots, s\} = A \cup A^c$ would satisfy $(\cup_{j\in A}\mathcal{Y}_j) \cap (\cup_{j\in A^c}\mathcal{Y}_j) = \emptyset$ (Cosslett 1981, Assumption 10), known as the connected case, $f(x,y)$ is identifiable from the sample, in the sense that the minus log likelihood

$$-\frac{1}{n}\sum_{i=1}^{n} \eta(X_i, Y_i) + \sum_{j=1}^{s} \frac{n_j}{n} \log \int_{\mathcal{X}} \int_{\mathcal{Y}_i} e^{\eta} \tag{6.37}$$

is strictly convex in $\eta = \eta_x + \eta_y + \eta_{x,y}$ that satisfies side conditions $A_x(\eta_x + \eta_{x,y}) = 0$, $\forall y$, and $A_y(\eta_y + \eta_{x,y}) = 0$, $\forall x$, where (X_i, Y_i) are the observed data and n_j are the sample sizes from the strata $\mathcal{Y}_j$, $\sum_{j=1}^{s} n_j = n$; see Problem 6.19. When there is a partition of $\{1, \cdots, s\} = A \cup A^c$ such that $(\cup_{j\in A}\mathcal{Y}_j) \cap (\cup_{j\in A^c}\mathcal{Y}_j) = \emptyset$, however, η_y is not identifiable although $\eta_x + \eta_{x,y}$ still is.

For the estimation of $f(x|y) = e^{\eta_x+\eta_{x,y}} / \int_{\mathcal{X}} e^{\eta_x+\eta_{x,y}}$, one can always cast the sampling scheme as separate sampling with $s = K$ and $\mathcal{Y}_j = \{y = j\}$, and the minus log conditional likelihood

$$-\frac{1}{n}\sum_{i=1}^{n} \eta(X_i, Y_i) + \sum_{j=1}^{K} \frac{n_j}{n} \log \int_{\mathcal{X}} e^{\eta(x,j)} \tag{6.38}$$

is strictly convex in $\eta = \eta_x + \eta_{x,y}$ that satisfies side conditions $A_x\eta = 0$, $\forall y$. Note that (6.38) is identical to (6.37) under separate sampling, with $\eta_y(j)$ in (6.37) canceling out.

6.8.2 Penalized Likelihood Estimation

The estimation of $f(x|y)$ has been treated in §6.7, so we only consider the connected case here. Write $\tilde{\pi}_j = n_j/n$. The joint density $f(x,y) = e^\eta / \int_{\mathcal{X}} \int_{\mathcal{Y}} e^\eta$ can be estimated through the minimization of

$$-\frac{1}{n}\sum_{i=1}^{n} \eta(X_i, Y_i) + \sum_{j=1}^{s} \tilde{\pi}_j \log \int_{\mathcal{X}} \int_{\mathcal{Y}_j} e^\eta + \frac{\lambda}{2} J(\eta), \tag{6.39}$$

for $\eta(x,y) = \eta_x + \eta_y + \eta_{x,y}$. The minimizer in $\mathcal{H} = \{f : J(f) < \infty\}$ is generally not computable, but that in $\mathcal{H}^* = \mathcal{N}_J \oplus \text{span}\{R_J((X_i, Y_i), \cdot)\}$ shares the same convergence rates; see §8.2. Define $\mu_f(g|j) = \int_{\mathcal{X}} \int_{\mathcal{Y}_j} g e^f / \int_{\mathcal{X}} \int_{\mathcal{Y}_j} e^f$ and $v_f(g,h|j) = \mu_f(gh|j) - \mu_f(g|j)\mu_f(h|x)$. The Newton updating formula (6.7) on page 181 again holds verbatim for the minimization of (6.39) in $\mathcal{H}^*$ with $\mu_f(g)$ and $V_f(g,h)$ modified as

$$\mu_f(g) = \sum_{j=1}^{K} \tilde{\pi}_j \mu_f(g|j), \qquad V_f(g,h) = \sum_{j=1}^{K} \tilde{\pi}_j v_f(g,h|j).$$

The Kullback-Leibler distance is now defined by

$$\text{KL}(\eta, \eta_\lambda) = \sum_{j=1}^{K} \pi_j \Big\{ \mu_\eta(\eta - \eta_\lambda | j) - \log \int_{\mathcal{X}} \int_{\mathcal{Y}_j} e^\eta + \log \int_{\mathcal{X}} \int_{\mathcal{Y}_j} e^{\eta_\lambda} \Big\},$$

with the relative Kullback-Leibler distance given by

$$\text{RKL}(\eta, \eta_\lambda) = \sum_{j=1}^{K} \pi_j \Big\{ \log \int_{\mathcal{X}} \int_{\mathcal{Y}_j} e^{\eta_\lambda} - \mu_\eta(\eta_\lambda | j) \Big\}.$$

The cross-validation and computation are again trivial to modify.

6.9 Bibliographic Notes

Sections 6.1 and 6.2

Penalized likelihood density estimation was pioneered by Good and Gaskins (1971), who used a square root transform for positivity and resorted to constrained optimization to enforce unity. The logistic density transform was introduced by Leonard (1978), and (6.12) was proposed by Silverman (1982) to ensure unity without numerically enforcing it. The early work was largely done in the univariate context, although the basic ideas are applicable in more general settings. Using B-spline basis with local support, O'Sullivan (1988a) developed a fast algorithm similar to that of §3.8.1 for the computation of Silverman's estimate.

The one-to-one logistic density transform through a side condition was introduced in Gu and Qiu (1993), where an asymptotic theory was developed that led to the computability of the estimate through $\mathcal{H}^*$ on generic domains. The estimation of the Poisson process and the link to Silverman's estimate was also noted by Gu and Qiu (1993).

Section 6.3

With a varying smoothing parameter λ in (6.7), a performance-oriented iteration similar to that in §5.2.1 was developed by Gu (1993b). This approach does not bode well with multiple smoothing parameters, however, as analytical derivatives similar to those behind Algorithm 3.2 are lacking. The direct cross-validation and its modifications presented here were developed in Gu and Wang (2001).

The simulated samples used in the empirical study were the ones used in Gu (1993b, §5) so that Figure 6.1 is comparable to Figure 5.1 of Gu (1993b); the samples were generated using FORTRAN routines `rnor` and `runi` from the CMLIB (`http://lib.stat.cmu.edu/cmlib`). The performances of the methods are visibly worse on samples generated from the random number generator in R, for reasons unknown to the author.

Section 6.4

The strategy for the handling of numerical singularity is similar to the one discussed in Gu (1993b, Appendix). The flop counts are largely taken from Golub and Van Loan (1989).

Section 6.5

The Buffalo snowfall data have been analyzed by numerous authors using various density estimation methods such as density-quantile autoregressive estimation (Parzen 1979), average shifted histograms (Scott 1985), and regression spline extended linear models (Stone, Hansen, Kooperberg, and Truong 1997). The estimates presented here differ slight from the ones shown in Gu (1993b), where a performance-oriented iteration was used to select the smoothing parameter.

The analysis of the CDC blood transfusion data presented here differ slightly from the one in Gu (1998b), where a performance-oriented iteration was used to select the smoothing parameters.

Section 6.6

An early reference on length-biased sampling and its applications is Cox (1969). The empirical distributions for data in the settings of Examples 6.5

and 6.6 were derived and their asymptotic properties studied by Vardi (1982, 1985) and Gill, Vardi, and Wellner (1988). The empirical distribution for truncated data of Example 6.7 was studied by Woodroofe (1985), Wang, Jewell, and Tsay (1986), Wang (1989), and Keiding and Gill (1990).

The smoothing of the empirical distribution for length-biased data of Example 6.4 through the kernel method was studied by Jones (1991). The general formulation of penalized likelihood density estimation for biased and truncated data as presented in this section is largely taken from an unpublished technical report by Gu (1992d).

Section 6.7

Most nonparametric regression techniques, such as the local polynomial methods (Cleveland 1979, Fan and Gijbels 1996), with the kernel methods as special cases, primarily concern the conditional mean. Work has also been done for the estimation of conditional quantiles (Koenker, Ng, and Portnoy 1994). Cole (1988) and Cole and Green (1992) considered a three-parameter model on the y axis in the form of Box-Cox transformation and estimated the three parameters as functions of x through penalized likelihood; the conditional mean and conditional quantiles could be easily obtained from the three-parameter model.

The materials of this section are largely taken from Gu (1995a). We estimate $f(y|x)$ as a "bivariate" function on generic domains $\mathcal{X}$ and $\mathcal{Y}$, where $\mathcal{X}$ and $\mathcal{Y}$ can both be multivariate. A similar approach to conditional density estimation can be found in Stone, Hansen, Kooperberg, and Truong (1997).

Regression with multinomial responses has also been studied by Kooperberg, Bose, and Stone (1997) and Lin (1998). For $\mathcal{Y} = \{1, 2\}^L$, a product of binary domains, Gao (1999) and Gao, Wahba, Klein, and Klein (2001) attempted a direct generalization of (5.1).

Section 6.8

The term response-based sampling was coined by Manski (1995). Parametric or partly parametric estimation of the odds ratio or the conditional density $f(y|x)$ under such a sampling scheme have been studied by Anderson (1972), Prentice and Pyke (1978), Cosslett (1981), and Scott and Wild (1986), among others. The empirical joint distribution based on enriched samples was derived by Morgenthaler and Vardi (1986) and was used as weights in their kernel estimate of $f(y|x)$. A version of penalized likelihood estimation adapted from Good and Gaskins (1971) was proposed by Anderson and Blair (1982) for the case of $K = 2$. The formulation of this section was largely taken from an unpublished technical report by Gu (1995b).

6.10 Problems

Section 6.1

6.1 Verify (6.5).

6.2 Verify (6.6) and (6.7).

Section 6.2

6.3 Verify (6.11).

6.4 Show that the minimizer $\tilde{\eta}^*$ of (6.12) satisfies the unity constraint $\int_{\mathcal{X}} e^{\tilde{\eta}^*} dx = 1$.

Section 6.3

6.5 For $L_{f,g}(\alpha) = \log \int_{\mathcal{X}} e^{f+\alpha g} dx$ as a function of α, verify that $\dot{L}_{f,g}(0) = \mu_f(g)$ and $\ddot{L}_{f,g}(0) = V_f(g)$.

6.6 Plugging (6.3) into (6.15), show that the minimizing coefficients satisfy (6.7).

6.7 Verify the cross-validation estimate given in (6.20) for prebinned data.

6.8 For $H = V + \text{diag}(\lambda Q, O)$, where $V = V_{\tilde{\eta}}(\breve{\boldsymbol{\xi}}, \breve{\boldsymbol{\xi}}^T)$, verify that

$$\left.\frac{dH^{-1}}{d\log\lambda}\right|_{\lambda=\tilde{\lambda}} = H^{-1} - H^{-1}VH^{-1}.$$

6.9 Verify (6.21), given the fact that the right-hand side of (6.6) is 0 for $\tilde{\eta} = \eta_{\tilde{\lambda}}$.

Section 6.4

6.10 Premultiply (6.23) by $\tilde{G}^{-T}$, and show that the linear system reduces to (6.24).

Section 6.6

6.11 Show that the Newton update for minimizing (6.25) satisfies (6.7), with $\mu_f(g)$ and $V_f(g, h)$ modified as in (6.26).

6.12 Derive the counterparts of (6.19) and (6.22) for use with (6.25).

6.13 Show that with (t, X) generated according to the scheme of Example 6.7, with $A = \{t < x\}$, $f(x)$ supported on $(0, a)$, and $g(t)$ uniform on $(0, a)$, X is length-biased from a density proportional to $xf(x)$.

Section 6.7

6.14 Consider $\mathcal{H}_{\langle x \rangle}$ on $\mathcal{X}$, with a reproducing kernel $R_{\langle x \rangle}(x_1, x_2)$ and an inner product $J_{\langle x \rangle}(f, g)$, and $\mathcal{H}_{\langle y \rangle}$ on $\mathcal{Y} = \{1, \ldots, K\}$, with the reproducing kernel $R_{\langle y \rangle}(y_1, y_2) = I_{[y_1 = y_2]} - (1/K)$ and the inner product $(f, g)_{\langle y \rangle} = f^T(I - \mathbf{1}\mathbf{1}^T/K)g$. Verify that in the tensor product space $\mathcal{H}_{\langle x \rangle} \otimes \mathcal{H}_{\langle y \rangle}$, with a reproducing kernel $R_{\langle x \rangle}(x_1, x_2)R_{\langle y \rangle}(y_1, y_2)$, the inner product is given by

$$J(f, g) = (1/K) \textstyle\sum_{y=1}^{K} J_{\langle x \rangle}((I - A_y)f, (I - A_y)g),$$

where $A_y f = \sum_{y=1}^{K} f(y)/K$.

6.15 Show that the Newton update for minimizing (6.29) satisfies (6.7), with $\mu_f(g)$ and $V_f(g, h)$ defined as in (6.30).

6.16 Show that (6.34) can be cast as (6.35) using the notation of §5.2.2, with $I_{[Y=1]}$ replaced by Y and $2\eta(x, 1)$ replaced by $\eta(x)$.

6.17 With $\mathcal{Y} = \{1, 2\}$, $\int_{\mathcal{Y}} f(y) = f(1) + f(2)$, $\eta(x, 1) = -\eta(x, 2)$, and $\eta_\lambda(x, 1) = -\eta_\lambda(x, 2)$, show that

$$\mathrm{KL}(\eta, \eta_\lambda) = \frac{1}{n} \sum_{i=1}^{n} \left\{ \mu_\eta(\eta - \eta_\lambda | x_i) - \log \int_{\mathcal{Y}} e^{\eta(x_i, y)} + \log \int_{\mathcal{Y}} e^{\eta_\lambda(x_i, y)} \right\}$$

coincides with (5.6) on page 155.

Section 6.8

6.18 Show that the odds ratio of (6.36) depends only on the interaction $\eta_{x,y}$.

6.19 Assuming the connected case and $n_j > 0$, $j = 1, \ldots, s$, show that the minus log likelihood of (6.37) is strictly convex in $\eta = \eta_x + \eta_y + \eta_{x,y}$.

7
Hazard Rate Estimation

For right-censored lifetime data with possible left-truncation, (1.6) of Example 1.3 defines penalized likelihood hazard estimation. Of interest are the selection of smoothing parameters, the computation of the estimates, and the asymptotic behavior of the estimates.

The existence and the computability of the penalized likelihood hazard estimates are discussed in §7.1, and it is shown that the numerical structure of hazard estimation parallels that of density estimation, as given in §6.1. In §7.2, a natural Kullback-Leibler loss is derived under the sampling mechanism, and a cross-validation scheme for smoothing parameter selection is developed to target the loss. It turns out that the algorithms for density estimation as developed in §6.3 and §6.4 are readily applicable to hazard estimation after trivial modifications. Real-data examples are given in §7.3. Missing from the discussion is user-friendly software, which is still under development, and the asymptotic convergence rates, which will be presented in Chapter 8.

Also of interest are the estimation of relative risk in a proportional hazard model through penalized partial likelihood (§7.4), which is shown to be isomorphic to density estimation under biased sampling, and models that are parametric in time (§7.5), which can be fitted following the lines of non-Gaussian regression, as discussed in Chapter 5.

7.1 Preliminaries

Let T be the lifetime of an item, Z be the left-truncation time at which the item enters the study, and C be the right-censoring time beyond which the item is dropped from surveillance, independent of each other. Let U be a covariate and $T|U$ follow a survival distribution with survival function $S(t,u) = P(T > t|U = u)$. Observing independent data $(Z_i, X_i, \delta_i, U_i)$, $i = 1, \dots, n$, where $X = \min(T, C)$, $\delta = I_{[T \leq C]}$, and $Z < X$, one is to estimate the hazard rate $\lambda(t,u) = -\partial \log S(t,u)/\partial t$.

When parametric models are assumed on the time axis, hazard estimation is not much different from non-Gaussian regression as treated in Chapter 5; see §7.5. Assuming a proportional hazard model $\lambda(t,u) = \lambda_0(t)\lambda_1(u)$, one may treat the base hazard $\lambda_0(t)$ as nuisance and estimate the "univariate" relative risk $\lambda_1(u)$ through penalized partial likelihood; see §7.4.

The main subject of this chapter is the estimation of the "bivariate" hazard function $\lambda(t,u) = e^{\eta(t,u)}$ through the minimization of the penalized likelihood functional

$$-\frac{1}{n}\sum_{i=1}^{n}\left\{\delta_i \eta(X_i, U_i) - \int_{Z_i}^{X_i} e^{\eta(t,U_i)} dt\right\} + \frac{\lambda}{2} J(\eta) \tag{7.1}$$

in a reproducing kernel Hilbert space $\mathcal{H} = \{f : J(f) < \infty\}$ of functions defined on the domain $\mathcal{T} \times \mathcal{U}$. With $\mathcal{U}$ a singleton and $\lambda = 0$, the nonparametric maximum likelihood yields a delta sum estimate of $\lambda(t)$ corresponding to the Kaplan-Meier estimate of the survival function; see Kaplan and Meier (1958). With $\lambda = \infty$, one fits a parametric model in the null space $\mathcal{N}_J = \{f : J(f) = 0\}$ of the penalty. The time domain $\mathcal{T}$ is understood to be $[0, T^*]$ for some T^* finite, which is not much of a constraint, as all observations are finite in practice.

Let $L(f) = -n^{-1}\sum_{i=1}^{n}\{\delta_i f(X_i, U_i) - \int_{Z_i}^{X_i} e^{f(t,U_i)} dt\}$ be the negative log likelihood. When the maximum likelihood estimate uniquely exists in the null space $\mathcal{N}_J$, the following lemmas establish the existence of the minimizer of (7.1) through Theorem 2.9.

Lemma 7.1
$L(f)$ is convex in f, and the convexity is strict if $f \in \mathcal{H}$ is uniquely determined by its restriction on $\cup_{i=1}^{n}\{(Z_i, X_i) \times \{U_i\}\}$.

Proof. For $\alpha, \beta > 0$, $\alpha + \beta = 1$,

$$\begin{aligned}
\int_Z^X e^{\alpha f(t,U) + \beta g(t,U)} dt &\leq \left\{\int_Z^X e^{f(t,U)} dt\right\}^{\alpha}\left\{\int_Z^X e^{g(t,U)} dt\right\}^{\beta} \\
&= \exp\left\{\alpha \log \int_Z^X e^{f(t,U)} dt + \beta \log \int_Z^X e^{g(t,U)} dt\right\} \\
&\leq \alpha \int_Z^X e^{f(t,U)} dt + \beta \int_Z^X e^{g(t,U)} dt,
\end{aligned}$$

where the first inequality (Hölder's) is strict unless $e^{f(t,U)} \propto e^{g(t,U)}$ on (Z, X) and the second is strict unless $\int_Z^X e^{f(t,U)}dt = \int_Z^X e^{g(t,U)}dt$. The lemma follows. □

Lemma 7.2
$L(f)$ is continuous in f if $f(t,u)$ is continuous in t, $\forall u \in \mathcal{U}$, $\forall f \in \mathcal{H}$.

Proof: The lemma follows from the continuity of evaluation in $\mathcal{H}$ and the Riemann sum approximations of $\int_Z^X e^{f(t,U)}dt$. □

A few examples follow.

Example 7.1 (Cubic spline with no covariate)
A singleton $\mathcal{U}$ characterizes the absence of the covariate. Take $\mathcal{T} = [0,1]$ and $J(\eta) = \int_0^1 \ddot{\eta}^2 dt$. One has $\mathcal{N}_J = \text{span}\{1, t\}$. □

Example 7.2 (Cubic spline with binary covariate)
Consider $\mathcal{U} = \{1, 2\}$. Take $\mathcal{T} = [0,1]$ and

$$J(\eta) = \theta_m^{-1} \int_0^1 (\ddot{\eta}(t,1) + \ddot{\eta}(t,2))^2 + \theta_c^{-1} \int_0^1 (\ddot{\eta}(t,1) - \ddot{\eta}(t,2))^2$$
$$= \theta_m^{-1} J_m(\eta) + \theta_c^{-1} J_c(\eta),$$

where $J_m(\eta)$ penalizes the mean log hazard and $J_c(\eta)$ penalizes the contrast. The null space is given by $\mathcal{N}_J = \text{span}\{I_{[u=1]}, I_{[u=2]}, tI_{[u=1]}, tI_{[u=2]}\}$. See Example 2.7 of §2.4.4.

Setting $\theta_c = 0$ and $\mathcal{N}_J = \text{span}\{I_{[u=1]}, I_{[u=2]}, t\}$, one obtains a proportional hazard model. The proportional hazard model can also be obtained from Example 7.1 using the partial spline technique of §4.1, by adding $I_{[u=2]}$ to $\mathcal{N}_J = \text{span}\{1, t\}$ in Example 7.1. □

Example 7.3 (Tensor product cubic spline)
Consider $\mathcal{U} = \mathcal{T} = [0,1]$. The tensor product cubic spline of Example 2.5 of §2.4.3 can be used in (7.1) for the estimation of the log hazard; see also Example 2.8 of §2.4.5. An additive model characterizes a proportional hazard model. □

Similar to the situation for density estimation, a minimizer η_λ of (7.1) in $\mathcal{H} = \{f : J(f) < \infty\}$ is, in general, not computable, but one may calculate a minimizer η_λ^* in a data-adaptive finite-dimensional space

$$\mathcal{H}^* = \mathcal{N}_J \oplus \text{span}\{R_J((X_i, U_i), \cdot), \delta_i = 1\}, \tag{7.2}$$

which shares the same asymptotic convergence rates as η_λ; see §8.3.

From now on, we shall focus on η_λ^* but drop the star from the notation. Denote by $N = \sum_{i=1}^n \delta_i$ the number of deaths and let $(T_j, \tilde{U}_j)$, $j = 1, \ldots, N$, be the observed lifetimes along with the associated covariates. Plugging the expression

$$\eta(t,u) = \sum_{\nu=1}^{m} d_\nu \phi_\nu(t,u) + \sum_{j=1}^{N} c_j R_J((T_j,\tilde{U}_j),(t,u)) = \boldsymbol{\phi}^T \mathbf{d} + \boldsymbol{\xi}^T \mathbf{c} \qquad (7.3)$$

into (7.1), the calculation of η_λ reduces to the minimization of

$$A_\lambda(\mathbf{c},\mathbf{d}) = -\frac{1}{n}\mathbf{1}^T(Q\mathbf{c}+S\mathbf{d}) + \frac{1}{n}\sum_{i=1}^{n}\int_{Z_i}^{X_i} \exp(\xi_i^T\mathbf{c}+\phi_i^T\mathbf{d})dt + \frac{\lambda}{2}\mathbf{c}^T Q\mathbf{c} \quad (7.4)$$

with respect to $\mathbf{c}$ and $\mathbf{d}$, where S is $N\times m$ with the (j,ν)th entry $\phi_\nu(T_j,\tilde{U}_j)$, Q is $N\times N$ with the (j,k)th entry $\xi_j(T_k,\tilde{U}_k) = R_J((T_j,\tilde{U}_j),(T_k,\tilde{U}_k))$, ξ_i is $N\times 1$ with the jth entry $\xi_j(t,U_i)$, and ϕ_i is $m\times 1$ with the νth entry $\phi_\nu(t,U_i)$.

Write $\mu_f(g) = (1/n)\sum_{i=1}^{n}\int_{Z_i}^{X_i} g(t,U_i)e^{f(t,U_i)}dt$ and $V_f(g,h) = \mu_f(gh)$. Taking derivatives of A_λ in (7.4) at $\tilde{\eta} = \boldsymbol{\xi}^T\tilde{\mathbf{c}} + \boldsymbol{\phi}^T\tilde{\mathbf{d}} \in \mathcal{H}^*$, one has

$$\begin{aligned}
\frac{\partial A_\lambda}{\partial \mathbf{c}} &= -Q\mathbf{1}/n + \mu_{\tilde{\eta}}(\boldsymbol{\xi}) + \lambda Q\tilde{\mathbf{c}} = -Q\mathbf{1}/n + \mu_\xi + \lambda Q\tilde{\mathbf{c}},\\
\frac{\partial A_\lambda}{\partial \mathbf{d}} &= -S^T\mathbf{1}/n + \mu_{\tilde{\eta}}(\boldsymbol{\phi}) = -S^T\mathbf{1}/n + \mu_\phi,\\
\frac{\partial^2 A_\lambda}{\partial \mathbf{c}\partial \mathbf{c}^T} &= V_{\tilde{\eta}}(\boldsymbol{\xi},\boldsymbol{\xi}^T) + \lambda Q = V_{\xi,\xi} + \lambda Q, \qquad (7.5)\\
\frac{\partial^2 A_\lambda}{\partial \mathbf{d}\partial \mathbf{d}^T} &= V_{\tilde{\eta}}(\boldsymbol{\phi},\boldsymbol{\phi}^T) = V_{\phi,\phi},\\
\frac{\partial^2 A_\lambda}{\partial \mathbf{c}\partial \mathbf{d}^T} &= V_{\tilde{\eta}}(\boldsymbol{\xi},\boldsymbol{\phi}^T) = V_{\xi,\phi},
\end{aligned}$$

which is a carbon copy of (6.5) on page 181; see Problem 7.1. With the altered definitions of $\mu_f(g)$, $V_f(g,h)$, Q, and S, the Newton updating equations (6.6) and (6.7) also hold verbatim for the minimization of $A_\lambda(\mathbf{c},\mathbf{d})$ in (7.4).

7.2 Smoothing Parameter Selection

Smoothing parameter selection for hazard estimation parallels that for density estimation. The performance-oriented iteration works fine when the covariate is absent, but it is numerically less efficient when the covariate is present. The direct cross-validation is as effective as the indirect one and is simpler to implement.

A Kullback-Leibler distance is derived for hazard estimation under the sampling mechanism, and a cross-validation score is derived to track the

Kullback-Leibler loss. The cross-validation procedure is nearly a carbon copy of the one derived for density estimation, so the modifications and the computation follow trivially. The effectiveness of the cross-validation scores is evaluated through simple simulation.

As in §3.2, §5.2, and §6.3, the dependence of entities on θ_β is suppressed in the notation.

7.2.1 Kullback-Leibler Loss and Cross-Validation

Denote by $N(t) = I_{[t \leq X, \delta=1]}$ the event process. Under independent censorship, the quantity $e^{\eta(t,u)}dt$ is the conditional probability that $N(t)$ makes a jump in $[t, t+dt)$ given that $t \leq X$ and $U = u$; see, e.g., Fleming and Harrington (1991, page 19). The Kullback-Leibler distance

$$e^{\eta}dt \log \frac{e^{\eta}dt}{e^{\eta_\lambda}dt} + (1 - e^{\eta}dt) \log \frac{1 - e^{\eta}dt}{1 - e^{\eta_\lambda}dt} \\ = \{(\eta - \eta_\lambda)e^{\eta} - (e^{\eta} - e^{\eta_\lambda})\}dt + O((dt)^2)$$

measures the proximity of the estimate $e^{\eta_\lambda}dt$ to the true "success" probability $e^{\eta}dt$. Weighting by the at-risk probability

$$\tilde{S}(t,u) = P(Z < t \leq X | U = u) = E[I_{[Z<t\leq X]} | U = u]$$

and accumulating over $\mathcal{T} \times \mathcal{U}$, one has a Kullback-Leibler distance

$$\mathrm{KL}(\eta, \eta_\lambda) = \int_{\mathcal{U}} m(u) \int_{\mathcal{T}} \{(\eta - \eta_\lambda)e^{\eta} - (e^{\eta} - e^{\eta_\lambda})\}\tilde{S}(t,u)dt \\ = E\Big[\int_{\mathcal{T}} Y(t)\{(\eta(t,U) - \eta_\lambda(t,U))e^{\eta(t,U)} - (e^{\eta(t,U)} - e^{\eta_\lambda(t,U)})\}dt\Big], \quad (7.6)$$

where $Y(t) = I_{[Z<t\leq X]}$ is the at-risk process, $m(u)$ is the density of U, and the expectation is with respect to Z, X, and U. Dropping terms that do not involve η_λ, one obtains a relative Kullback-Leibler distance

$$\mathrm{RKL}(\eta, \eta_\lambda) = E\Big[\int_{\mathcal{T}} Y(t)\{e^{\eta_\lambda(t,U)} - \eta_\lambda(t,U)e^{\eta(t,U)}\}dt\Big],$$

which can be estimated by

$$\frac{1}{n}\sum_{i=1}^{n}\int_{Z_i}^{X_i} e^{\eta_\lambda(t,U_i)}dt - \frac{1}{n}\sum_{i=1}^{n}\int_{Z_i}^{X_i} \eta_\lambda(t,U_i)e^{\eta(t,U_i)}dt. \quad (7.7)$$

The first term of (7.7) is readily computable, but the second term $\mu_\eta(\eta_\lambda)$ involves the unknown $\eta(t,u)$.

Write $A(t) = \int_0^t Y(s)e^{\eta_0(s,U)}ds$. Conditioning on Z and U, $M(t) = N(t) - A(t)$ is a martingale; see, e.g., Fleming and Harrington (1991,

§1.3). For predictable function $h(t)$, the Stieltjes integral $\int_0^t h(s)dM(s)$ is also a martingale; see, e.g., Fleming and Harrington (1991, §2.4). A deterministic (meaning independent of $M(t)$) continuous function is predictable. For $h(t,u)$ continuous in t, $\forall u \in \mathcal{U}$, and independent of Z and X, $E[\int_{\mathcal{T}} h(t,U)dM(t)] = 0$, where $M(t)$ depends on Z, X, and U. "Estimating" 0 by the sample mean $n^{-1}\sum_{i=1}^n \int_{\mathcal{T}} h(t,U_i)dM_i(t)$, one has

$$0 \approx \frac{1}{n}\sum_{i=1}^n \left\{ \int_{\mathcal{T}} h(t,U_i)dN_i(t) - \int_{\mathcal{T}} h(t,U_i)I_{[Z_i<t\leq X_i]}e^{\eta(t,U_i)}dt \right\}$$
$$= \frac{1}{n}\sum_{i=1}^n \left\{ \delta_i h(X_i,U_i) - \int_{Z_i}^{X_i} h(t,U_i)e^{\eta(t,U_i)}dt \right\},$$

which, upon setting $h(t,U_i) = \eta_{\lambda,\tilde{\eta}}^{[i]}(t,U_i)$, yields

$$\tilde{\mu}_\eta(\eta_\lambda) = \frac{1}{n}\sum_{i=1}^n \int_{Z_i}^{X_i} \eta_{\lambda,\tilde{\eta}}^{[i]}(t,U_i)e^{\eta(t,U_i)}dt \approx \frac{1}{n}\sum_{i=1}^n \delta_i \eta_{\lambda,\tilde{\eta}}^{[i]}(X_i,U_i), \qquad (7.8)$$

where $\eta_{\lambda,\tilde{\eta}}^{[i]}$ minimizes the delete-one version of the quadratic approximation of (7.1) at $\tilde{\eta} = \eta_\lambda$. The derivation of the quadratic approximation is left as an exercise (Problem 7.2).

Write $\breve{\boldsymbol{\xi}} = (\boldsymbol{\xi}^T, \boldsymbol{\phi}^T)^T$. Define $\breve{Q} = (\breve{\boldsymbol{\xi}}(T_1,\tilde{U}_1), \ldots, \breve{\boldsymbol{\xi}}(T_N,\tilde{U}_N))$ and $H = V_{\tilde{\eta}}(\breve{\boldsymbol{\xi}}, \breve{\boldsymbol{\xi}}^T) + \text{diag}(\lambda Q, O)$. Similar to (6.17) on page 185,

$$\eta_{\lambda,\tilde{\eta}}^{[i]}(X_i,U_i) = \eta_\lambda(X_i,U_i) - \frac{1}{n-1}\breve{\boldsymbol{\xi}}(X_i,U_i)^T H^{-1}(\delta_i\breve{\boldsymbol{\xi}}(X_i,U_i) - \breve{Q}\mathbf{1}/n), \qquad (7.9)$$

where $\sum_{i=1}^n \delta_i\breve{\boldsymbol{\xi}}(X_i,U_i) = \breve{Q}\mathbf{1}$; see Problem 7.3. It follows that

$$\tilde{\mu}_\eta(\eta_\lambda) = \frac{1}{n}\sum_{i=1}^n \delta_i\eta_\lambda(X_i,U_i) - \frac{\text{tr}(\breve{Q}^T H^{-1}\breve{Q})}{n(n-1)} + \frac{\mathbf{1}^T\breve{Q}^T H^{-1}\breve{Q}\mathbf{1}}{n^2(n-1)}. \qquad (7.10)$$

Substituting (7.10) for the second term in (7.7), one gets a cross-validation estimate of the relative Kullback-Leibler distance,

$$-\frac{1}{n}\sum_{i=1}^n \left\{ \delta_i\eta_\lambda(X_i,U_i) - \int_{Z_i}^{X_i} e^{\eta_\lambda(t,U_i)}dt \right\}$$
$$+ \left\{ \frac{\text{tr}(\breve{Q}^T H^{-1}\breve{Q})}{n(n-1)} - \frac{\mathbf{1}^T\breve{Q}^T H^{-1}\breve{Q}\mathbf{1}}{n^2(n-1)} \right\}, \qquad (7.11)$$

where the first term is the minus log likelihood of η_λ. Modifications of (7.11) are trivial, following the lines of §6.3.2, and the computation of the cross-validated hazard estimates requires little change to the algorithms developed for density estimation.

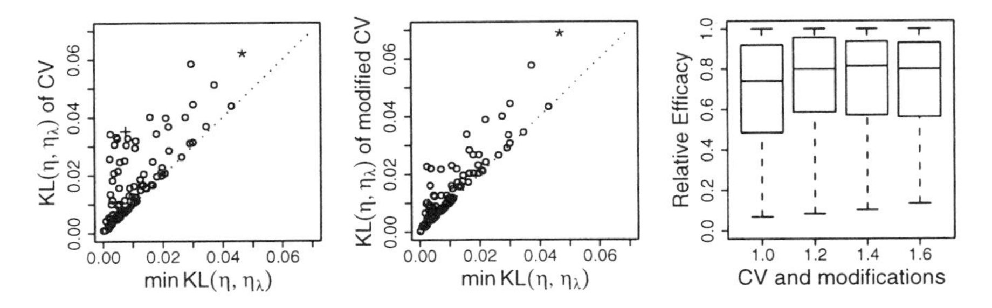

FIGURE 7.1. Performance of Cross-Validation and Modifications Thereof for Hazard Estimation. Left: Loss achieved by unmodified cross-validation. Center: Loss achieved by modified cross-validation with $\alpha = 1.4$. Right: Relative efficacy of cross-validation with $\alpha = 1, 1.2, 1.4, 1.6$.

7.2.2 Empirical Performance

We now illustrate the effectiveness of cross-validation through some simple simulation. Take a singleton $\mathcal{U}$ and a test hazard

$$\lambda(t) = e^{\eta(t)} = 24(t - 0.35)^2 + 2.$$

T_i were generated from the test hazard, C_i from a truncated exponential distribution with survival function $P(C > c) = I_{[c \leq 1]} e^{-4c/3}$, and Z_i from an exponential distribution with survival function $P(Z > z) = e^{-5z}$. Generated were 100 replicates with sample size $n = 150$, and the observed number of failures $N = \sum_{i=1}^{150} \delta_i$ ranged from 89 to 118; the overall empirical censoring rate was 4683/15000=31.22%.

Take $\mathcal{T} = [0, 1]$. Using the cubic spline of Example 7.1, estimates η_λ were calculated on the grid $\log_{10} \lambda = (-6)(0.1)(-2)$. Recorded for each of the estimates were the cross-validation scores and the Kullback-Leibler distance

$$\mathrm{KL}(\eta, \eta_\lambda) = \frac{1}{n} \sum_{i=1}^{n} \int_{Z_i}^{X_i} \{(\eta(t) - \eta_\lambda(t)) e^{\eta(t)} - (e^{\eta(t)} - e^{\eta_\lambda(t)})\} dt.$$

Note that this definition of $\mathrm{KL}(\eta, \eta_\lambda)$ depends on Z_i and X_i. The counterpart of Figure 6.1 of §6.3.3 is given in Figure 7.1. The fits for two replicates corresponding to the plus and the star in Figure 7.1 are plotted in Figure 7.2, where the solid lines are the fits from the modified ($\alpha = 1.4$) cross-validation, the dashed lines are the fits from the unmodified one, and the dotted lines are the test hazard. The plus is not visible in the middle frames of Figure 7.1, as it falls into the crowd. The data are also superimposed, where the histograms plot the empirical hazards of discretized data and the dashed curves hanging from above plot the risk set size $\sum_{i=1}^{n} I_{[Z_i < t \leq X_i]}$.

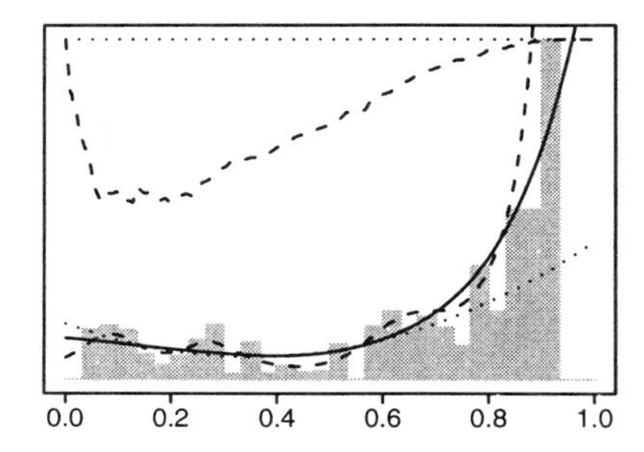

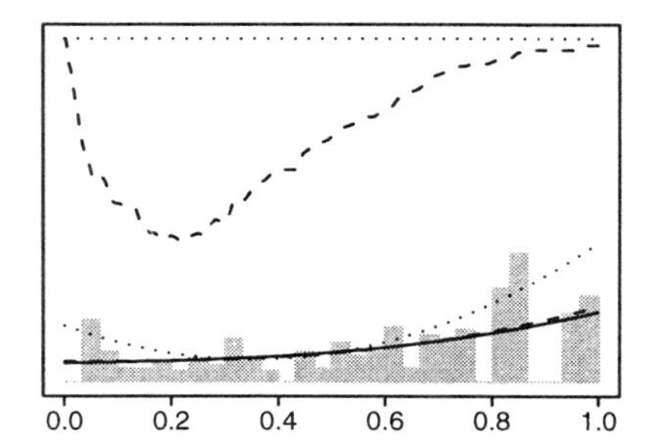

FIGURE 7.2. Selected Hazard Fits from Simulation. Left: The plus in Figure 7.1. Right: The star in Figure 7.1. The solid lines are the fits with modified cross-validation ($\alpha = 1.4$), the dashed lines those with unmodified cross-validation, and the dotted lines the test hazard. The histograms plot the empirical hazards of discretized data. The dashed curves hanging from above plot the risk set size.

7.3 Case Studies

We now apply the techniques developed so far to analyze a few real data sets.

7.3.1 Treatments of Gastric Cancer

The survival times of 90 gastric cancer patients are listed in Moreau et al. (1985). Half of the patients were treated by chemotherapy ($u = 1$), the other half by chemotherapy combined with radiotherapy ($u = 2$). There were 37 recorded deaths and 8 censorings in each of the treatment groups. The follow-up times ranged from 1 to 1519 days.

To estimate the hazard $\lambda(t, u) = e^{\eta(t,u)}$ via (7.1), we employed the formulation of Example 7.2 after mapping $\mathcal{T} = [0, 1550]$ onto $[0, 1]$. Modified cross-validation with $\alpha = 1.4$ was used to select the smoothing parameters. Plotted in the left frame of Figure 7.3 are the estimated $\lambda(t, u)$ indicated by the solid line ($u = 1$) and the dashed line ($u = 2$), with their geometric mean drawn as the dotted line. The follow-up times are superimposed as the rugs on the bottom ($u = 1$) and the top ($u = 2$). Separate estimates of $\lambda(t, 1)$ and $\lambda(t, 2)$ using Example 7.1 are also superimposed as faded lines. The separate estimate of $\lambda(t, 2)$ is virtually in the null space $\text{span}\{1, t\}$ and the separate estimate of $\lambda(t, 1)$ has a bump, and the smoothness of the joint estimate appears to be somewhere in between. Decomposing the log hazard as $\eta(t, u) = \eta_\emptyset + \eta_t + \eta_u + \eta_{t,u}$, where $\int_0^{1550} \eta_t(t)dt = \int_0^{1550} \eta_{t,u}(t, u)dt = 0$ and $\eta_u(1) + \eta_u(2) = \eta_{t,u}(t, 1) + \eta_{t,u}(t, 2) = 0$, the mean log hazard is given by $\eta_\emptyset + \eta_t$ and the contrast is given by $\eta_u + \eta_{t,u}$. Plotted in the right frame of Figure 7.3 are $\eta_t(t)$ and $\eta_{t,u}(t, 1)$, which demonstrate the time trend of

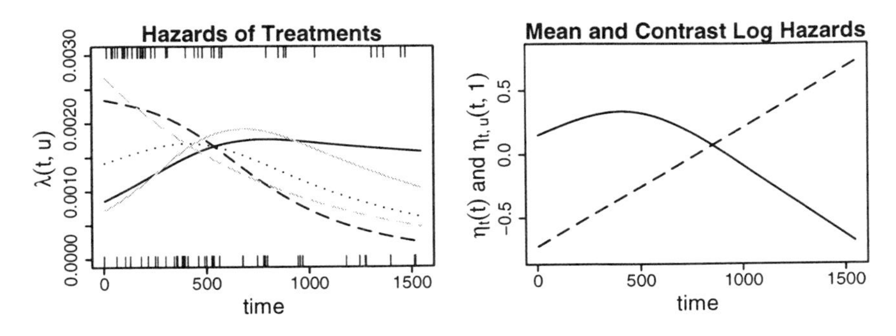

FIGURE 7.3. Treatments of Gastric Cancer. Left: The solid line plots $\lambda(t,1)$, the dashed line $\lambda(t,2)$, and the dotted line their geometric mean; the rugs on the bottom ($u = 1$) and the top ($u = 2$) mark the follow-up times, and the faded lines plot separate estimates of $\lambda(t,1)$ and $\lambda(t,2)$. Right: The solid line plots $\eta_t(t)$ and the dashed line plots $\eta_{t,u}(t,1)$.

the mean and contrast log hazards, respectively. The contrast log hazard is virtually linear in t.

The combined therapy appeared to take a heavier toll than chemotherapy alone in the early going, but for those who survived beyond about 500 days, the comparison was reversed. This, however, does not necessarily mean that the radiation would eventually benefit. A small group of stronger patients would probably survive a long time anyway, regardless of the therapy, but for the rest of the patients, radiation seemed to kill most of them before long, and after about 500 days, there simply were not many left.

7.3.2 Survival After Heart Transplant

We shall now fill in more details concerning the analysis of the Stanford heart transplant data previewed in §1.4.3. The survival or censoring times after transplant were between 0 and 3695 days, and the ages of patients at transplant were between 12 and 64. As mentioned in §1.4.3, a square root transform $t^* = \sqrt{t}$ was applied on the time axis to spread out the data. The domain $\mathcal{T}$ (after the transform) was taken as $[0, 62]$ and $\mathcal{U}$ was taken as $[11, 65]$. The product domain $\mathcal{T} \times \mathcal{U}$ was mapped onto $[0,1]^2$ and the tensor product cubic spline of Example 7.3 was used in the estimation. The smoothing parameters were selected using modified cross-validation with $\alpha = 1.4$.

Two fits were calculated, one with the interaction and one without. Plotted in Figure 7.4 are the contours of the estimated $\tilde{\eta}(t^*, u) = \log \tilde{\lambda}(t^*, u)$, from which the hazard rates on the original time scale can be obtained through $\lambda(t,u) = \tilde{\lambda}(\sqrt{t}, u)/(2\sqrt{t})$. The two fits are visually very close, especially in data-dense areas, so a proportional hazard model appears ade-

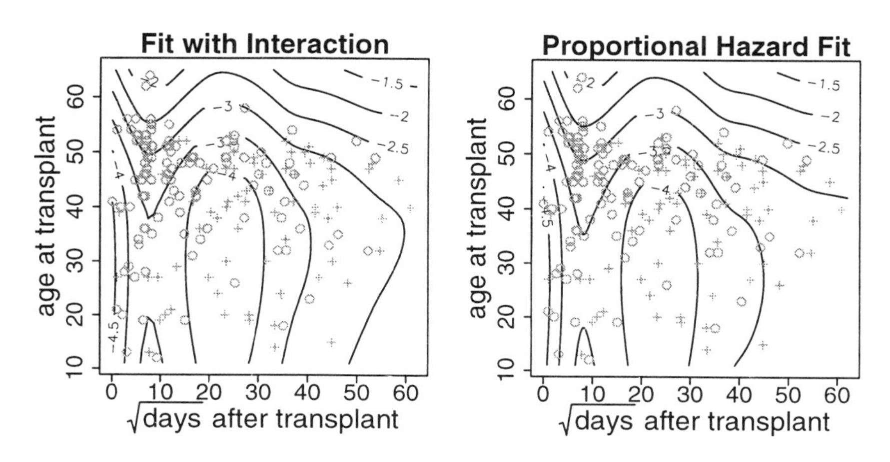

FIGURE 7.4. Hazard After Heart Transplant. The contours are the estimated $\log \tilde{\lambda}(\sqrt{t}, u)$, with deceased (circles) and censored (pluses) patients superimposed.

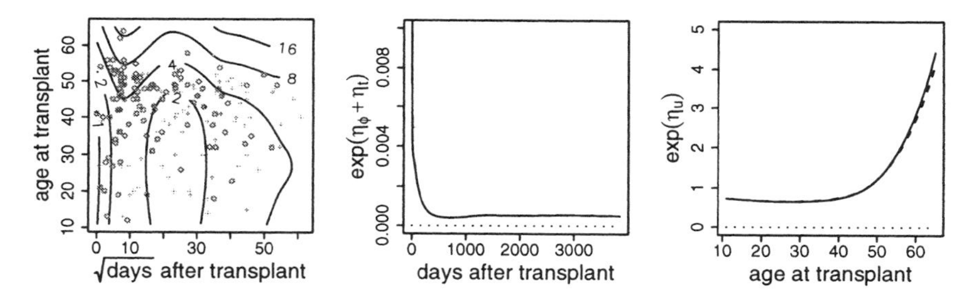

FIGURE 7.5. Hazard After Heart Transplant: Proportional Hazard Fit. Left: Contours of $100\tilde{\lambda}(\sqrt{t}, u)$, with deceased (circles) and censored (pluses) patients superimposed. Center: Base hazard $e^{\eta_\emptyset + \eta_t}$ on the original time scale. Right: Age effect e^{η_u}, with an estimate through penalized partial likelihood superimposed as the dashed line.

quate. Diagnostic tools for the practical significance of the interaction are still under development.

For the proportional hazard fit, the contours of $\tilde{\lambda}(t^*, u)$, the base hazard $e^{\eta_\emptyset + \eta_t}$ on the original time scale, and the relative risk e^{η_u} as shown in Figure 1.4 are reproduced in Figure 7.5. The dashed line superimposed in the right frame is a penalized partial likelihood estimate of the relative risk, to be discussed in the next section.

It is seen that once a patient survived the initial shock, the hazard rate would remain stable over extended time period. The relative risk was flat for younger patients up to about 40 years of age, then quickly took off for older patients.

7.4 Penalized Partial Likelihood

Assume a proportional hazard model $\lambda(t,u) = \lambda_0(t)\lambda_1(u)$. Treating the base hazard $\lambda_0(t)$ as a nuisance parameter, one may estimate the relative risk $\lambda_1(u)$ using penalized partial likelihood.

The estimation of relative risk through penalized partial likelihood is isomorphic to density estimation under biased sampling, as treated in §6.6, so no new tool is needed here. The technique is illustrated on the Stanford heart transplant data of §1.4.3 and §7.3.2.

7.4.1 Partial Likelihood and Biased Sampling

Let $Y_i(t) = I_{[Z_i<t\leq X_i]}$ be the at-risk process of the ith observation. For the estimation of the relative risk $\lambda_1(u) = e^{\eta(u)}$, Cox (1972) proposed working with the partial likelihood,

$$\prod_{i=1}^{n}\left(\frac{e^{\eta(U_i)}}{\sum_{k=1}^{n} Y_k(X_i)e^{\eta(U_k)}}\right)^{\delta_i} = \prod_{j=1}^{N}\left(\frac{e^{\eta(\tilde{U}_j)}}{\sum_{k=1}^{n} Y_k(T_j)e^{\eta(U_k)}}\right), \tag{7.12}$$

where $(T_j, \tilde{U}_j)$ are the observed lifetimes and the corresponding covariates. Note that the relative risk is defined only up to a multiplicative constant, so a side condition $A\eta = 0$ on the log relative risk would be needed to pin down the function to be estimated; see the related discussion on logistic density transform in §6.1.

Writing $\int f = \sum_{k=1}^{n} f(U_k)$, $e^{\eta}/\int e^{\eta}$ defines a probability density on the discrete domain $\{U_k, k = 1,\ldots,n\}$. One may write

$$\frac{e^{\eta(\tilde{U}_j)}}{\sum_{k=1}^{n} Y_k(T_j)e^{\eta(U_k)}} = \frac{w_j(\tilde{U}_j)e^{\eta(\tilde{U}_j)}}{\int w_j(u)e^{\eta(u)}},$$

where $w_j(u)$ is defined by $w_j(U_k) = Y_k(T_j)$. Hence, the partial likelihood of (7.12) can be cast as a likelihood for density estimation under biased sampling. The estimation of relative risk through penalized partial likelihood can thus be conducted using the tools developed in §6.6. Further details are left as exercises (Problems 7.4 and 7.5).

7.4.2 Case Study: Survival After Heart Transplant

Consider a proportional hazard model for the Stanford heart transplant data of §1.4.3 and §7.3.2. Estimating the age effect through penalized partial likelihood using the cubic spline of Example 6.1, one gets the dashed line in the right frame of Figure 7.5, which nearly coincides with the estimate through penalized full likelihood. The side condition was $\int_{11}^{65} \eta(u)du = 0$, the same as the one on η_u in §7.3.2. The smoothing parameter was selected using modified cross-validation with $\alpha = 1.4$.

7.5 Models Parametric in Time

When parametric models are assumed on the time axis, one usually needs to estimate a parameter of the survival distribution as a function of the covariate. The problem is similar to non-Gaussian regression as treated in Chapter 5, although the response likelihood may not belong to an exponential family.

We discuss the accelerated life models through location-scale families for the log lifetime, and spell out details concerning the Weibull family, the log normal family, and the log logistic family. Software tools are provided through the `gssanova` suite in the R package `gss`.

7.5.1 Location-Scale Families and Accelerated Life Models

Let $F(z)$ be a cumulative distribution function on $(-\infty, \infty)$ and $f(z)$ be its density. A location-scale family is defined by $P(X \leq x|\mu, \sigma) = F((x-\mu)/\sigma)$, where μ is the location parameter and $\sigma > 0$ is the scale parameter.

Assume a location-scale family for $\log T$. The survival function and the hazard function are easily seen to be

$$S(t) = 1 - F(z), \qquad \lambda(t) = \frac{1}{\sigma t}\frac{f(z)}{1 - F(z)}, \tag{7.13}$$

where $z = (\log t - \mu)/\sigma$.

Let σ be a constant and μ be a function of a covariate u with $\mu(u_0) = 0$ at a "control" point u_0. It follows that

$$S(t|u) = 1 - F((\log t - \mu(u))/\sigma) = 1 - F(\log(te^{-\mu(u)})/\sigma) = S(te^{-\mu(u)}|u_0),$$

so the covariate is effectively rescaling the time axis. Such models are known as accelerated life models.

Example 7.4 (Extreme value and Weibull distributions)
Setting $F(z) = 1 - e^{-w}$ with $f(z) = we^{-w}$, where $w = e^z$, one has the extreme value distribution. When $\log T$ follows the extreme value distribution, T follows the Weibull distribution with the survival function and the hazard function

$$\begin{aligned} S(t) &= \exp\{-e^{(\log t - \mu)/\sigma}\} = \exp\{-(t/e^{\mu})^{1/\sigma}\} = \exp\{-(t/\beta)^{\alpha}\}, \\ \lambda(t) &= \frac{1}{\sigma t}e^{(\log t - \mu)/\sigma} = \frac{1}{\sigma t}\left(\frac{t}{e^{\mu}}\right)^{1/\sigma} = \frac{\alpha}{t}\left(\frac{t}{\beta}\right)^{\alpha}, \end{aligned} \tag{7.14}$$

where $\alpha = 1/\sigma$ is called the shape parameter and $\beta = e^{\mu}$ is called the scale parameter. When $\alpha = 1$, the Weibull distribution reduces to the exponential distribution. □

Example 7.5 (Normal and log normal distributions)
Setting $F(z) = \Phi(z)$, the cumulative distribution function of the standard normal with $f(z) = \phi(z) = e^{-z^2/2}/\sqrt{2\pi}$, one has the normal distribution. When $\log T$ follows a normal distribution, T is log normal with the survival function and the hazard function

$$S(t) = 1 - \Phi(z), \qquad \lambda(t) = \frac{1}{\sigma t} \frac{\phi(z)}{1 - \Phi(z)}, \tag{7.15}$$

where $z = (\log t - \mu)/\sigma$. □

Example 7.6 (Logistic and log logistic distributions)
Setting $F(z) = w/(1 + w)$ with $f(z) = w/(1 + w)^2$, where $w = e^z$, one has the logistic distribution. When $\log T$ follows a logistic distribution, T follows a log logistic distribution with the survival function and the hazard function

$$S(t) = \frac{1}{1 + e^z}, \qquad \lambda(t) = \frac{1}{\sigma t} \frac{e^z}{1 + e^z}, \tag{7.16}$$

where $z = (\log t - \mu)/\sigma$. □

Now let us take a closer look at the accelerated life model. Observing $(Z_i, X_i, \delta_i, U_i)$, $i = 1, \ldots, n$, the minus log likelihood is seen to be

$$-\sum_{i=1}^{n} \left\{ \delta_i \log \lambda(X_i; \mu_i, \sigma) - \int_{Z_i}^{X_i} \lambda(t; \mu_i, \sigma) dt \right\}, \tag{7.17}$$

where $\lambda(t; \mu, \sigma)$ spells out the dependence of $\lambda(t)$ on the parameters μ and σ, and $\mu_i = \mu(U_i)$; see Problem 1.2. Fix σ and write

$$h_1(t; \mu) = -\frac{\partial \log \lambda(t; \mu, \sigma)}{\partial \mu}, \qquad h_2(t; \mu) = \frac{\partial h_1(t; \mu)}{\partial \mu}.$$

The first and second derivatives of (7.17) with respect to μ_i are given by

$$\begin{aligned} \tilde{u}_i &= \delta_i h_1(X_i; \mu_i) - \int_{Z_i}^{X_i} h_1(t; \mu_i) \lambda(t; \mu_i, \sigma) dt = \int h_1(t; \mu_i) dM_i(t), \\ \tilde{w}_i &= \delta_i h_2(X_i; \mu_i) - \int_{Z_i}^{X_i} h_2(t; \mu_i) \lambda(t; \mu_i, \sigma) dt + \int_{Z_i}^{X_i} h_1^2(t; \mu_i) \lambda(t; \mu_i, \sigma) dt \\ &= \int h_2(t; \mu_i) dM_i(t) + \int h_1^2(t; \mu_i) dA_i(t), \end{aligned}$$

where $M_i(t) = N_i(t) - A_i(t)$ is a martingale, $N_i(t) = I_{[t \leq X_i, \delta_i = 1]}$, and $A_i(t) = \int_0^t I_{[Z_i < s \leq X_i]} \lambda(s; \mu_i, \sigma) ds$, as discussed in §7.2.1. By martingale properties, one has $E[\tilde{u}_i] = 0$ and $E[\tilde{u}_i^2] = E[\tilde{w}_i]$; see, e.g., Fleming and Harrington (1991, §2.7). See also §8.3.1. These derivatives may be used in (5.3) on page 151 to calculate a penalized likelihood estimate of $\mu(u)$ through iterated penalized weighted least squares. Since $\int h_2(t; \mu_i) dM_i(t)$ can be negative, one may set it to zero and use only the second term $\int h_1^2(t; \mu_i) dA_i(t)$ as $\tilde{w}_i$, which is always positive.

7.5.2 Weibull Family

For the Weibull family of Example 7.4, the minus log likelihood of μ and $\alpha = 1/\sigma$ is seen to be

$$-\sum_{i=1}^{n}\{\delta_i(\alpha(\log X_i - \mu_i) + \log\alpha) - (X_i^\alpha - Z_i^\alpha)e^{-\alpha\mu_i}\}; \tag{7.18}$$

see Problem 7.6. Fixing α, the first and second derivatives of (7.18) with respect to η_i are given by

$$\tilde{u}_i = \alpha(\delta_i - (X_i^\alpha - Z_i^\alpha)e^{-\alpha\eta_i}),$$
$$\tilde{w}_i = \alpha^2(X_i^\alpha - Z_i^\alpha)e^{-\alpha\eta_i}.$$

Note that $\log\lambda(t,u) = \alpha(\log t - \mu(u)) + \log(\alpha/t)$, so the Weibull model is also a proportional hazard model, with the relative risk proportional to $e^{-\alpha\mu(u)}$. It is easily seen that $h_1(t;\mu_i) = \alpha$ is a constant and $h_2 = 0$.

In the implementation of Weibull regression with censored data through (5.3), the α that minimizes (7.18) is calculated after each iteration, which is then used in $\tilde{u}_i$ and $\tilde{w}_i$ for the updating of $\tilde{Y}_i = \tilde{\mu}_i - \tilde{u}_i/\tilde{w}_i$.

Example 7.7 (Weibull regression)
The following R commands generate some synthetic data and fit a cubic spline Weibull regression model. The fit is plotted in Figure 7.6 with the data and the test function superimposed.

```
# generate the data
set.seed(5732)
test <- function(x)
        {(1e6*(x^11*(1-x)^6)+1e4*(x^3*(1-x)^10))+.1}
x <- (0:150)/150
y <- NULL
for (i in 1:151) {
    mu <- test(x[i])
    repeat {
        t.wk <- rweibull(1,2,mu)
        c.wk <- rweibull(1,1,2*mu)
        x.wk <- min(t.wk,c.wk)
        delta.wk <- t.wk<=c.wk
        z.wk <- rweibull(1,1,mu/2)
        if (x.wk>z.wk) break
    }
    y <- rbind(y,c(x.wk,delta.wk,z.wk))
}
# calculate the fit
weibull.fit <- gssanova(y~x,family="weibull")
```

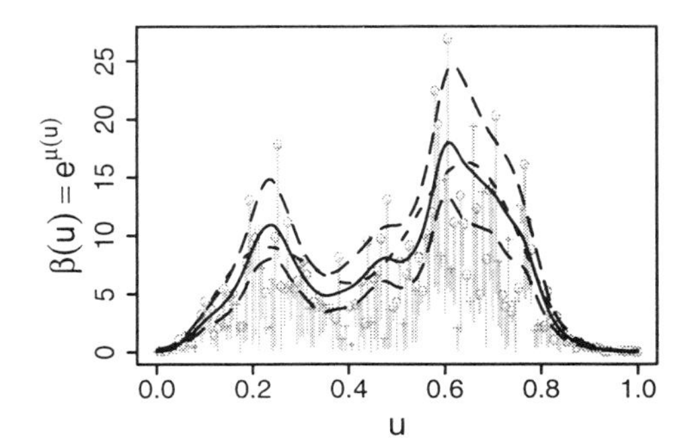

FIGURE 7.6. Cubic Spline Weibull Regression with Censored Data. The estimated $e^{\mu(u)}$ is indicated by the solid line, the 95% Bayesian confidence intervals are indicated by long dashed lines, and the test function is indicated by the short dashed line. The data are superimposed as circles (failures) or pluses (censorings). The vertical lines extended from the data mark the at-risk processes $I_{[Z_i<t\leq X_i]}$.

```
# predict
est <- predict(weibull.fit,data.frame(x=x),se=TRUE)
# plot the fit
plot(x,y[,1],type="n")
for (i in 1:151) {
    lines(c(x[i],x[i]),c(y[i,1],y[i,3]),col=5)
}
points(x,y[,1],pch=c("+","o")[y[,2]+1])
lines(x,exp(est$fit))
lines(x,exp(est$fit+1.96*est$se),lty=5)
lines(x,exp(est$fit-1.96*est$se),lty=5)
lines(x,test(x),lty=2)
```

Note that the response is entered as a matrix with three columns in the order X_i, δ_i, and Z_i, where the third column can be omitted if $Z_i = 0, \forall i$. By default, the performance-oriented iteration is driven by $U_w(\lambda)$ with the dispersion set to 1 for the Weibull family. □

7.5.3 Log Normal Family

For the log normal family of Example 7.5, the minus log likelihood of μ and $\alpha = 1/\sigma$ is seen to be

$$-\sum_{i=1}^{n}\left\{\delta_i(-z_i^2/2 - \log(1-\Phi(z_i)) + \log\alpha) - \log\frac{1-\Phi(\tilde{z}_i)}{1-\Phi(z_i)}\right\}, \qquad (7.19)$$

where $z_i = \alpha(\log X_i - \mu_i)$ and $\tilde{z}_i = \alpha(\log Z_i - \mu_i)$; see Problem 7.7. The first derivative of (7.19) with respect to η_i is given by

$$\tilde{u}_i = \alpha\, \delta_i \left(\frac{\phi(z_i)}{1 - \Phi(z_i)} - z_i \right) - \alpha \left(\frac{\phi(z_i)}{1 - \Phi(z_i)} - \frac{\phi(\tilde{z}_i)}{1 - \Phi(\tilde{z}_i)} \right).$$

Since $h_1(t; \mu_i) = \alpha(\phi(z)/(1 - \Phi(z)) - z)$, where $z = \alpha(\log t - \mu_i)$,

$$\begin{aligned}
&\int_{Z_i}^{X_i} h_1^2(t; \mu_i)\lambda(t; \mu_i, \alpha)dt \\
&\quad = \int_{Z_i}^{X_i} \alpha^2 \left(\frac{\phi(z)}{1 - \Phi(z)} - z \right)^2 \frac{\phi(z)}{1 - \Phi(z)} \frac{\alpha dt}{t} \\
&\quad = \alpha^2 \int_{\tilde{z}_i}^{z_i} \left(\frac{\phi(z)}{1 - \Phi(z)} - z \right) d \left(\frac{\phi(z)}{1 - \Phi(z)} \right) \\
&\quad = \alpha^2 \left\{ \left(\frac{\phi(z)}{1 - \Phi(z)} - z \right) \left(\frac{\phi(z)}{1 - \Phi(z)} \right) \Bigg|_{\tilde{z}_i}^{z_i} \right. \\
&\qquad \left. - \int_{\tilde{z}_i}^{z_i} \frac{\phi(z)}{1 - \Phi(z)} d \left(\frac{\phi(z)}{1 - \Phi(z)} - z \right) \right\} \\
&\quad = \alpha^2 \left\{ \left(\frac{1}{2} \left(\frac{\phi(z_i)}{1 - \Phi(z_i)} \right)^2 - \frac{z_i \phi(z_i)}{1 - \Phi(z_i)} - \log(1 - \Phi(z_i)) \right) \right. \\
&\qquad \left. - \left(\frac{1}{2} \left(\frac{\phi(\tilde{z}_i)}{1 - \Phi(\tilde{z}_i)} \right)^2 - \frac{\tilde{z}_i \phi(\tilde{z}_i)}{1 - \Phi(\tilde{z}_i)} - \log(1 - \Phi(\tilde{z}_i)) \right) \right\},
\end{aligned}$$

which is to be used as $\tilde{w}_i$ in (5.3).

Log normal regression with censored data has been implemented through `gssanova` as the `lognorm` family, in the same manner as Weibull regression. The parameter α is updated in each iteration via the minimization of (7.19), before the calculation of $\tilde{u}_i$ and $\tilde{w}_i$.

7.5.4 Log Logistic Family

For the log logistic family of Example 7.6, the minus log likelihood of μ and $\alpha = 1/\sigma$ is seen to be

$$-\sum_{i=1}^{n} \{\delta_i(z_i - \log(1 + e^{z_i}) + \log \alpha) - (\log(1 + e^{z_i}) - \log(1 + e^{\tilde{z}_i}))\}, \quad (7.20)$$

where $z_i = \alpha(\log X_i - \mu_i)$ and $\tilde{z}_i = \alpha(\log Z_i - \mu_i)$; see Problem 7.8. The first derivative of (7.20) with respect to μ_i is given by

$$\tilde{u}_i = \frac{\alpha \delta_i}{1 + e^{z_i}} - \alpha \left(\frac{1}{1 + e^{\tilde{z}_i}} - \frac{1}{1 + e^{z_i}} \right).$$

Since $h_1(t;\mu_i) = \alpha/(1+e^z)$, where $z = \alpha(\log t - \mu_i)$,

$$\int_{Z_i}^{X_i} h_1^2(t;\mu_i)\lambda(t;\mu_i,\alpha)dt = \int_{Z_i}^{X_i} \frac{\alpha^2}{(1+e^z)^2}\frac{e^z}{1+e^z}\frac{\alpha dt}{t}$$
$$= \alpha^2 \int_{\tilde{z}_i}^{z_i} \frac{de^z}{(1+e^z)^3}$$
$$= \frac{\alpha^2}{2}\left(\frac{1}{(1+e^{\tilde{z}_i})^2} - \frac{1}{(1+e^{z_i})^2}\right),$$

which is to be used as $\tilde{w}_i$ in (5.3).

Log logistic regression with censored data has been implemented through `gssanova` as the `loglogis` family, in the same manner as Weibull regression. The parameter α is updated in each iteration via the minimization of (7.20), before the calculation of $\tilde{u}_i$ and $\tilde{w}_i$.

7.5.5 Case Study: Survival After Heart Transplant

The Stanford heart transplant data are read into an R data frame `stan` with components `time`, `status`, and `age`. We now fit a Weibull model to the data.

```
stan.fit <- gssanova(cbind(time+.01,status)~age,
                     data=stan,family="weibull",
                     ext=1/(64-12))
```

The follow-up times in the records were rounded to whole days and there was a recorded death at 0, and we chose to add 0.01 to the follow-up times instead of deleting the 0. The same data set is found in the `stanford2` data frame in the R package `survival`, but the 0 has been replaced by 0.5 there. The ages ranged from 12 to 64, and the argument `ext=1/(64-12)` sets the domain to be [11, 65]; see discussion in §3.5 and the online documentation of `gssanova`. One may evaluate the `age` effect $\mu_u(u)$ of the estimated $\mu(u)$ on a regular grid, which satisfies $\int_{11}^{65} \mu_u(u)du = 0$, and obtain the standard errors.

```
uu <- seq(11,65,length=201)
est <- predict(stan.fit,data.frame(age=uu),
               se=TRUE,include="age")
```

The following commands plot $e^{\mu_u(u)}$ along with the 95% Bayesian confidence intervals, as shown in the left frame of Figure 7.7, with the data superimposed. Note that $\mu(x) = \mu_\emptyset + \mu_u(u)$, so the follow-up times were divided by $e^{\mu_\emptyset}$ to be brought on to the same scale as $e^{\mu_u(u)}$.

```
const <- predict(stan.fit,stan,inc="1")[1]
plot(stan$age,stan$time/exp(const),type="n")
```

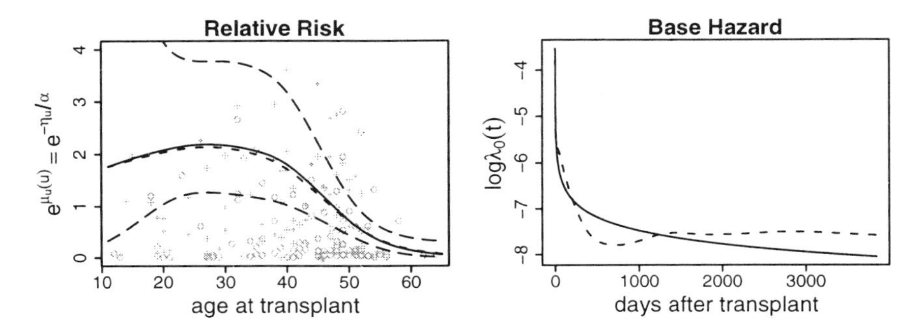

FIGURE 7.7. Hazard After Heart Transplant: Weibull Fit. Left: The solid line plots the estimated $e^{\mu_u(u)}$ and the long dashed lines the 95% Bayesian confidence intervals; the circles (failures) or pluses (censorings) mark the data, and the short dashed line corresponds to the solid line in the right frame of Figure 7.5. Right: The solid line plots the estimated log base hazard, and the dashed line corresponds to the line in the middle frame of Figure 7.5.

```
points(stan$age,stan$time/exp(const),
       pch=c("+","o")[stan$status+1])
lines(uu,exp(est$fit))
lines(uu,exp(est$fit+1.96*est$se),lty=5)
lines(uu,exp(est$fit-1.96*est$se),lty=5)
```

The Weibull model is a proportional hazard model with the relative risk proportional to $e^{-\alpha\mu_u(u)}$, where α was estimated to be 0.5733; check the value of `stan.fit$alpha`. For comparison, the relative risk e^{η_u} estimated through penalized full likelihood in §7.3.2, which was plotted as the solid line in the right frame of Figure 7.5, is converted to $e^{-\eta_u/\alpha}$ and superimposed as the short dashed line in the left frame of Figure 7.7; remember that η_u satisfies the same side condition $\int_{11}^{65}\eta_u(u)du = 0$. Plotted in the right frame of Figure 7.7 is the estimated log base hazard $\log\lambda_0(t) = \alpha(\log t - \mu_\emptyset) + \log(\alpha/t)$ (solid line), with the logarithm of the base hazard plotted in the middle frame of Figure 7.5 superimposed (dashed line).

7.6 Bibliographic Notes

Section 7.1

Absent of the covariate, penalized likelihood hazard estimation was studied by Anderson and Senthilselvan (1980), Bartoszyński, Brown, McBride, and Thompson (1981), O'Sullivan (1988a), Antoniadis (1989), and Gu (1994). With the covariate, the estimation of the "bivariate" hazard function through penalized full likelihood was formulated and studied by Gu (1996, 1998b).

Section 7.2

A performance-oriented iteration similar to that in §5.2.1 and in Gu (1993b) was proposed and illustrated by Gu (1994) for $\mathcal{U}$ a singleton, where a martingale moment estimate similar to (7.8) was used to derive an indirect cross-validation score. The direct cross-validation score presented here is adapted from §6.3 and has not appeared previously.

A comprehensive treatment of the counting process approach to survival analysis and the related martingale structure has been given by Fleming and Harrington (1991). A technically less demanding exposition can be found in Gill (1984).

Section 7.3

The gastric cancer data were used as an example by Moreau, O'Quigley, and Mesbah (1985) to illustrate their goodness-of-fit test for the proportional hazard model; the p-value calculated on the data was between 0.01 and 0.02, indicating the inadequacy of the proportional hazard model.

The analysis of the Stanford heart transplant data presented here differ slightly from the one in Gu (1998b), where a performance-oriented iteration was used to select the smoothing parameters.

Section 7.4

Partial likelihood was proposed by Cox (1972) based on a conditioning argument, and the maximum partial likelihood has become the golden standard for the parametric estimation of relative risk. Penalized partial likelihood was studied by O'Sullivan (1988b); see also Hastie and Tibshirani (1986) and Gray (1992). The isomorphism between the partial likelihood and the likelihood under biased sampling has its root in Cox's conditioning argument. The induced estimation procedure discussed here appears to be new.

Zucker and Karr (1990) considered a generalization of the proportional hazard model of the form $\lambda(t,u) = \lambda_0(t)\lambda_1(\beta(t),u)$, where $\lambda_1(\beta(t),u)$ was parametric in u with a time-varying parameter $\beta(t)$, and $\beta(t)$ was estimated via the penalized partial likelihood.

Section 7.5

Accelerated life models are among classical tools in reliability and survival analysis; see, e.g., Kalbfleisch and Prentice (1980, §2.3). Basic properties of the Weibull, the log normal, and the log logistic distributions can be found in Kalbfleisch and Prentice (1980, §2.2) along with properties of other lifetime distributions. Parametric linear models for $\mu(u)$ have been implemented by Terry Therneau in his `survival` package, ported to R from the Splus original by Thomas Lumley.

The tools developed in this section for the penalized likelihood estimation of $\mu(u)$ appear to be new.

7.7 Problems

Section 7.1

7.1 Verify (7.5).

Section 7.2

7.2 Using the calculus leading to (6.14) on page 184, one can obtain the quadratic approximation of (7.1).

(a) Define $L_{f,g}(\alpha) = (1/n)\sum_{i=1}^{n}\int_{Z_i}^{X_i} e^{(f+\alpha g)(t,U_i)}dt$, where f and g are functions and α is real. Calculate $\dot{L}_{f,g}(0)$ and $\ddot{L}_{f,g}(0)$.

(b) Obtain the quadratic approximation of (7.1) at $\tilde{\eta}$.

7.3 Verify (7.9).

Section 7.4

7.4 Formally formulate the penalized partial likelihood functional for the estimation of relative risk. Discuss basic properties such as the existence and uniqueness of the estimate.

7.5 Applying the tools developed in §6.6 to the estimation of relative risk via the penalized partial likelihood, characterize the Kullback-Leibler loss that is targeted by the cross-validation scheme.

Section 7.5

7.6 Verify the minus log likelihood (7.18) for the Weibull family.

7.7 Verify the minus log likelihood (7.19) for the log normal family.

7.8 Verify the minus log likelihood (7.20) for the log logistic family.

8
Asymptotic Convergence

In this chapter, we develop an asymptotic theory concerning the rates of convergence of penalized likelihood estimates to the target functions as the sample size goes to infinity. The rates are calculated in terms of problem-specific loss functions derived from the respective stochastic settings.

The primary tool used in the development is the eigenvalue analysis in a Hilbert space, of which a brief introduction is given in §8.1. Convergence rates are established in §8.2 for the density estimates of Chapter 6, in §8.3 for the hazard estimates of §7.1—7.3, and in §8.4 for the regression estimates of Chapters 3 and 5 and §7.5. For density estimation and hazard estimation, an important part of the theory is the justification of the semi-parametric estimates in the space $\mathcal{H}^*$, defined for density estimation in (6.2) and for hazard estimation in (7.2), respectively. For regression, the theory is developed in a setting more general than that of §5.1.

When an estimate is sought in a space $\mathcal{H}$ for the target function $\eta_0 \notin \mathcal{H}$, the estimate converges to a Kullback-Leibler projection η_0^* of η_0 in $\mathcal{H}$, at the same rates as established for the convergence to $\eta_0 \in \mathcal{H}$.

8.1 Preliminaries

Let $V(f)$ be a quadratic functional that defines a statistically interpretable metric so that a small $V(\hat{\eta} - \eta)$ indicates a good estimate $\hat{\eta}$ of η. The asymptotic convergence rates of penalized likelihood estimates can be characterized through an eigenvalue analysis of $J(f)$ with respect to $V(f)$, to be discussed below. Following the convention of §2.1.1, abstract concepts are set in boldface at the point of definition and are followed by simple examples set in italic.

A quadratic functional B is said to be **completely continuous** with respect to another quadratic functional A, if for any $\epsilon > 0$, there exist a finite number of linear functionals $L_1, \ldots, L_k$ such that $L_j f = 0$, $j = 1, \ldots, k$, implies that $B(f) \leq \epsilon A(f)$; see Weinberger (1974, §3.3).

Consider the space $\mathcal{P}[0,1]$ of periodic functions permitting the Fourier series expansion (4.3) on page 113. Define $B(f) = 2\int_0^1 f^2 dx$, $A(f) = 2\int_0^1 (f^{(m)})^2 dx$, and

$$L_{2\mu} f = \int_0^1 f(x) \sin 2\pi\mu x \, dx,$$
$$L_{2\mu+1} f = \int_0^1 f(x) \cos 2\pi\mu x \, dx, \qquad \mu = 0, 1, \ldots.$$

A function f satisfying $L_j f = 0$, $j = 1, \ldots, 2k-1$, has an expression

$$f(x) = \sum_{\mu=k}^{\infty} (a_\mu \cos 2\pi\mu x + b_\mu \sin 2\pi\mu x)$$

and, consequently,

$$B(f) = \sum_{\mu=k}^{\infty} (a_\mu^2 + b_\mu^2) \leq \frac{1}{(2\pi k)^{2m}} \sum_{\mu=k}^{\infty} (a_\mu^2 + b_\mu^2)(2\pi\mu)^{2m} = \frac{1}{(2\pi k)^{2m}} A(f).$$

Hence, B is completely continuous with respect to A.

When B is completely continuous with respect to A and, hence, to $A+B$, there exist **eigenvalues** λ_ν and the associated **eigenfunctions** ψ_ν such that

$$B(\psi_\nu, \psi_\mu) = \lambda_\nu \delta_{\nu,\mu}, \qquad (A+B)(\psi_\nu, \psi_\mu) = \delta_{\nu,\mu},$$

where $\delta_{\nu,\mu}$ is the Kronecker delta and $1 \geq \lambda_\nu \downarrow 0$; see Theorem 3.1 of Weinberger (1974, page 52). Write $\phi_\nu = \lambda_\nu^{-1/2} \psi_\nu$. It follows that

$$B(\phi_\nu, \phi_\mu) = \delta_{\nu,\mu}, \qquad A(\phi_\nu, \phi_\mu) = \rho_\nu \delta_{\nu,\mu},$$

where $0 \leq \rho_\nu = \lambda_\nu^{-1} - 1 \uparrow \infty$. We refer to ρ_ν as the eigenvalues of A with respect to B and to ϕ_ν as the associated eigenfunctions. Functions satisfying $A(f) < \infty$ can be expressed as a **Fourier series expansion** $f = \sum_\nu f_\nu \phi_\nu$, where $f_\nu = B(f, \phi_\nu)$ are the **Fourier coefficients**.

Take $\phi_{2\mu} = \sin 2\pi\mu x$, $\phi_{2\mu+1} = \cos 2\pi\mu x$, $\mu = 0, 1, \ldots$, in the periodic function example given above. It is easy to see that

$$B(\phi_\nu, \phi_\mu) = \delta_{\nu,\mu}, \quad A(\phi_\nu, \phi_\mu) = \frac{(2\pi\lfloor \nu/2 \rfloor)^{2m}}{2} \delta_{\nu,\mu}, \qquad \nu, \mu = 1, 2, \ldots,$$

where $\lfloor \nu/2 \rfloor$ *is the integer part of* $\nu/2$. *The eigenvalues* $\rho_\nu = (2\pi\lfloor \nu/2 \rfloor)^{2m}/2$ *grow at a rate* $O(\nu^{2m})$. *The Fourier coefficients are given by* $f_{2\mu} = b_\mu$, $f_{2\mu+1} = a_\mu$, $\mu = 0, 1, \ldots$.

To possibly achieve noise reduction in estimation, the effective dimension of the model space has to be kept finite, and to make the procedure non-restrictive, the dimension has to be expandable when more data become available. When V is completely continuous with respect to J, this can be achieved through constraints of the form $J(f) \leq \rho$ with $\rho \to \infty$ as $n \to \infty$ or, equivalently, by Theorem 2.12, through penalized likelihood with $\lambda \to 0$ as $n \to \infty$. The growth rate of the eigenvalues ρ_ν of J with respect to V, which typically is of the form $O(\nu^r)$, dictates how fast λ should approach 0, as will be seen in the sections to follow.

A few examples are given in the rest of the section.

Example 8.1 (Polynomial splines)
Consider $J(f) = \int_0^1 (f^{(m)})^2 dx$ and $V(f) = \int_0^1 f^2 w(x) dx$ on $\mathcal{X} = [0, 1]$, where $w(x)$ satisfies $0 < c_1 < w(x) < c_2 < \infty$ for some c_1, c_2. V is known to be completely continuous with respect to J, and it can be shown that for ν sufficiently large, $\beta_1 \nu^{2m} < \rho_\nu < \beta_2 \nu^{2m}$ for some $0 < \beta_1 < \beta_2 < \infty$. See, e.g., Utreras (1981).

For $J(f) = \int_0^1 (Lf)^2 dx$ with L given in (4.39) on page 126, the same results hold as $\int_0^1 (Lf)^2 dx$ is equivalent to $\int_0^1 (f^{(m)})^2 dx$. □

Let $\{\varphi_\nu\}$ be a sequence of functions on $[0, 1]$ satisfying $\int_0^1 \varphi_\nu \varphi_\mu dx = \delta_{\nu,\mu}$ and $\int_0^1 \ddot{\varphi}_\nu \ddot{\varphi}_\mu dx = \sigma_\nu \delta_{\nu,\mu}$, where $O(\nu^4) = \sigma_\nu \uparrow \infty$. The first two entries are $\varphi_1 = 1$ and $\varphi_2 = \sqrt{12}(\cdot - 0.5)$, with $\sigma_1 = \sigma_2 = 0$.

Example 8.2 (Tensor product cubic spline)
Consider $\mathcal{X} = [0, 1]^2$. Write $\tilde{V}(f) = \int_0^1 \int_0^1 f^2 dx_{\langle 1 \rangle} dx_{\langle 2 \rangle}$ and

$$\tilde{J}(f) = J_{1,00}(f) + J_{00,1}(f) + J_{1,01}(f) + J_{01,1}(f) + J_{1,1}(f),$$

where

$$J_{1,00}(f) = \int_0^1 \left\{ \int_0^1 \ddot{f}_{11} dx_{\langle 2 \rangle} \right\}^2 dx_{\langle 1 \rangle},$$

$$J_{00,1}(f) = \int_0^1 \left\{ \int_0^1 \ddot{f}_{22} dx_{\langle 1 \rangle} \right\}^2 dx_{\langle 2 \rangle},$$

$$J_{1,01}(f) = \int_0^1 \left\{ \int_0^1 f^{(3)}_{112} dx_{\langle 2 \rangle} \right\}^2 dx_{\langle 1 \rangle},$$

$$J_{01,1}(f) = \int_0^1 \left\{ \int_0^1 f^{(3)}_{122} dx_{\langle 1 \rangle} \right\}^2 dx_{\langle 2 \rangle},$$

$$J_{1,1}(f) = \int_0^1 \int_0^1 (f^{(4)}_{1122})^2 dx_{\langle 1 \rangle} dx_{\langle 2 \rangle}.$$

The sequence $\{\varphi_\nu(x_{\langle 1\rangle})\varphi_\mu(x_{\langle 2\rangle})\}$ are orthonormal with respect to $\tilde{V}(f,g)$ and are orthogonal with respect to $\tilde{J}(f,g)$. More precisely, $J_\beta(f)$ define square norms in $\mathcal{H}_\beta$, where

$$\begin{aligned}
\mathcal{H}_{1,00} &= \{\varphi_\nu(x_{\langle 1\rangle})\varphi_1(x_{\langle 2\rangle})\}_{\nu\geq 3},\\
\mathcal{H}_{00,1} &= \{\varphi_1(x_{\langle 1\rangle})\varphi_\nu(x_{\langle 2\rangle})\}_{\nu\geq 3},\\
\mathcal{H}_{1,01} &= \{\varphi_\nu(x_{\langle 1\rangle})\varphi_2(x_{\langle 2\rangle})\}_{\nu\geq 3},\\
\mathcal{H}_{01,1} &= \{\varphi_2(x_{\langle 1\rangle})\varphi_\nu(x_{\langle 2\rangle})\}_{\nu\geq 3},\\
\mathcal{H}_{1,1} &= \{\varphi_\nu(x_{\langle 1\rangle})\varphi_\mu(x_{\langle 2\rangle})\}_{\nu,\mu\geq 3}.
\end{aligned}$$

The null space of $\tilde{J}(f)$ is given by $\mathcal{N}_{\tilde{J}} = \{\varphi_\nu(x_{\langle 1\rangle})\varphi_\mu(x_{\langle 2\rangle})\}_{\nu,\mu=1,2}$. Putting $\{\sigma_\nu\sigma_\mu\}_{\nu,\mu\geq 3}$ in an increasing order as $\{\tilde{\sigma}_\nu\}$, it can be shown that $\tilde{\sigma}_\nu$ grow at a rate faster than $(\nu/\log\nu)^4$ but slower than ν^4; see, e.g., Wahba (1990, §12.1).

When $w(x)$ is bounded away from 0 and ∞, $V(f) = \int_0^1\int_0^1 wf^2dx_{\langle 1\rangle}dx_{\langle 2\rangle}$ is equivalent to $\tilde{V}(f)$. For $\theta_\beta > 0$, $\beta = \{1,00\}, \{00,1\}, \{1,01\}, \{01,1\}$, and $\{1,1\}$, $J(f) = \sum_\beta \theta_\beta J_\beta(f)$ is equivalent to $\tilde{J}(f)$. V is thus completely continuous with respect to J, and the eigenvalues ρ_ν of J with respect to V satisfy $\beta_1\nu^{4-\epsilon} < \rho_\nu < \beta_2\nu^4$ for some $0 < \beta_1 < \beta_2 < \infty$ and ν sufficiently large, $\forall\epsilon > 0$. If $\mathcal{H}_{1,1}$ is eliminated with $\theta_{1,1} = 0$, ϵ can be set to 0. □

Example 8.3 (Thin-plate splines)
For the thin-plate splines of §4.4, $J_m^d(f)$ in (4.48) on page 136 is defined on the unbounded domain $(-\infty,\infty)^d$, on which the usual L_2 norm is not defined.

Consider a bounded domain Ω satisfying certain boundary conditions. Let $J(f)$ be the integral of (4.48) restricted to Ω and $V(f) = \int_\Omega f^2dx$. It can be shown that V is completely continuous with respect to J and that the eigenvalues ρ_ν of J with respect to V satisfy $\beta_1\nu^{2m/d} < \rho_\nu < \beta_2\nu^{2m/d}$, for some $0 < \beta_1 < \beta_2 < \infty$ and ν sufficiently large; see Cox (1984) and Utreras (1988). This does not address the thin-plate splines directly, but appears to be as close as one can get. □

8.2 Rates for Density Estimates

Denote by $e^{\eta_0}/\int_\mathcal{X} e^{\eta_0}$ the density to be estimated and by $e^{\hat{\eta}}/\int_\mathcal{X} e^{\hat{\eta}}$ the estimate through the minimization of (6.1). We shall establish the asymptotic convergence rates in terms of the symmetrized Kullback-Leibler distance

$$\mathrm{SKL}(\eta_0,\hat{\eta}) = \mu_{\eta_0}(\eta_0-\hat{\eta}) + \mu_{\hat{\eta}}(\hat{\eta}-\eta_0),$$

where $\mu_\eta(f) = \int_\mathcal{X} f e^\eta / \int_\mathcal{X} e^\eta$, and in terms of $V(\hat{\eta} - \eta_0) = V_{\eta_0}(\hat{\eta} - \eta_0)$, where $V_\eta(f) = \mu_\eta(f^2) - \mu_\eta^2(f)$.

The rates are first established for the minimizer $\tilde{\eta}$ of the quadratic approximation of (6.1) at η_0, then extended to $\hat{\eta}$ by bounding the magnitude of $\hat{\eta} - \tilde{\eta}$. The rates are further extended to the minimizer $\hat{\eta}^*$ of (6.1) in a finite-dimensional space $\mathcal{H}^*$, given in (6.2), by bounding the magnitudes of $\hat{\eta} - \eta^*$ and $\eta^* - \hat{\eta}^*$, where η^* is the projection of $\hat{\eta}$ in $\mathcal{H}^*$. The geometry in the spaces and the Fourier series expansion provide convenient tools throughout the analysis.

When $\eta_0 \notin \mathcal{H}$, the estimates are seen to converge to a Kullback-Leibler projection of η_0 in $\mathcal{H}$ at the same rates. The theory can also be easily adapted for the analysis of conditional density estimates and of estimates based on samples that are subject to selection bias.

8.2.1 Linear Approximation

Take $V(f) = V_{\eta_0}(f)$. The following conditions are needed in our analysis.

Condition 8.2.1 V is completely continuous with respect to J.

Condition 8.2.2 For ν sufficiently large and some $\beta > 0$, the eigenvalues ρ_ν of J with respect to V satisfy $\rho_\nu > \beta\nu^r$, where $r > 1$.

Consider the quadratic approximation of (6.1) at η_0, which is given by

$$-\frac{1}{n}\sum_{i=1}^{n}\eta(X_i) + \mu_{\eta_0}(\eta) + \frac{1}{2}V(\eta - \eta_0) + \frac{\lambda}{2}J(\eta); \tag{8.1}$$

see (6.14) on page 184. Plugging the Fourier series expansions $\eta = \sum_\nu \eta_\nu\phi_\nu$ and $\eta_0 = \sum_\nu \eta_{\nu,0}\phi_\nu$ into (8.1), one has

$$-\sum_\nu\left\{\eta_\nu\left(\frac{1}{n}\sum_{i=1}^{n}\phi_\nu(X_i) - \mu_{\eta_0}(\phi_\nu)\right) + \frac{1}{2}(\eta_\nu - \eta_{\nu,0})^2 + \frac{\lambda}{2}\rho_\nu\eta_\nu^2\right\}. \tag{8.2}$$

Write $\beta_\nu = n^{-1}\sum_{i=1}^{n}\phi_\nu(X_i) - \mu_{\eta_0}(\phi_\nu)$. The Fourier coefficients that minimize (8.2) are given by

$$\tilde{\eta}_\nu = (\beta_\nu + \eta_{\nu,0})/(1 + \lambda\rho_\nu).$$

The minimizer $\tilde{\eta} = \sum_\nu \tilde{\eta}_\nu\phi_\nu$ of (8.1) is called a linear approximation of $\hat{\eta}$ since it is linear in $\phi_\nu(X_i)$. Straightforward calculation yields

$$V(\tilde{\eta} - \eta_0) = \sum_\nu(\tilde{\eta}_\nu - \eta_{\nu,0})^2 = \sum_\nu\frac{\beta_\nu^2 - 2\beta_\nu\lambda\rho_\nu\eta_{\nu,0} + \lambda^2\rho_\nu^2\eta_{\nu,0}^2}{(1 + \lambda\rho_\nu)^2},$$

$$\lambda J(\tilde{\eta} - \eta_0) = \sum_\nu\lambda\rho_\nu(\tilde{\eta}_\nu - \eta_{\nu,0})^2 = \sum_\nu\lambda\rho_\nu\frac{\beta_\nu^2 - 2\beta_\nu\lambda\rho_\nu\eta_{\nu,0} + \lambda^2\rho_\nu^2\eta_{\nu,0}^2}{(1 + \lambda\rho_\nu)^2}.$$

Note that $E[\beta_\nu] = 0$ and $E[\beta_\nu^2] = n^{-1}$. It follows that

$$\begin{aligned} E[V(\tilde{\eta} - \eta_0)] &= \frac{1}{n} \sum_\nu \frac{1}{(1 + \lambda\rho_\nu)^2} + \lambda \sum_\nu \frac{\lambda\rho_\nu}{(1 + \lambda\rho_\nu)^2} \rho_\nu \eta_{\nu,0}^2, \\ E[\lambda J(\tilde{\eta} - \eta_0)] &= \frac{1}{n} \sum_\nu \frac{\lambda\rho_\nu}{(1 + \lambda\rho_\nu)^2} + \lambda \sum_\nu \frac{(\lambda\rho_\nu)^2}{(1 + \lambda\rho_\nu)^2} \rho_\nu \eta_{\nu,0}^2. \end{aligned} \tag{8.3}$$

These quantities can be bounded with the help of the following lemma.

Lemma 8.1
Under Condition 8.2.2, as $\lambda \to 0$, *one has*

$$\begin{aligned} \sum_\nu \frac{\lambda\rho_\nu}{(1 + \lambda\rho_\nu)^2} &= O(\lambda^{-1/r}), \\ \sum_\nu \frac{1}{(1 + \lambda\rho_\nu)^2} &= O(\lambda^{-1/r}), \\ \sum_\nu \frac{1}{1 + \lambda\rho_\nu} &= O(\lambda^{-1/r}). \end{aligned}$$

Proof: We prove the first equation.

$$\begin{aligned} \sum_\nu \frac{\lambda\rho_\nu}{(1 + \lambda\rho_\nu)^2} &= \Big(\sum_{\nu < \lambda^{-1/r}} + \sum_{\nu \geq \lambda^{-1/r}} \Big) \frac{\lambda\rho_\nu}{(1 + \lambda\rho_\nu)^2} \\ &= O(\lambda^{-1/r}) + O\Big(\int_{\lambda^{-1/r}}^\infty \frac{\lambda x^r}{(1 + \lambda x^r)^2} dx \Big) \\ &= O(\lambda^{-1/r}) + \lambda^{-1/r} O\Big(\int_1^\infty \frac{x^r}{(1 + x^r)^2} dx \Big) \\ &= O(\lambda^{-1/r}). \end{aligned}$$

The other two follow similar arguments. □

Theorem 8.2
Assume $J(\eta_0) < \infty$. *Under Conditions 8.2.1 and 8.2.2, as* $n \to \infty$ *and* $\lambda \to 0$,

$$(V + \lambda J)(\tilde{\eta} - \eta_0) = O(n^{-1}\lambda^{-1/r} + \lambda).$$

Proof: Note that $\sum_\nu \rho_\nu \eta_{\nu,0}^2 = J(\eta_0) < \infty$. The theorem follows from (8.3) and Lemma 8.1. □

When η_0 is "supersmooth," in the sense that $\sum_\nu \rho_\nu^p \eta_{\nu,0}^2 < \infty$ for some $p > 1$, the rates can be improved to $O(n^{-1}\lambda^{-1/r} + \lambda^p)$, for p up to 2; see Problem 8.1.

8.2.2 Approximation Error and Main Results

We now turn to the approximation error $\hat{\eta} - \tilde{\eta}$. Define

$$A_{f,g}(\alpha) = -\frac{1}{n}\sum_{i=1}^{n}(f+\alpha g)(X_i) + \log\int_{\mathcal{X}} e^{f+\alpha g} + \frac{\lambda}{2}J(f+\alpha g),$$

$$B_{f,g}(\alpha) = -\frac{1}{n}\sum_{i=1}^{n}(f+\alpha g)(X_i) + \mu_{\eta_0}(f+\alpha g) + \frac{1}{2}V(f+\alpha g-\eta_0) + \frac{\lambda}{2}J(f+\alpha g).$$

It is easy to verify that (Problem 8.2)

$$\dot{A}_{f,g}(0) = -\frac{1}{n}\sum_{i=1}^{n} g(X_i) + \mu_f(g) + \lambda J(f,g), \tag{8.4}$$

$$\dot{B}_{f,g}(0) = -\frac{1}{n}\sum_{i=1}^{n} g(X_i) + \mu_{\eta_0}(g) + V(f-\eta_0, g) + \lambda J(f,g). \tag{8.5}$$

Setting $f = \hat{\eta}$ and $g = \hat{\eta} - \tilde{\eta}$ in (8.4), one has

$$-\frac{1}{n}\sum_{i=1}^{n}(\hat{\eta}-\tilde{\eta})(X_i) + \mu_{\hat{\eta}}(\hat{\eta}-\tilde{\eta}) + \lambda J(\hat{\eta}, \hat{\eta}-\tilde{\eta}) = 0, \tag{8.6}$$

and setting $f = \tilde{\eta}$ and $g = \hat{\eta} - \tilde{\eta}$ in (8.5) yields

$$-\frac{1}{n}\sum_{i=1}^{n}(\hat{\eta}-\tilde{\eta})(X_i) + \mu_{\eta_0}(\hat{\eta}-\tilde{\eta}) + V(\tilde{\eta}-\eta_0, \hat{\eta}-\tilde{\eta}) + \lambda J(\tilde{\eta}, \hat{\eta}-\tilde{\eta}) = 0. \tag{8.7}$$

Combining (8.6) and (8.7), it follows that

$$\begin{aligned}\mu_{\hat{\eta}}(\hat{\eta}-\tilde{\eta}) - \mu_{\tilde{\eta}}(\hat{\eta}-\tilde{\eta}) + \lambda J(\hat{\eta}-\tilde{\eta}) \\ = V(\tilde{\eta}-\eta_0, \hat{\eta}-\tilde{\eta}) + \mu_{\eta_0}(\hat{\eta}-\tilde{\eta}) - \mu_{\tilde{\eta}}(\hat{\eta}-\tilde{\eta}).\end{aligned} \tag{8.8}$$

Now, define

$$C(\alpha) = \mu_{\eta_0+\alpha(\tilde{\eta}-\eta_0)/\sigma}(\hat{\eta}-\tilde{\eta}) - \mu_{\eta_0}(\hat{\eta}-\tilde{\eta}),$$

where $\sigma = \{V(\tilde{\eta}-\eta_0)\}^{1/2} = o_p(1)$. A Taylor expansion gives $C(\alpha) = \alpha(1+o(1))V(\tilde{\eta}-\eta_0, \hat{\eta}-\tilde{\eta})/\sigma$, where $o(1)$ is with respect to $\alpha \to 0$. This leads to

$$\mu_{\tilde{\eta}}(\hat{\eta}-\tilde{\eta}) - \mu_{\eta_0}(\hat{\eta}-\tilde{\eta}) = C(\sigma) = V(\tilde{\eta}-\eta_0, \hat{\eta}-\tilde{\eta})(1+o_p(1)), \tag{8.9}$$

as $\lambda \to 0$ and $n\lambda^{1/r} \to \infty$. Now, define $D(\alpha) = \mu_{\tilde{\eta}+\alpha(\hat{\eta}-\tilde{\eta})}(\hat{\eta}-\tilde{\eta})$. It can be shown that $\dot{D}(\alpha) = V_{\tilde{\eta}+\alpha(\hat{\eta}-\tilde{\eta})}(\hat{\eta}-\tilde{\eta})$. By the mean value theorem,

$$\mu_{\hat{\eta}}(\hat{\eta}-\tilde{\eta}) - \mu_{\tilde{\eta}}(\hat{\eta}-\tilde{\eta}) = D(1) - D(0) = \dot{D}(\alpha) = V_{\tilde{\eta}+\alpha(\hat{\eta}-\tilde{\eta})}(\hat{\eta}-\tilde{\eta}), \tag{8.10}$$

for some $\alpha \in [0,1]$. The following condition is needed to proceed.

Condition 8.2.3 For η in a convex set B_0 around η_0 containing $\hat{\eta}$ and $\tilde{\eta}$, $c_1 V(f) \leq V_\eta(f)$ holds uniformly for some $c_1 > 0$.

Condition 8.2.3 is satisfied when the members of B_0 have uniform upper and lower bounds on domain $\mathcal{X}$.

Theorem 8.3
Assume $\sum_\nu \rho_\nu^p \eta_{\nu,0}^2 < \infty$ *for some* $p \in [1,2]$. *Under Conditions 8.2.1—8.2.3, as* $\lambda \to 0$ *and* $n\lambda^{1/r} \to \infty$,

$$(V + \lambda J)(\hat{\eta} - \tilde{\eta}) = o_p(n^{-1}\lambda^{-1/r} + \lambda^p).$$

Consequently,

$$(V + \lambda J)(\hat{\eta} - \eta_0) = O_p(n^{-1}\lambda^{-1/r} + \lambda^p).$$

Proof: From (8.8), (8.9), (8.10), and Condition 8.2.3,

$$\begin{aligned} c_1 V(\hat{\eta} - \tilde{\eta}) + \lambda J(\hat{\eta} - \tilde{\eta}) &\leq o_p(V(\tilde{\eta} - \eta_0, \hat{\eta} - \tilde{\eta})) \\ &= o_p(\{V(\hat{\eta} - \tilde{\eta})V(\tilde{\eta} - \eta_0)\}^{1/2}). \end{aligned}$$

The theorem follows from Theorem 8.2 after trivial manipulation. □

Theorem 8.4
Assume $\sum_\nu \rho_\nu^p \eta_{\nu,0}^2 < \infty$ *for some* $p \in [1,2]$. *Under Conditions 8.2.1—8.2.3, as* $\lambda \to 0$ *and* $n\lambda^{1/r} \to \infty$,

$$\text{SKL}(\eta_0, \hat{\eta}) = O_p(n^{-1}\lambda^{-1/r} + \lambda^p).$$

Proof: Setting $f = \hat{\eta}$ and $g = \hat{\eta} - \eta_0$ in (8.4), one has

$$\begin{aligned} &\mu_{\eta_0}(\eta_0 - \hat{\eta}) + \mu_{\hat{\eta}}(\hat{\eta} - \eta_0) \\ &\quad = \left\{ \frac{1}{n} \sum_{i=1}^n (\hat{\eta} - \eta_0)(X_i) - \mu_{\eta_0}(\hat{\eta} - \eta_0) \right\} - \lambda J(\hat{\eta}, \hat{\eta} - \eta_0) \qquad (8.11) \end{aligned}$$

For the first term on the right-hand side of (8.11), write

$$\frac{1}{n} \sum_{i=1}^n (\hat{\eta} - \eta_0)(X_i) - \mu_{\eta_0}(\hat{\eta} - \eta_0) = \sum_\nu (\hat{\eta}_\nu - \eta_{\nu,0})\beta_\nu,$$

where $\hat{\eta}_\nu$ are the Fourier coefficients of $\hat{\eta}$ and $\beta_\nu = n^{-1}\sum_{i=1}^n \phi_\nu(X_i) - \mu_{\eta_0}(\phi_\nu)$. By the Cauchy-Schwartz inequality,

$$\sum_\nu |(\hat{\eta}_\nu - \eta_{\nu,0})\beta_\nu| \leq \left(\sum_\nu \alpha_\nu^2 (\hat{\eta}_\nu - \eta_{\nu,0})^2\right)^{1/2} \left(\sum_\nu \alpha_\nu^{-2}\beta_\nu^2\right)^{1/2}, \qquad (8.12)$$

for some sequence α_ν. Setting $\alpha_\nu^2 = 1 + \lambda\rho_\nu$, one has

$$\sum_\nu (1+\lambda\rho_\nu)(\hat{\eta}_\nu - \eta_{\nu,0})^2 = (V+\lambda J)(\hat{\eta}-\eta_0) = O_p(n^{-1}\lambda^{-1/r} + \lambda^p)$$

by Theorem 8.3, and $E[\sum_\nu (1+\lambda\rho_\nu)^{-1}\beta_\nu^2] = O(n^{-1}\lambda^{-1/r})$ by Lemma 8.1 and the fact that $E[\beta_\nu^2] = n^{-1}$. Substituting these into (8.12), one has

$$\sum_\nu |(\hat{\eta}_\nu - \eta_{\nu,0})\beta_\nu| = O_p(n^{-1}\lambda^{-1/r} + n^{-1/2}\lambda^{-1/2r+p/2}). \tag{8.13}$$

Similarly, $\lambda J(\hat{\eta}, \hat{\eta}-\eta_0) = \lambda J(\hat{\eta}-\eta_0) + \lambda J(\eta_0, \hat{\eta}-\eta_0)$, where

$$\begin{aligned}\lambda J(\eta_0, \hat{\eta}-\eta_0) &= \sum_\nu \lambda\rho_\nu\eta_{\nu,0}(\hat{\eta}_\nu - \eta_{\nu,0}) \\ &\le \Big\{\sum_\nu (1+\lambda\rho_\nu)(\hat{\eta}_\nu-\eta_{\nu,0})^2\Big\}^{1/2} \\ &\quad \times \Big\{\lambda^p \sum_\nu \frac{(\lambda\rho_\nu)^{2-p}}{1+\lambda\rho_\nu}\rho_\nu^p\eta_{\nu,0}^2\Big\}^{1/2} \\ &= \{(V+\lambda J)(\hat{\eta}-\eta_0)\}^{1/2} O(\lambda^{p/2}).\end{aligned}$$

By Theorem 8.3, $\lambda J(\hat{\eta}, \hat{\eta}-\eta_0) = O_p(n^{-1}\lambda^{-1/r} + \lambda^p)$. Combining this with (8.13), the theorem follows. □

8.2.3 Semiparametric Approximation

As was noted in §6.1, the minimizer $\hat{\eta}$ of (6.1) in $\mathcal{H}$ is, in general, not computable. The minimizer $\hat{\eta}^*$ in the space

$$\mathcal{H}^* = \mathcal{N}_J \oplus \text{span}\{R_J(X_i,\cdot), i=1,\dots,n\}$$

was computed instead. We shall now establish the same convergence rates for $\hat{\eta}^*$ under an extra condition.

Condition 8.2.4 $V(\phi_\nu\phi_\mu) \le c_2$ holds uniformly for some $c_2 > 0$, $\forall \nu, \mu$.

Condition 8.2.4 virtually calls for uniformly bounded fourth moments of $\phi_\nu(X)$. The condition appears mild, as ϕ_ν typically grow in roughness but not necessarily in magnitude, but since ϕ_ν are generally not available in explicit forms, the condition is extremely difficult to verify from more primitive conditions, if at all possible.

Lemma 8.5
Under Conditions 8.2.1, 8.2.2, and 8.2.4, as $\lambda \to 0$ and $n\lambda^{2/r} \to \infty$, $V(h) = o_p(\lambda J(h))$, $\forall h \in \mathcal{H} \ominus \mathcal{H}^*$.

Note that when $\lambda \to 0$ and $n\lambda^{2/r} \to \infty$, $n\lambda^{1/r} \to \infty$. With the optimal rate of $\lambda = O(n^{-r/(pr+1)})$, $n\lambda^{2/r} = O(n^{(pr-1)/(pr+1)}) \to \infty$, for $r > 1$ and $p \geq 1$.

Proof of Lemma 8.5: For $h \in \mathcal{H} \ominus \mathcal{H}^*$, since $h(X_i) = J(R_J(X_i, \cdot), h) = 0$, $\sum_{i=1}^n h^2(X_i) = 0$. Write $h = \sum_\nu h_\nu \phi_\nu$. It follows that

$$\begin{aligned}
V(h) \leq \mu_{\eta_0}(h^2) &= \sum_\nu \sum_\mu h_\nu h_\mu \mu_{\eta_0}(\phi_\nu \phi_\mu) \\
&= \sum_\nu \sum_\mu h_\nu h_\mu \Big\{ \mu_{\eta_0}(\phi_\nu \phi_\mu) - \frac{1}{n} \sum_{i=1}^n \phi_\nu(X_i)\phi_\mu(X_i) \Big\} \\
&\leq \Big\{ \sum_\nu \sum_\mu \frac{1}{1+\lambda\rho_\nu} \frac{1}{1+\lambda\rho_\mu} \\
&\qquad \times \Big\{ \frac{1}{n} \sum_{i=1}^n \phi_\nu(X_i)\phi_\mu(X_i) - \mu_{\eta_0}(\phi_\nu\phi_\mu) \Big\}^2 \Big\}^{1/2} \\
&\quad \times \Big\{ \sum_\nu \sum_\mu (1+\lambda\rho_\nu)(1+\lambda\rho_\mu) h_\nu^2 h_\mu^2 \Big\}^{1/2} \\
&= O_p(n^{-1/2}\lambda^{-1/r})(V + \lambda J)(h),
\end{aligned}$$

where Lemma 8.1 and the fact that

$$E\Big[\frac{1}{n} \sum_{i=1}^n \phi_\nu(X_i)\phi_\mu(X_i) - \mu_{g_0}(\phi_\nu\phi_\mu) \Big]^2 \leq \frac{c_2}{n}$$

are used. The lemma follows. □

Let η^* be the projection of $\hat{\eta}$ in $\mathcal{H}^*$. Setting $f = \hat{\eta}$ and $g = \hat{\eta} - \eta^*$ in (8.4), one has

$$\mu_{\hat{\eta}}(\hat{\eta} - \eta^*) + \lambda J(\hat{\eta} - \eta^*) = 0, \tag{8.14}$$

remember that $(\hat{\eta} - \eta^*)(X_i) = 0$ and $J(\eta^*, \hat{\eta} - \eta^*) = 0$. Similar to (8.9), one has

$$\mu_{\hat{\eta}}(\hat{\eta} - \eta^*) - \mu_{\eta_0}(\hat{\eta} - \eta^*) = V(\hat{\eta} - \eta_0, \hat{\eta} - \eta^*)(1 + o_p(1)), \tag{8.15}$$

and similar to (8.13), it can be shown that

$$\mu_{\eta_0}(\hat{\eta} - \eta^*) = O_p(n^{-1/2}\lambda^{-1/2r})\{(V + \lambda J)(\hat{\eta} - \eta^*)\}^{1/2}. \tag{8.16}$$

Theorem 8.6
Assume $\sum_\nu \rho_\nu^p \eta_{\nu,0}^2 < \infty$ *for some* $p \in [1, 2]$. *Under Conditions 8.2.1—8.2.4, as* $\lambda \to 0$ *and* $n\lambda^{2/r} \to \infty$,

$$\begin{aligned}
\lambda J(\hat{\eta} - \eta^*) &= O_p(n^{-1}\lambda^{-1/r} + \lambda^p), \\
V(\hat{\eta} - \eta^*) &= o_p(n^{-1}\lambda^{-1/r} + \lambda^p).
\end{aligned}$$

Proof: Combining (8.14), (8.15), and (8.16) and applying Theorem 8.3, one has

$$\lambda J(\hat{\eta} - \eta^*) = O_p(n^{-1/2}\lambda^{-1/2r} + \lambda^{p/2})\{(V + \lambda J)(\hat{\eta} - \eta^*)\}^{1/2}.$$

The theorem follows from Lemma 8.5. □

We can now obtain the rates for $(V + \lambda J)(\hat{\eta}^* - \eta^*)$ and, in turn, for $(V + \lambda J)(\hat{\eta}^* - \hat{\eta})$. Condition 8.2.3 needs to be modified to include $\hat{\eta}^*$ and η^* in the convex set B_0.

Theorem 8.7
Assume $\sum_\nu \rho_\nu^p \eta_{\nu,0}^2 < \infty$ *for some* $p \in [1, 2]$. *Under Conditions 8.2.1—8.2.4, as* $\lambda \to 0$ *and* $n\lambda^{2/r} \to \infty$,

$$(V + \lambda J)(\hat{\eta}^* - \eta^*) = O_p(n^{-1}\lambda^{-1/r} + \lambda^p),$$
$$(V + \lambda J)(\hat{\eta} - \hat{\eta}^*) = O_p(n^{-1}\lambda^{-1/r} + \lambda^p).$$

Proof: Setting $f = \hat{\eta}^*$ and $g = \hat{\eta}^* - \eta^* \in \mathcal{H}^*$ in (8.4), one has

$$-\frac{1}{n}\sum_{i=1}^{n}(\hat{\eta}^* - \eta^*)(X_i) + \mu_{\hat{\eta}^*}(\hat{\eta}^* - \eta^*) + \lambda J(\hat{\eta}^*, \hat{\eta}^* - \eta^*) = 0. \quad (8.17)$$

Setting $f = \hat{\eta}$ and $g = \hat{\eta} - \hat{\eta}^*$ in (8.4), one gets

$$-\frac{1}{n}\sum_{i=1}^{n}(\hat{\eta} - \hat{\eta}^*)(X_i) + \mu_{\hat{\eta}}(\hat{\eta} - \hat{\eta}^*) + \lambda J(\hat{\eta}, \hat{\eta} - \hat{\eta}^*) = 0. \quad (8.18)$$

Note that $(\hat{\eta} - \eta^*)(X_i) = 0$ and $J(\hat{\eta} - \eta^*, \eta^*) = J(\hat{\eta} - \eta^*, \hat{\eta}^*) = 0$. Equation (8.18) leads to

$$-\frac{1}{n}\sum_{i=1}^{n}(\eta^* - \hat{\eta}^*)(X_i) + \mu_{\hat{\eta}}(\hat{\eta} - \hat{\eta}^*) + \lambda J(\hat{\eta} - \eta^*) + \lambda J(\eta^*, \eta^* - \hat{\eta}^*) = 0. \quad (8.19)$$

Adding (8.17) and (8.19), some algebra yields

$$\begin{aligned}\mu_{\hat{\eta}^*}(\hat{\eta}^* - \eta^*) - \mu_{\eta^*}(\hat{\eta}^* - \eta^*) + \lambda J(\hat{\eta}^* - \eta^*) + \lambda J(\hat{\eta} - \eta^*) \\ = \mu_{\hat{\eta}}(\eta^* - \hat{\eta}) + \mu_{\hat{\eta}}(\hat{\eta}^* - \eta^*) - \mu_{\eta^*}(\hat{\eta}^* - \eta^*). \quad (8.20)\end{aligned}$$

Now, by Condition 8.2.3,

$$\mu_{\hat{\eta}^*}(\hat{\eta}^* - \eta^*) - \mu_{\eta^*}(\hat{\eta}^* - \eta^*) \geq c_1 V(\hat{\eta}^* - \eta^*).$$

From (8.15), (8.16), and Theorems 8.3 and 8.6, one has $\mu_{\hat{\eta}}(\eta^* - \hat{\eta}) = O_p(n^{-1}\lambda^{-1/r} + \lambda^p)$ and

$$\begin{aligned}\mu_{\hat{\eta}}&(\hat{\eta}^* - \eta^*) - \mu_{\eta^*}(\hat{\eta}^* - \eta^*) \\ &= V(\hat{\eta} - \eta^*, \hat{\eta}^* - \eta^*)(1 + o_p(1)) \\ &= o_p(n^{-1/2}\lambda^{-1/2r} + \lambda^{p/2})\{V(\hat{\eta}^* - \eta^*)\}^{1/2}. \quad (8.21)\end{aligned}$$

Hence, (8.20) becomes

$$\begin{aligned} &c_1 V(\hat{\eta}^* - \eta^*) + \lambda J(\hat{\eta}^* - \eta^*) + \lambda J(\hat{\eta} - \eta^*) \\ &\quad = O_p(n^{-1}\lambda^{-1/r} + \lambda^p) + o_p(n^{-1/2}\lambda^{-1/2r} + \lambda^{p/2})\{V(\hat{\eta}^* - \eta^*)\}^{1/2}. \end{aligned}$$

The theorem follows. □

Theorem 8.8
Assume $\sum_\nu \rho_\nu^p \eta_{\nu,0}^2 < \infty$ *for some* $p \in [1,2]$. *Under Conditions 8.2.1—8.2.4, as* $\lambda \to 0$ *and* $n\lambda^{2/r} \to \infty$,

$$\begin{aligned} (V + \lambda J)(\hat{\eta}^* - \eta_0) &= O_p(n^{-1}\lambda^{-1/r} + \lambda^p), \\ \text{SKL}(\eta_0, \hat{\eta}^*) &= O_p(n^{-1}\lambda^{-1/r} + \lambda^p). \end{aligned}$$

Proof: The first part of the theorem follows from Theorems 8.3, 8.6, and 8.7. To prove the second part, set $f = \hat{\eta}$ and $h = \hat{\eta}^* - \eta_0$ in (8.4). This yields

$$-\frac{1}{n}\sum_{i=1}^n (\hat{\eta}^* - \eta_0)(X_i) + \mu_{\hat{\eta}}(\hat{\eta}^* - \eta_0) + \lambda J(\hat{\eta}, \hat{\eta}^* - \eta_0) = 0.$$

Hence,

$$\begin{aligned} &\mu_{\eta_0}(\eta_0 - \hat{\eta}^*) + \mu_{\hat{\eta}^*}(\hat{\eta}^* - \eta_0) \\ &\qquad = \mu_{\eta_0}(\eta_0 - \hat{\eta}^*) + \mu_{\hat{\eta}^*}(\hat{\eta}^* - \eta_0) \\ &\qquad\quad + \frac{1}{n}\sum_{i=1}^n (\hat{\eta}^* - \eta_0)(X_i) - \mu_{\hat{\eta}}(\hat{\eta}^* - \eta_0) + \lambda J(\hat{\eta}, \eta_0 - \hat{\eta}^*) \\ &\qquad = \lambda J(\hat{\eta}, \eta_0 - \hat{\eta}^*) + \left\{\frac{1}{n}\sum_{i=1}^n (\hat{\eta}^* - \eta_0)(X_i) - \mu_{\eta_0}(\hat{\eta}^* - \eta_0)\right\} \\ &\qquad\quad + [\mu_{\hat{\eta}^*}(\hat{\eta}^* - \eta_0) - \mu_{\hat{\eta}}(\hat{\eta}^* - \eta_0)]. \end{aligned}$$

By the same arguments used to prove Theorem 8.4, the first two terms above can be shown to be of the order $O_p(n^{-1}\lambda^{-1/r} + \lambda^p)$. Similar to (8.21), the third term is easily seen to be of the same order. □

8.2.4 Convergence Under Incorrect Model

It has been implicitly assumed thus far that $\eta_0 \in \mathcal{H}$. In the case $\eta_0 \notin \mathcal{H}$, say an additive model is fitted while the interaction is present in η_0, modifications are needed in the problem formulation. The convergence rates remain valid under the modified formulation, however.

Write $\text{RKL}(\eta|\eta_0) = \log \int_{\mathcal{X}} e^\eta - \mu_{\eta_0}(\eta)$, the relative Kullback-Leibler distance of η from η_0; $\text{RKL}(\eta|\eta_0)$ is strictly convex in $\eta \in \mathcal{H} \not\supset \{1\}$. When

$\eta_0 \notin \mathcal{H}$, the minimizer η_0^* of $\text{RKL}(\eta|\eta_0)$ in $\mathcal{H}$, when it exists, is probably the best proxy of η_0 that one can hope to estimate in the context. Such an η_0^* will be referred to as the Kullback-Leibler projection of η_0 in $\mathcal{H}$.

Lemma 8.9
Assume that the Kullback-Leibler projection η_0^* of η_0 exists in $\mathcal{H}$. It follows that $\mu_{\eta_0^*}(h) = \mu_{\eta_0}(h)$, $\forall h \in \mathcal{H}$.

Proof: Write $A_{g,h}(\alpha) = \text{RKL}(g + \alpha h|\eta_0)$, where $g, h \in \mathcal{H}$, α real. It is easy to check that $\dot{A}_{g,h}(0) = \mu_g(h) - \mu_{\eta_0}(h)$. The lemma follows the fact that $\dot{A}_{\eta_0^*,h}(0) = 0$, $\forall h \in \mathcal{H}$. □

Substituting η_0^* for η_0 everywhere in §§8.2.1—8.2.3, all results and arguments remain valid if one can show that

$$E\left[\frac{1}{n}\sum_{i=1}^{n}\phi_\nu(X_i) - \mu_{\eta_0^*}(\phi_\nu)\right]^2 = \frac{1}{n},$$

$$E\left[\frac{1}{n}\sum_{i=1}^{n}\phi_\nu(X_i)\phi_\mu(X_i) - \mu_{\eta_0^*}(\phi_\nu\phi_\mu)\right]^2 \leq \frac{c_2}{n}.$$

These equations follow from Lemma 8.9 under an extra condition.

Condition 8.2.0 $gh - C_{gh} \in \mathcal{H}$ for some constant C_{gh}, $\forall g, h \in \mathcal{H}$.

Note that if $\mu_{\eta_0}(gh - C_{gh}) = \mu_{\eta_0^*}(gh - C_{gh})$, then $\mu_{\eta_0}(gh) = \mu_{\eta_0^*}(gh)$. The key requirement here is that $J(gh) < \infty$ whenever $J(g) < \infty$, $J(h) < \infty$; the constant C_{gh} takes care of the side condition on log density. Condition 8.2.0 is satisfied by all the spaces appearing in the examples in Chapter 6.

8.2.5 Estimation Under Biased Sampling

Now, consider the setting of §6.6. Observations (t_i, X_i) are taken from $\mathcal{T} \times \mathcal{X}$ with $X|t \sim w(t,x)e^{\eta_0(x)}/\int_{\mathcal{X}} w(t,x)e^{\eta_0(x)}$, and the density estimate $e^{\hat{\eta}}/\int_{\mathcal{X}} e^{\hat{\eta}}$ is obtained via the minimization of (6.25). The theory developed in the proceeding sections remain valid with due modifications, although some of the intermediate o_p rates might have to be replaced by the respective O_p rates.

Let $m(t)$ be the limiting density of t_i on $\mathcal{T}$. Write

$$\mu_\eta(f|t) = \frac{\int_{\mathcal{X}} f(x)w(t,x)e^{\eta(x)}}{\int_{\mathcal{X}} w(t,x)e^{\eta(x)}}, \quad v_\eta(f|t) = \mu_\eta(f^2|t) - \mu_\eta^2(f|t),$$

and define

$$\mu_\eta(f) = \int_{\mathcal{T}} m(t)\mu_\eta(f|t), \quad V_\eta(f) = \int_{\mathcal{T}} m(t)v_\eta(f|t).$$

The convergence rates are given in terms of

$$\mathrm{SKL}(\eta_0, \hat{\eta}) = \int_{\mathcal{T}} m(t)\{\mu_{\eta_0}(\eta_0 - \hat{\eta}|t) + \mu_{\hat{\eta}}(\hat{\eta} - \eta_0|t)\}$$

and $V(\hat{\eta} - \eta_0)$, where $V(f) = V_{\eta_0}(f)$.

For the theory of §§8.2.1—8.2.3 to hold in this setting, Conditions 8.2.1 and 8.2.2 need little change except for the definition of V. Conditions 8.2.3 and 8.2.4 shall be modified as follows.

Condition 8.2.3.b For η in a convex set B_0 around η_0 containing $\tilde{\eta}$, $\hat{\eta}$, η^*, and $\hat{\eta}^*$, $c_1 v_{\eta_0}(f|t) \leq v_\eta(f|t) \leq c_2 v_{\eta_0}(f|t)$ holds uniformly for some $0 < c_1 < c_2 < \infty$, $\forall f \in \mathcal{H}$, $\forall t \in \mathcal{T}$.

Condition 8.2.4.b $\int_{\mathcal{T}} m(t)\{v_{\eta_0}(\phi_\nu, \phi_\mu|t)\}^2 \leq c_3$ holds uniformly for some $c_3 < \infty$, $\forall \nu, \mu$.

To apply the arguments of §8.2.4, the relative Kullback-Leibler distance shall be modified as $\mathrm{RKL}(\eta|\eta_0) = \int_{\mathcal{T}} m(t)\{\log \int_{\mathcal{X}} w(t,x)e^{\eta(x)} - \mu_{\eta_0}(\eta|t)\}$. Details are straightforward to work out and are left as an exercise (Problem 8.3).

8.2.6 Estimation of Conditional Density

For the estimation of the conditional density $f(y|x) = e^{\eta_0(x,y)} / \int_{\mathcal{Y}} e^{\eta_0(x,y)}$ via the minimization of (6.29), the theory is also easy to modify.

Let $f(x)$ be the marginal density of X on $\mathcal{X}$. Write

$$\mu_\eta(g|x) = \frac{\int_{\mathcal{Y}} g(x,y)e^{\eta(x,y)}}{\int_{\mathcal{Y}} e^{\eta(x,y)}}, \quad v_\eta(g|x) = \mu_\eta(g^2|x) - \mu_\eta^2(g|x)$$

and define

$$\mu_\eta(g) = \int_{\mathcal{X}} f(x)\mu_\eta(g|x), \quad V_\eta(g) = \int_{\mathcal{X}} f(x)v_\eta(g|x).$$

The convergence rates are given in terms of

$$\mathrm{SKL}(\eta_0, \hat{\eta}) = \int_{\mathcal{X}} f(x)\{\mu_{\eta_0}(\eta_0 - \hat{\eta}|x) + \mu_{\hat{\eta}}(\hat{\eta} - \eta_0|x)\}$$

and $V(\hat{\eta} - \eta_0)$, where $V(g) = V_{\eta_0}(g)$.

For the theory of §§8.2.1—8.2.3 to hold for conditional density estimates, Conditions 8.2.1 and 8.2.2 need little change except for the definition of V. Conditions 8.2.3 and 8.2.4 shall be modified as follows.

Condition 8.2.3.c For η in a convex set B_0 around η_0 containing $\tilde{\eta}$, $\hat{\eta}$, η^*, and $\hat{\eta}^*$, $c_1 v_{\eta_0}(g|x) \leq v_\eta(g|x) \leq c_2 v_{\eta_0}(g|x)$ holds uniformly for some $0 < c_1 < c_2 < \infty$, $\forall g \in \mathcal{H}$, $\forall x \in \mathcal{X}$.

Condition 8.2.4.c There exist $c_3, c_4, c_5 < \infty$, such that

$$\int_{\mathcal{X}} f(x)\{v_{\eta_0}(\phi_\nu, \phi_\mu|x)\}^2 \leq c_3,$$
$$\int_{\mathcal{X}} f(x) v_{\eta_0}(\phi_\nu\phi_\mu, \phi_\nu\phi_\mu|x) \leq c_4,$$
$$\int_{\mathcal{X}} f(x)\{\mu_{\eta_0}(\phi_\nu\phi_\mu|x) - \mu_{\eta_0}(\phi_\nu\phi_\mu)\}^2 \leq c_5,$$

hold uniformly, $\forall \nu, \mu$,

To apply the arguments of §8.2.4, the relative Kullback-Leibler distance shall be modified as $\text{RKL}(\eta|\eta_0) = \int_{\mathcal{X}} f(x)\{\log \int_{\mathcal{Y}} e^\eta - \mu_{\eta_0}(\eta|x)\}$, and the constant C_{gh} in Condition 8.2.0 may be a function of x. Details are left as an exercise (Problem 8.4).

8.2.7 Estimation Under Response-Based Sampling

Consider the connected case in the setting of §6.8, where the strata $\mathcal{Y}_j$ are sampled with probability π_j, and the samples $(X, Y)|\mathcal{Y}_j$ are taken from $e^{\eta_0(x,y)} / \int_{\mathcal{X}\times\mathcal{Y}_j} e^{\eta_0(x,y)}$. Write

$$\mu_\eta(f|j) = \frac{\int_{\mathcal{X}\times\mathcal{Y}_j} fe^\eta}{\int_{\mathcal{X}\times\mathcal{Y}_j} e^\eta}, \quad v_\eta(f|j) = \mu_\eta(f^2|j) - \mu_\eta^2(f|j)$$

and define

$$\mu_\eta(f) = \sum_{j=1}^{s} \pi_j \mu_\eta(f|j), \quad V_\eta(f) = \sum_{j=1}^{s} \pi_j v_\eta(f|j).$$

The rates for the minimizers of (6.39) can be derived in terms of

$$\text{SKL}(\eta_0, \hat{\eta}) = \sum_{j=1}^{s} \pi_j\{\mu_{\eta_0}(\eta_0 - \hat{\eta}|j) + \mu_{\hat{\eta}}(\hat{\eta} - \eta_0|j)\}$$

and $V(\hat{\eta} - \eta_0)$, where $V(f) = V_{\eta_0}(f)$. The conditions needed are similar to those for conditional density estimates. The relative Kullback-Leibler distance is defined by $\text{RKL}(\eta|\eta_0) = \sum_{j=1}^{s} \pi_j\{\log \int_{\mathcal{X}\times\mathcal{Y}_j} e^\eta - \mu_{\eta_0}(\eta|j)\}$. Further details are left as an exercise (Problem 8.5).

8.3 Rates for Hazard Estimates

The convergence rates for the minimizers of (7.1) are to be established in this section. The martingale structure of censored lifetime data, which was mentioned in §7.2.1 and §7.5.1, serves as the primary tool for the stochastic calculations involved.

Some basic facts concerning the martingale structure are summarized, and a quadratic functional V is derived under the sampling structure. The rates are given in terms of $V(\hat{\eta} - \eta_0)$ and in terms of the symmetrized version of $\text{KL}(\eta_0, \hat{\eta})$ as defined in (7.6). The analysis parallels that in §8.2.

8.3.1 Martingale Structure

Write $N(t) = I_{[X \leq t, \delta=1]}$, $Y(t) = I_{[Z<t\leq X]}$, and $A(t) = \int_0^t Y(s)e^{\eta_0(s,U)}ds$, as in §7.2.1. Under independent censorship, $M(t) = N(t) - A(t)$ is a martingale conditional on U and Z. We shall now summarize some martingale properties needed in the asymptotic analysis. The results are quoted from Fleming and Harrington (1991, §2.7) and Gill (1984).

First of all, one has

$$E[M(t)|U,Z] = 0,$$
$$E[M^2(t)|U,Z] = E[A(t)|U,Z] = \int_0^t e^{\eta_0(s,U)}E[Y(s)|U,Z]ds.$$

For any deterministic function $h(t,u)$ continuous in t, $\forall u$ (so it is locally bounded predictable), the Stieltjes integral

$$\int_0^t h(s,U)dM(s)$$

is also a martingale as long as $\int_{\mathcal{T}} h^2(t,U)e^{\eta_0(t,U)}E[Y(t)|U,Z]dt < \infty$. It follows that

$$E\big[\int_0^t h(s,U)dM(s)|U,Z\big] = 0,$$
$$E\big[\{\int_0^t h(s,U)dM(s)\}^2|U,Z\big] = \int_0^t h^2(s,U)e^{\eta_0(s,U)}E[Y(s)|U,Z]ds.$$

This yields

$$E\Big[\int_0^t h\,dN(s)\Big] - \int_{\mathcal{U}} m(u)\int_0^t h\,e^{\eta_0}\tilde{S}ds = E\Big[\int_0^t h\,dM(s)\Big] = 0, \quad (8.22)$$
$$E\Big[\Big\{\int_0^t h\,dM(s)\Big\}^2\Big] = E\Big[\int_0^t h^2dA(s)\Big] = \int_{\mathcal{U}} m(u)\int_0^t h^2e^{\eta_0}\tilde{S}ds, \quad (8.23)$$

where $\tilde{S}(t,u) = E[Y(t)|U=u] = P(Z<t\leq X|U=u)$. Furthermore,

$$\begin{aligned}
&E\Big[\Big\{\int_0^t h\,dN(s) - \int_{\mathcal{U}} m(u)\int_0^t h\,e^{\eta_0}\tilde{S}ds\Big\}^2\Big] \\
&\quad = E\Big[\Big\{\int_0^t h\,dM(s) + \int_0^t h\,e^{\eta_0}Y(s)ds - \int_{\mathcal{U}} m(u)\int_0^t h\,e^{\eta_0}\tilde{S}ds\Big\}^2\Big] \\
&\quad = E\Big[\Big\{\int_0^t h\,dM(s)\Big\}^2\Big] \\
&\qquad + E\Big[\Big\{\int_0^t h\,e^{\eta_0}Y(s)ds - \int_{\mathcal{U}} m(u)\int_0^t h\,e^{\eta_0}\tilde{S}ds\Big\}^2\Big], \quad (8.24)
\end{aligned}$$

where $E[\int_0^t h\,dM(s)(\int_0^t h\,e^{\eta_0}Y(s)ds - \int_{\mathcal{U}} m(u)\int_0^t h\,e^{\eta_0}\tilde{S}ds)|U,Z] = 0$ because $\int_0^t h\,e^{\eta_0}Y(s)ds - \int_{\mathcal{U}} m(u)\int_0^t h\,e^{\eta_0}\tilde{S}ds$ is predictable.

Note that $\delta\eta(X,U) = \int_{\mathcal{T}} \eta(t,U)dN(t)$, $\int_Z^X e^{\eta(t,U)}dt = \int_{\mathcal{T}} Y(t)e^{\eta(t,U)}dt$. The penalized likelihood functional (7.1) on page 212 shall be written as

$$-\frac{1}{n}\sum_{i=1}^{n}\Big\{\int_{\mathcal{T}}\eta_i dN_i(t) - \int_{\mathcal{T}} Y_i e^{\eta_i}dt\Big\} + \frac{\lambda}{2}J(\eta), \tag{8.25}$$

where $\eta_i(t) = \eta(t,U_i)$. Define

$$V(f) = \int_{\mathcal{U}} m(u)\int_{\mathcal{T}} f^2(t,u)e^{\eta_0(t,u)}\tilde{S}(t,u)dt. \tag{8.26}$$

Convergence rates for the minimizer $\hat{\eta}$ of (8.25) shall be established in terms of $V(\hat{\eta}-\eta_0)$ and

$$\mathrm{SKL}(\eta_0,\hat{\eta}) = \int_{\mathcal{U}} m(u)\int_{\mathcal{T}}(e^{\hat{\eta}(t,u)} - e^{\eta_0(t,u)})(\hat{\eta}(t,u)-\eta_0(t,u))\tilde{S}(t,u)dt,$$

which is the symmetrized version of $\mathrm{KL}(\eta_0,\hat{\eta})$ defined in (7.6) on page 215.

8.3.2 Linear Approximation

The following conditions are needed in our analysis, which are carbon copies of Conditions 8.2.1 and 8.2.2 but with V as defined in (8.26).

Condition 8.3.1 V is completely continuous with respect to J.

Condition 8.3.2 For ν sufficiently large and some $\beta > 0$, the eigenvalues ρ_ν of J with respect to V satisfy $\rho_\nu > \beta\nu^r$, where $r > 1$.

Consider the quadratic functional

$$-\frac{1}{n}\sum_{i=1}^{n}\Big\{\int_{\mathcal{T}}\eta_i dN_i(t) - \int_{\mathcal{T}}\eta_i Y_i e^{\eta_{0,i}}dt\Big\} + \frac{1}{2}V(\eta-\eta_0) + \frac{\lambda}{2}J(\eta), \tag{8.27}$$

where $\eta_{0,i}(t) = \eta_0(t,U_i)$. Plugging the Fourier expansions $\eta = \sum_\nu \eta_\nu\phi_\nu$ and $\eta_0 = \sum_\nu \eta_{\nu,0}\phi_\nu$ into (8.27), the minimizer $\tilde{\eta}$ of (8.27) has Fourier coefficients

$$\tilde{\eta}_\nu = (\beta_\nu + \eta_{\nu,0})/(1+\lambda\rho_\nu),$$

where $\beta_\nu = n^{-1}\sum_{i=1}^n \int_{\mathcal{T}}\phi_{\nu,i}dM_i(t)$ with $\phi_{\nu,i}(t) = \phi_\nu(t,U_i)$. From (8.22), (8.23), and the fact that $\int_{\mathcal{U}} m(u)\int_{\mathcal{T}}\phi_\nu^2 e^{\eta_0}\tilde{S}dt = V(\phi_\nu) = 1$, it is easy to see that $E[\beta_\nu] = 0$ and $E[\beta_\nu^2] = n^{-1}$. See Problem 8.6.

Theorem 8.10
Assume $\sum_\nu \rho_\nu^p\eta_{\nu,0}^2 < \infty$ for some $p \in [1,2]$. Under Conditions 8.3.1 and 8.3.2, as $n\to\infty$ and $\lambda\to 0$,

$$(V+\lambda J)(\tilde{\eta}-\eta_0) = O_p(n^{-1}\lambda^{-1/r} + \lambda^p).$$

Proof: See the proof of Theorem 8.2. □

8.3.3 Approximation Error and Main Result

We now turn to the approximation error $\hat{\eta} - \tilde{\eta}$. Define

$$\begin{aligned} A_{f,g}(\alpha) &= -\frac{1}{n}\sum_{i=1}^{n}\left\{\int_{\mathcal{T}}(f+\alpha g)_i dN_i(t) - \int_{\mathcal{T}} Y_i e^{(f+\alpha g)_i} dt\right\} \\ &\quad + \frac{\lambda}{2}J(f+\alpha g), \\ B_{f,g}(\alpha) &= -\frac{1}{n}\sum_{i=1}^{n}\left\{\int_{\mathcal{T}}(f+\alpha g)_i dN_i(t) - \int_{\mathcal{T}}(f+\alpha g)_i Y_i e^{\eta_{0,i}} dt\right\} \\ &\quad + \frac{1}{2}V(f+\alpha g-\eta_0) + \frac{\lambda}{2}J(f+\alpha g). \end{aligned}$$

It can be shown that

$$\dot{A}_{f,g}(0) = -\frac{1}{n}\sum_{i=1}^{n}\left\{\int_{\mathcal{T}} g_i dN_i(t) - \int_{\mathcal{T}} g_i Y_i e^{f_i} dt\right\} + \lambda J(f,g), \tag{8.28}$$

$$\begin{aligned} \dot{B}_{f,g}(0) &= -\frac{1}{n}\sum_{i=1}^{n}\left\{\int_{\mathcal{T}} g_i dN_i(t) - \int_{\mathcal{T}} g_i Y_i e^{\eta_{0,i}} dt\right\} \\ &\quad + V(f-\eta_0, g) + \lambda J(f,g). \end{aligned} \tag{8.29}$$

Setting $f = \hat{\eta}$ and $g = \hat{\eta} - \tilde{\eta}$ in (8.28), one has

$$-\frac{1}{n}\sum_{i=1}^{n}\left\{\int_{\mathcal{T}}(\hat{\eta}-\tilde{\eta})_i dN_i(t) - \int_{\mathcal{T}}(\hat{\eta}-\tilde{\eta})_i Y_i e^{\hat{\eta}_i} dt\right\} + \lambda J(\hat{\eta}, \hat{\eta}-\tilde{\eta}) = 0, \tag{8.30}$$

and setting $f = \tilde{\eta}$ and $g = \hat{\eta} - \tilde{\eta}$ in (8.29), one gets

$$\begin{aligned} -\frac{1}{n}\sum_{i=1}^{n}\left\{\int_{\mathcal{T}}(\hat{\eta}-\tilde{\eta})_i dN_i(t) - \int_{\mathcal{T}}(\hat{\eta}-\tilde{\eta})_i Y_i e^{\eta_{0,i}} dt\right\} \\ + V(\tilde{\eta}-\eta_0, \hat{\eta}-\tilde{\eta}) + \lambda J(\tilde{\eta}, \hat{\eta}-\tilde{\eta}) = 0. \end{aligned} \tag{8.31}$$

Subtracting (8.31) from (8.30), some algebra yields

$$\begin{aligned} \frac{1}{n}\sum_{i=1}^{n}\int_{\mathcal{T}}(\hat{\eta}-\tilde{\eta})_i(e^{\hat{\eta}} - e^{\tilde{\eta}})_i Y_i dt + \lambda J(\hat{\eta}-\tilde{\eta}) \\ = V(\tilde{\eta}-\eta_0, \hat{\eta}-\tilde{\eta}) - \frac{1}{n}\sum_{i=1}^{n}\int_{\mathcal{T}}(\hat{\eta}-\tilde{\eta})_i(e^{\tilde{\eta}} - e^{\eta_0})_i Y_i dt. \end{aligned} \tag{8.32}$$

One needs the following conditions in addition to Conditions 8.3.1 and 8.3.2 to proceed.

Condition 8.3.3 For η in a convex set B_0 around η_0 containing $\hat{\eta}$ and $\tilde{\eta}$, $c_1 e^{\eta_0(t,u)} \leq e^{\eta(t,u)} \leq c_2 e^{\eta_0(t,u)}$ holds uniformly for some $c_1 > 0$ and $c_2 < \infty$ on $\{(t,u) : \tilde{S}(t,u) > 0\}$.

Condition 8.3.4 $\int_{\mathcal{U}} m(u) \int_{\mathcal{T}} \phi_\nu^2 \phi_\mu^2 e^{k\eta_0} \tilde{S} dt \leq c_3$ for some $c_3 < \infty$, $k = 1, 2$, $\forall \nu, \mu$.

By the mean value theorem, Condition 8.3.3 implies the equivalence of $V(\eta - \eta_0)$ and $\text{SKL}(\eta_0, \eta)$ for η in B_0. When η_0 is bounded, Condition 8.3.4 essentially asks for a uniform bound on the fourth moments of ϕ_ν.

Lemma 8.11
Under Conditions 8.3.1, 8.3.2, and 8.3.4, as $\lambda \to 0$ and $n\lambda^{2/r} \to \infty$,

$$\frac{1}{n} \sum_{i=1}^{n} \int_{\mathcal{T}} f_i^2 e^{\eta_{0,i}} Y_i dt = V(f) + o_p((V + \lambda J)(f)),$$

where $f_i = f(t, U_i)$. Similarly,

$$\frac{1}{n} \sum_{i=1}^{n} \int_{\mathcal{T}} f_i g_i e^{\eta_{0,i}} Y_i dt = V(f, g) + o_p(\{(V + \lambda J)(f)(V + \lambda J)(g)\}^{1/2}).$$

Proof: We only prove the first statement. The same arguments apply to the second. Write $\tau(f) = \int_{\mathcal{U}} m(u) \int_{\mathcal{T}} f e^{\eta_0} \tilde{S} dt$. Using the Fourier series expansion $f = \sum_\nu f_\nu \phi_\nu$, one has

$$\begin{aligned}
&\left| \frac{1}{n} \sum_{i=1}^{n} \int_{\mathcal{T}} f_i^2 e^{\eta_{0,i}} Y_i dt - V(f) \right| \\
&\quad = \left| \sum_\nu \sum_\mu f_\nu f_\mu \Big\{ \frac{1}{n} \sum_{i=1}^{n} \int_{\mathcal{T}} \phi_{\nu,i} \phi_{\mu,i} e^{\eta_{0,i}} Y_i dt - \tau(\phi_\nu \phi_\mu) \Big\} \right| \\
&\quad \leq \left\{ \sum_\nu \sum_\mu \frac{1}{1 + \lambda\rho_\nu} \frac{1}{1 + \lambda\rho_\mu} \right. \\
&\qquad \left. \times \left\{ \frac{1}{n} \sum_{i=1}^{n} \int_{\mathcal{T}} \phi_{\nu,i} \phi_{\mu,i} e^{\eta_{0,i}} Y_i dt - \tau(\phi_\nu \phi_\mu) \right\}^2 \right\}^{1/2} \\
&\qquad \times \left\{ \sum_\nu \sum_\mu (1 + \lambda\rho_\nu)(1 + \lambda\rho_\mu) f_\nu^2 f_\mu^2 \right\}^{1/2} \\
&\quad = O_p(n^{-1/2} \lambda^{-1/r})(V + \lambda J)(f),
\end{aligned}$$

where the Cauchy-Schwartz inequality, Lemma 8.1, and the fact that

$$E\left[\left\{ \frac{1}{n} \sum_{i=1}^{n} \int_{\mathcal{T}} \phi_{\nu,i} \phi_{\mu,i} e^{\eta_{0,i}} Y_i - \tau(\phi_\nu \phi_\mu) \right\}^2 \right] = O(n^{-1}) \tag{8.33}$$

are used. To see (8.33), note that

$$\begin{aligned}
E&\Big[\Big\{\int_{\mathcal{T}}\phi_\nu\phi_\mu e^{\eta_0}Y\,dt-\int_{\mathcal{U}}m(u)\int_{\mathcal{T}}\phi_\nu\phi_\mu e^{\eta_0}\tilde{S}dt\Big\}^2\Big]\\
&=E\Big[\Big\{\int_{\mathcal{T}}\phi_\nu\phi_\mu e^{\eta_0}(Y-\tilde{S})dt\Big\}^2\Big]\\
&\quad+E\Big[\Big\{\int_{\mathcal{T}}\phi_\nu\phi_\mu e^{\eta_0}\tilde{S}dt-\int_{\mathcal{U}}m(u)\int_{\mathcal{T}}\phi_\nu\phi_\mu e^{\eta_0}\tilde{S}dt\Big\}^2\Big]\\
&\leq E\Big[\Big(\int_{\mathcal{T}}|\phi_\nu\phi_\mu|e^{\eta_0}\tilde{S}^{1/2}dt\Big)\Big(\int_{\mathcal{T}}|\phi_\nu\phi_\mu|e^{\eta_0}\tilde{S}^{-1/2}E[(Y-\tilde{S})^2|U]dt\Big)\Big]\\
&\quad+E\Big[\Big\{\int_{\mathcal{T}}\phi_\nu\phi_\mu e^{\eta_0}\tilde{S}dt\Big\}^2\Big]\\
&\leq E\Big[\Big\{\int_{\mathcal{T}}|\phi_\nu\phi_\mu|e^{\eta_0}\tilde{S}^{1/2}dt\Big\}^2\Big]+\int_{\mathcal{U}}m(u)\int_{\mathcal{T}}\phi_\nu^2\phi_\mu^2e^{2\eta_0}\tilde{S}^2dt\\
&\leq 2c_3.
\end{aligned}$$

This completes the proof. □

Theorem 8.12
Assume $\sum_\nu \rho_\nu^p\eta_{\nu,0}^2<\infty$ for some $p\in[1,2]$. Under Conditions 8.3.1—8.3.4, as $\lambda\to 0$ and $n\lambda^{2/r}\to\infty$,

$$(V+\lambda J)(\hat{\eta}-\tilde{\eta})=O_p(n^{-1}\lambda^{-1/r}+\lambda^p).$$

Consequently,

$$\begin{aligned}
(V+\lambda J)(\hat{\eta}-\eta_0)&=O_p(n^{-1}\lambda^{-1/r}+\lambda^p),\\
\mathrm{SKL}(\eta_0,\hat{\eta})&=O_p(n^{-1}\lambda^{-1/r}+\lambda^p).
\end{aligned}$$

Proof: By the mean value theorem, Condition 8.3.3, and Lemma 8.11, (8.32) leads to

$$\begin{aligned}
(c_1V&+\lambda J)(\hat{\eta}-\tilde{\eta})(1+o_p(1))\\
&\leq\{(|1-c|V+\lambda J)(\hat{\eta}-\tilde{\eta})\}^{1/2}O_p(\{(|1-c|V+\lambda J)(\tilde{\eta}-\eta_0)\}^{1/2})
\end{aligned}$$

for some $c\in[c_1,c_2]$. The theorem follows Theorem 8.10. □

8.3.4 Semiparametric Approximation

As was noted in §7.1, the minimizer $\hat{\eta}$ of (7.1) in $\mathcal{H}$ is, in general, not computable. The minimizer $\hat{\eta}^*$ in the space

$$\mathcal{H}^*=\mathcal{N}_J\oplus\mathrm{span}\{R_J((X_i,U_i),\cdot),\delta_i=1\}$$

was computed instead. We now establish the convergence rates for $\hat{\eta}^*$.

For $h \in \mathcal{H} \ominus \mathcal{H}^*$, one has $\delta_i h(X_i, U_i) = \delta_i J(R_J((X_i, U_i), \cdot), h) = 0$, so $\sum_{i=1}^n \int_{\mathcal{T}} h_i^2 dN_i(t) = \sum_{i=1}^n \delta_i h^2(X_i, U_i) = 0$, where $h_i(t) = h(t, U_i)$.

Lemma 8.13
Under Conditions 8.3.1, 8.3.2, and 8.3.4, as $\lambda \to 0$ and $n\lambda^{2/r} \to \infty$, $V(h) = o_p(\lambda J(h))$, $\forall h \in \mathcal{H} \ominus \mathcal{H}^*$.

Proof: Define $\tau(f) = \int_{\mathcal{U}} m(u) \int_{\mathcal{T}} f e^{\eta_0} \tilde{S} dt$. From (8.22)—(8.24), Condition 8.3.4, and the proof of (8.33), one has

$$
\begin{aligned}
E&\Big[\Big\{\int_{\mathcal{T}} \phi_\nu \phi_\mu dN(t) - \tau(\phi_\nu\phi_\mu)\Big\}^2\Big] \\
&= E\Big[\Big\{\int_{\mathcal{T}} \phi_\nu \phi_\mu dM(t)\Big\}^2\Big] + E\Big[\Big\{\int_{\mathcal{T}} \phi_\nu \phi_\mu e^{\eta_0} Y dt - \tau(\phi_\nu\phi_\mu)\Big\}^2\Big] \\
&\leq \tau(\phi_\nu^2\phi_\mu^2) + 2c_3 \leq 3c_3.
\end{aligned}
$$

By the same arguments used in the proof of Lemma 8.11,

$$
V(h) = \left|\frac{1}{n}\sum_{i=1}^n \int_{\mathcal{T}} h_i^2 dN_i(t) - V(h)\right| = O_p(n^{-1/2}\lambda^{-1/r})(V + \lambda J)(h).
$$

The lemma follows. □

Theorem 8.14
Let η^* be the projection of $\hat{\eta}$ in $\mathcal{H}^*$. Assume $\sum_\nu \rho_\nu^p \eta_{\nu,0}^2 < \infty$ for some $p \in [1, 2]$. Under Conditions 8.3.1—8.3.4, as $\lambda \to 0$ and $n\lambda^{2/r} \to \infty$,

$$
\begin{aligned}
\lambda J(\hat{\eta} - \eta^*) &= O_p(n^{-1}\lambda^{-1/r} + \lambda^p), \\
V(\hat{\eta} - \eta^*) &= o_p(n^{-1}\lambda^{-1/r} + \lambda^p).
\end{aligned}
$$

Proof: Setting $f = \hat{\eta}$ and $g = \hat{\eta} - \eta^*$ in (8.28), some algebra yields

$$
\begin{aligned}
\lambda J(\hat{\eta} - \eta^*) = &\frac{1}{n}\sum_{i=1}^n \int_{\mathcal{T}} (\hat{\eta} - \eta^*)_i dM_i(t) \\
&- \frac{1}{n}\sum_{i=1}^n \int_{\mathcal{T}} (\hat{\eta} - \eta^*)_i (e^{\hat{\eta}} - e^{\eta_0})_i Y_i dt; \qquad (8.34)
\end{aligned}
$$

remember that $J(\eta^*, \hat{\eta} - \eta^*) = 0$. Now, with $\beta_\nu = n^{-1}\sum_{i=1}^n \int_{\mathcal{T}} \phi_{\nu,i} dM_i(t)$,

$$
\begin{aligned}
\left|\frac{1}{n}\sum_{i=1}^n \int_{\mathcal{T}} (\hat{\eta} - \eta^*)_i dM_i(t)\right| &= \left|\sum_\nu (\hat{\eta}_\nu - \eta_\nu^*)\beta_\nu\right| \\
&= \Big\{\sum_\nu (1 + \lambda\rho_\nu)(\hat{\eta}_\nu - \eta_\nu^*)^2\Big\}^{1/2} \Big\{\sum_\nu (1 + \lambda\rho_\nu)^{-1}\beta_\nu^2\Big\}^{1/2} \\
&= \{(V + \lambda J)(\hat{\eta} - \eta^*)\}^{1/2} O_p(n^{-1/2}\lambda^{-1/2r}). \qquad (8.35)
\end{aligned}
$$

By the mean value theorem, Condition 8.3.3, and Lemmas 8.11 and 8.13,

$$\left|\frac{1}{n}\sum_{i=1}^{n}\int_{\mathcal{T}}(\hat{\eta}-\eta^*)_i(e^{\hat{\eta}}-e^{\eta_0})_iY_i dt\right| \\ = o_p(\{\lambda J(\hat{\eta}-\eta^*)(V+\lambda J)(\hat{\eta}-\eta_0)\}^{1/2}); \quad (8.36)$$

see Problem 8.7. Plugging (8.35) and (8.36) into (8.34) and applying Theorem 8.12 and Lemma 8.13, one has

$$\lambda J(\hat{\eta}-\eta^*) = \{\lambda J(\hat{\eta}-\eta^*)\}^{1/2}\{O_p(n^{-1/2}\lambda^{-1/2r}) + o_p(\lambda^{p/2})\}.$$

The theorem follows. □

We shall now calculate $(V+\lambda J)(\hat{\eta}^*-\eta^*)$. Setting $f=\hat{\eta}^*$ and $g=\hat{\eta}^*-\eta^*$ in (8.28), one has

$$-\frac{1}{n}\sum_{i=1}^{n}\left\{\int_{\mathcal{T}}(\hat{\eta}^*-\eta^*)_i dN_i(t) - \int_{\mathcal{T}}(\hat{\eta}^*-\eta^*)_iY_ie^{\hat{\eta}_i^*}dt\right\} \\ + \lambda J(\hat{\eta}^*,\hat{\eta}^*-\eta^*) = 0. \quad (8.37)$$

Setting $f=\hat{\eta}$ and $g=\hat{\eta}-\hat{\eta}^*$ in (8.28), one gets

$$-\frac{1}{n}\sum_{i=1}^{n}\left\{\int_{\mathcal{T}}(\hat{\eta}-\hat{\eta}^*)_i dN_i(t) - \int_{\mathcal{T}}(\hat{\eta}-\hat{\eta}^*)_iY_ie^{\hat{\eta}_i}dt\right\} + \lambda J(\hat{\eta},\hat{\eta}-\hat{\eta}^*) = 0. \quad (8.38)$$

Adding (8.37) and (8.38) and noting that $J(\hat{\eta}-\eta^*,\eta^*) = J(\hat{\eta}-\eta^*,\hat{\eta}^*) = 0$ so $J(\hat{\eta},\hat{\eta}-\hat{\eta}^*) = J(\hat{\eta}-\eta^*) + J(\eta^*,\eta^*-\hat{\eta}^*)$, some algebra yields

$$\frac{1}{n}\sum_{i=1}^{n}\int_{\mathcal{T}}(\hat{\eta}^*-\eta^*)_i(e^{\hat{\eta}^*}-e^{\eta^*})_iY_i dt + \lambda J(\hat{\eta}^*-\eta^*) + \lambda J(\hat{\eta}-\eta^*) \\ = \frac{1}{n}\sum_{i=1}^{n}\int_{\mathcal{T}}(\hat{\eta}-\eta^*)_i dM_i(t) + \frac{1}{n}\sum_{i=1}^{n}\int_{\mathcal{T}}(\hat{\eta}-\eta^*)_i(e^{\eta_0}-e^{\hat{\eta}})_iY_i dt \\ + \frac{1}{n}\sum_{i=1}^{n}\int_{\mathcal{T}}(\hat{\eta}^*-\eta^*)_i(e^{\hat{\eta}}-e^{\eta^*})_iY_i dt; \quad (8.39)$$

see Problem 8.8. Condition 8.3.3 has to be modified to include η^* and $\hat{\eta}^*$ in the convex set B_0.

Theorem 8.15
Assume $\sum_\nu \rho_\nu^p \eta_{\nu,0}^2 < \infty$ for some $p\in[1,2]$. Under Conditions 8.3.1—8.3.4, as $\lambda\to 0$ and $n\lambda^{2/r}\to\infty$,

$$(V+\lambda J)(\hat{\eta}^*-\eta^*) = O_p(n^{-1}\lambda^{-1/r}+\lambda^p).$$

Consequently,

$$(V + \lambda J)(\hat{\eta}^* - \eta_0) = O_p(n^{-1}\lambda^{-1/r} + \lambda^p),$$
$$\mathrm{SKL}(\eta_0, \hat{\eta}^*) = O_p(n^{-1}\lambda^{-1/r} + \lambda^p).$$

Proof: By the mean value theorem, Condition 8.3.3, and Lemma 8.11, the left-hand side of (8.39) is bounded from below by

$$(c_1 V + \lambda J)(\hat{\eta}^* - \eta^*)(1 + o_p(1)) + \lambda J(\hat{\eta} - \eta^*).$$

For the right-hand side, the first term is of the order $O_p(n^{-1}\lambda^{-1/r} + \lambda^p)$ by (8.35) and Theorem 8.14, the second term is of the order $o_p(n^{-1}\lambda^{-1/r} + \lambda^p)$ by (8.36) and Theorems 8.12 and 8.14, and the third term is of the order

$$\{(V + \lambda J)(\hat{\eta}^* - \eta^*)\}^{1/2} o_p(\{\lambda J(\hat{\eta} - \eta^*)\}^{1/2})$$

by Condition 8.3.3 and Lemmas 8.11 and 8.13. Putting things together, one obtains the rates stated in the theorem. The latter part follows this, Theorems 8.12 and 8.14, and Condition 8.3.3. □

8.3.5 Convergence Under Incorrect Model

For $\eta_0 \notin \mathcal{H}$, one defines the relative Kullback-Leibler distance as

$$\mathrm{RKL}(\eta|\eta_0) = \int_{\mathcal{U}} m(u) \int_{\mathcal{T}} \{e^{\eta(t,u)} - \eta(t,u)e^{\eta_0(t,u)}\}\tilde{S}(t)dt.$$

The minimizer η_0^* of $\mathrm{RKL}(\eta|\eta_0)$ in $\mathcal{H}$, when it exists, satisfies

$$\int_{\mathcal{U}} m(u) \int_{\mathcal{T}} f(t,u)\{e^{\eta_0^*(t,u)} - e^{\eta_0(t,u)}\}\tilde{S}(t)dt, \quad \forall f \in \mathcal{H}.$$

Substituting η_0^* for η_0 in §§8.3.1—8.3.4, the analysis remain valid under a couple of extra conditions.

Condition 8.3.0 $fg \in \mathcal{H}$, $\forall f, g \in \mathcal{H}$.

Condition 8.3.5 $\int_{\mathcal{U}} m(u) \int_{\mathcal{T}} \phi_\nu^2 (e^{\eta_0} - e^{\eta_0^*})^2 \tilde{S} dt < c_4$ holds uniformly for some $c_4 < \infty$, $\forall \nu$.

Further details are left as an exercise (Problem 8.9).

8.4 Rates for Regression Estimates

We now establish convergence rates for regression estimates, which include those discussed in Chapters 3 and 5 and §7.5.

A formulation more general than (5.1) is presented, and a quadratic functional V is defined in the general setting. Rates are established in terms of $V(\hat{\eta} - \eta_0)$. The first step is, again, the analysis of a linear approximation $\tilde{\eta}$, but semiparametric approximation is no longer needed since $\hat{\eta}$ is computable.

8.4.1 General Formulation

Denote by $l(\eta; y)$ a minus log likelihood of η with observation y. We shall consider the penalized likelihood functional

$$\frac{1}{n}\sum_{i=1}^{n} l(\eta(x_i); Y_i) + \frac{\lambda}{2}J(\eta). \tag{8.40}$$

When η is the canonical parameter of an exponential family distribution, (8.40) reduces to (5.1) on page 150. The general formulation of (8.40) covers the noncanonical links used in the gamma family, the inverse Gaussian family, and the negative binomial family of §5.4. It also covers the log likelihoods of §7.5, where $\eta(x)$ was written as $\mu(u)$ and y consisted of several components. The dispersion parameter of an exponential family distribution can be absorbed into λ, known or unknown, but the α parameter in the negative binomial family of §5.4 or in the accelerated life models of §7.5 is assumed to be known.

Write $u(\eta; y) = dl/d\eta$ and $w(\eta; y) = d^2l/d\eta^2$; it is assumed that

$$E[u(\eta_0(x); Y)] = 0, \quad E[u^2(\eta_0(x); Y)] = \sigma^2 E[w(\eta_0(x); Y)], \tag{8.41}$$

which hold for all the log likelihoods appearing in §5.4 and §7.5, where σ^2 is a constant. Let $f(x)$ be the limiting density of x_i. Write $v_\eta(x) = E[w(\eta(x); Y)]$ and define

$$V(g) = \int_{\mathcal{X}} g^2(x) v_{\eta_0}(x) f(x) dx. \tag{8.42}$$

The specific forms of V for the families of §5.4 and §7.5 are easy to work out; see Problem 8.10. Convergence rates for the minimizer $\hat{\eta}$ of (8.40) shall be established in terms of $V(\hat{\eta} - \eta_0)$.

8.4.2 Linear Approximation

The following conditions are needed in our analysis, which are carbon copies of Conditions 8.2.1 and 8.2.2 but with V as defined in (8.42) in the regression setting.

Condition 8.4.1 V is completely continuous with respect to J.

Condition 8.4.2 For ν sufficiently large and some $\beta > 0$, the eigenvalues ρ_ν of J with respect to V satisfy $\rho_\nu > \beta\nu^r$, where $r > 1$.

Consider the quadratic functional

$$\frac{1}{n}\sum_{i=1}^{n} u(\eta_0(x_i); Y_i)\eta(x_i) + \frac{1}{2}V(\eta - \eta_0) + \frac{\lambda}{2}J(\eta). \tag{8.43}$$

Plugging the Fourier series expansions $\eta = \sum_\nu \eta_\nu \phi_\nu$ and $\eta_0 = \sum_\nu \eta_{\nu,0}\phi_\nu$ into (8.43), it is easy to show that the minimizer $\tilde{\eta}$ of (8.43) has Fourier coefficients

$$\tilde{\eta}_\nu = (\beta_\nu + \eta_{\nu,0})/(1 + \lambda\rho_\nu),$$

which are linear in $\beta_\nu = -n^{-1}\sum_{i=1}^n u(\eta_0(x_i); Y_i)\phi_\nu(x_i)$; see Problem 8.11. Note that $E[\beta_\nu] = 0$ and $E[\beta_\nu^2] = \sigma^2/n$. The following theorem can be easily proved parallel to Theorem 8.2.

Theorem 8.16
Assume $\sum_\nu \rho_\nu^p \eta_{\nu,0}^2 < \infty$ *for some* $p \in [1,2]$. *Under Conditions 8.4.1 and 8.4.2, as* $n \to \infty$ *and* $\lambda \to 0$,

$$(V + \lambda J)(\tilde{\eta} - \eta_0) = O_p(n^{-1}\lambda^{-1/r} + \lambda^p).$$

8.4.3 Approximation Error and Main Result

We now turn to the approximation error $\hat{\eta} - \tilde{\eta}$. Define

$$\begin{aligned} A_{g,h}(\alpha) &= \frac{1}{n}\sum_{i=1}^n l((g+\alpha h)(x_i); Y_i) + \frac{\lambda}{2}J(g+\alpha h), \\ B_{g,h}(\alpha) &= \frac{1}{n}\sum_{i=1}^n u(\eta_0(x_i); Y_i)(g+\alpha h)(x_i) \\ &\quad + \frac{1}{2}V(g+\alpha h - \eta_0) + \frac{\lambda}{2}J(g+\alpha h). \end{aligned}$$

It can be easily shown that

$$\dot{A}_{g,h}(0) = \frac{1}{n}\sum_{i=1}^n u(g(x_i); Y_i)h(x_i) + \lambda J(g,h), \tag{8.44}$$

$$\dot{B}_{g,h}(0) = \frac{1}{n}\sum_{i=1}^n u(\eta_0(x_i); Y_i)h(x_i) + V(g - \eta_0, h) + \lambda J(g,h). \tag{8.45}$$

Setting $g = \hat{\eta}$ and $h = \hat{\eta} - \tilde{\eta}$ in (8.44), one has

$$\frac{1}{n}\sum_{i=1}^n u(\hat{\eta}(x_i); Y_i)(\hat{\eta} - \tilde{\eta})(x_i) + \lambda J(\hat{\eta}, \hat{\eta} - \tilde{\eta}) = 0, \tag{8.46}$$

and setting $g = \tilde{\eta}$ and $h = \hat{\eta} - \tilde{\eta}$ in (8.45), one gets

$$\frac{1}{n}\sum_{i=1}^n u(\eta_0(x_i); Y_i)(\hat{\eta} - \tilde{\eta})(x_i) + V(\tilde{\eta} - \eta_0, \hat{\eta} - \tilde{\eta}) + \lambda J(\tilde{\eta}, \hat{\eta} - \tilde{\eta}) = 0. \tag{8.47}$$

Subtracting (8.47) from (8.46), some algebra yields

$$\frac{1}{n}\sum_{i=1}^{n}\{u(\hat\eta(x_i);Y_i)-u(\tilde\eta(x_i);Y_i)\}(\hat\eta-\tilde\eta)(x_i)+\lambda J(\hat\eta-\tilde\eta)$$
$$=V(\tilde\eta-\eta_0,\hat\eta-\tilde\eta)-\frac{1}{n}\sum_{i=1}^{n}\{u(\tilde\eta(x_i);Y_i)-u(\eta_0(x_i);Y_i)\}(\hat\eta-\tilde\eta)(x_i). \tag{8.48}$$

By the mean value theorem,

$$\begin{aligned} u(\hat\eta(x_i);Y_i)-u(\tilde\eta(x_i);Y_i)&=w(\eta_1(x_i);Y_i)(\hat\eta-\tilde\eta)(x_i),\\ u(\tilde\eta(x_i);Y_i)-u(\eta_0(x_i);Y_i)&=w(\eta_2(x_i);Y_i)(\tilde\eta-\eta_0)(x_i), \end{aligned} \tag{8.49}$$

where η_1 is a convex combination of $\hat\eta$ and $\tilde\eta$, and η_2 is that of $\tilde\eta$ and η_0.

To proceed, one needs the following conditions in addition to Conditions 8.4.1 and 8.4.2.

Condition 8.4.3 For η in a convex set B_0 around η_0 containing $\hat\eta$ and $\tilde\eta$, $c_1 w(\eta_0(x);Y)\le w(\eta(x);Y)\le c_2 w(\eta_0(x);Y)$ holds uniformly for some $0<c_1<c_2<\infty$, $\forall x\in\mathcal{X}$, $\forall Y$.

Condition 8.4.4 $\mathrm{Var}[\phi_\nu(X)\phi_\mu(X)w(\eta_0(X),Y)]\le c_3$ for some $c_3<\infty$, $\forall\nu,\mu$.

To understand the practical meanings of these conditions, one needs to work out their specific forms for the families of §5.4 and §7.5 (Problem 8.12). Roughly speaking, Condition 8.4.3 concerns the equivalence of the information in B_0 and Condition 8.4.4 asks for a uniform bound for the fourth moments of $\phi_\nu(X)$.

Lemma 8.17
Under Conditions 8.4.1, 8.4.2, and 8.4.4, as $\lambda\to 0$ and $n\lambda^{2/r}\to\infty$,

$$\frac{1}{n}\sum_{i=1}^{n}g(x_i)h(x_i)w(\eta_0(x_i);Y_i)$$
$$=V(g,h)+o_p(\{(V+\lambda J)(g)(V+\lambda J)(h)\}^{1/2}).$$

Proof: Write $\tau(g)=\int_{\mathcal{X}}g(x)v_{\eta_0}(x)f(x)$. Under Condition 8.4.4,

$$\frac{1}{n}\sum_{i=1}^{n}\phi_\nu(x_i)\phi_\mu(x_i)w(\eta_0(x_i);Y_i)-\tau(\phi_\nu\phi_\mu)\le\frac{c_3}{n}.$$

Write $g = \sum_\nu g_\nu \phi_\nu$ and $h = \sum_\nu h_\nu \phi_\nu$. Similar to the proofs of Lemmas 8.5 and 8.11, as $n\lambda^{2/r} \to \infty$,

$$\begin{aligned}
&\left|\frac{1}{n}\sum_{i=1}^n g(x_i)h(x_i)w(\eta_0(x_i);Y_i) - V(g,h)\right| \\
&\quad =\left|\sum_\nu\sum_\mu g_\nu h_\mu \left\{\frac{1}{n}\sum_{i=1}^n \phi_\nu(x_i)\phi_\mu(x_i)w(\eta_0(x_i);Y_i) - \tau(\phi_\nu\phi_\mu)\right\}\right| \\
&\quad \le\left\{\sum_\nu\sum_\mu \frac{1}{1+\lambda\rho_\nu}\frac{1}{1+\lambda\rho_\mu}\right. \\
&\qquad\quad \left.\times\left\{\frac{1}{n}\sum_{i=1}^n \phi_\nu(x_i)\phi_\mu(x_i)w(\eta_0(x_i);Y_i) - \tau(\phi_\nu\phi_\mu)\right\}^2\right\}^{1/2} \\
&\qquad \times\left\{\sum_\nu\sum_\mu (1+\lambda\rho_\nu)(1+\lambda\rho_\mu)g_\nu^2 h_\mu^2\right\}^{1/2} \\
&\quad =\{(V+\lambda J)(g)(V+\lambda J)(h)\}^{1/2} O_p(n^{-1/2}\lambda^{-1/r}) \\
&\quad =o_p(\{(V+\lambda J)(g)(V+\lambda J)(h)\}^{1/2})
\end{aligned}$$

where the Cauchy-Schwartz inequality and Lemma 8.1 are used. □

Theorem 8.18
Assume $\sum_\nu \rho_\nu^p \eta_{\nu,0}^2 < \infty$ for some $p \in [1,2]$. Under Conditions 8.4.1—8.4.4, as $\lambda \to 0$ and $n\lambda^{2/r} \to \infty$,

$$(V+\lambda J)(\hat\eta - \tilde\eta) = O_p(n^{-1}\lambda^{-1/r} + \lambda^p).$$

Consequently,

$$(V+\lambda J)(\hat\eta - \eta_0) = O_p(n^{-1}\lambda^{-1/r} + \lambda^p).$$

Proof: Plug (8.49) into (8.48). Applying Condition 8.4.3 and Lemma 8.17, one has

$$\begin{aligned}
&(c_1 V + \lambda J)(\hat\eta - \tilde\eta)(1 + o_p(1)) \\
&\qquad \le \{(|1-c|V+\lambda J)(\hat\eta - \tilde\eta)\}^{1/2} O_p(\{(|1-c|V+\lambda J)(\tilde\eta - \eta_0)\}^{1/2}),
\end{aligned}$$

for some $c \in [c_1, c_2]$. The theorem follows Theorem 8.16. □

8.4.4 Convergence Under Incorrect Model

For $\eta_0 \notin \mathcal{H}$, one may define the relative Kullback-Leibler distance as

$$\mathrm{RKL}(\eta|\eta_0) = \int_{\mathcal{X}} E[l(\eta(x);Y)]f(x).$$

The minimizer η_0^* of $\text{RKL}(\eta|\eta_0)$ in $\mathcal{H}$, when it exists, satisfies

$$\int_{\mathcal{X}} g(x)E[u(\eta_0^*(x);Y)]f(x) = 0, \quad \forall g \in \mathcal{H}.$$

Substituting η_0^* for η_0 in (8.43) but not in the definition of V, Theorem 8.16 may be proved under some extra condition that assures uniformly bounded $E[\beta_\nu^2]$. For Theorem 8.18 to hold, further conditions are also needed to ensure the uniform boundedness of

$$E[\{\phi_\nu(X)\phi_\mu(X)w(\eta_0^*(X);Y) - \tau(\phi_\nu\phi_\mu)\}^2];$$

details are tedious. It would be easier to work with specific families than with the general setting; see Problem 8.13.

8.5 Bibliographic Notes

Section 8.1

An general theory of eigenvalue analysis can be found in Weinberger (1974). Results on eigenvalues related to smoothing splines can be found in, e.g., Cox (1984, 1988) and Utreras (1981, 1983, 1988), among others. Example 8.2 is taken from Gu (1996).

Section 8.2

An asymptotic theory was developed by Silverman (1982) for the minimizer of (6.12), which laid the groundwork for later analysis. Cox and O'Sullivan (1990) developed a general asymptotic theory for penalized likelihood estimates, of which the estimate of Silverman (1982) was listed as an example. The materials of §§8.2.1—8.2.3 represent a refinement of the analysis found in Gu and Qiu (1993, §5, §6), where the semiparametric approximation was first proposed and studied. The analysis of §8.2.4 was noted by Gu (1998a). The adaptations of §§8.2.5—8.2.7 are found in Gu (1992d, 1995a, 1995b).

Section 8.3

The materials of this section are refined versions of the analysis found in Gu (1996). For $\mathcal{U}$ a singleton, the analyses of Antoniadis (1989) and Cox and O'Sullivan (1990) apply, although not to the semiparametric approximation.

Section 8.4

The analysis in the general setting as presented is adapted from that of Gu and Qiu (1994), where η was taken as the canonical parameter of an exponential family distribution, as in §5.1. The analysis of Cox and O'Sullivan (1990) also applies in the setting of §5.1.

Convergence rates for penalized least squares estimates have been studied extensively in the literature. For results on multidimensional domains, see Cox (1984), Utreras (1988), Chen (1991), and Lin (2000).

8.6 Problems

Section 8.2

8.1 Assume $\sum_\nu \nu^{pr}\eta_{\nu,0}^2 < \infty$ for some $p > 1$. Show that the rates in Theorem 8.2 can be improved to $O(n^{-1}\lambda^{-1/r} + \lambda^p)$, with p up to 2.

8.2 Verify (8.4) and (8.5).

8.3 In the setting of §8.2.5, state and prove the counterparts of all the lemmas and theorems appearing in §§8.2.1–8.2.3.

8.4 In the setting of §8.2.6, state and prove the counterparts of all the lemmas and theorems appearing in §§8.2.1–8.2.3.

8.5 In the setting of §8.2.7, state and prove the counterparts of all the lemmas and theorems appearing in §§8.2.1–8.2.3.

Section 8.3

8.6 Show that the minimizer $\tilde{\eta}$ of (8.27) has Fourier coefficients

$$\tilde{\eta}_\nu = (\beta_\nu + \eta_{\nu,0})/(1 + \lambda\rho_\nu),$$

where

$$\beta_\nu = \frac{1}{n}\sum_{i=1}^n \int_{\mathcal{T}} \phi_{\nu,i} dM_i(t),$$

with $\phi_{\nu,i}(t) = \phi_\nu(t, U_i)$, satisfy $E[\beta_\nu] = 0$ and $E[\beta_\nu^2] = n^{-1}$.

8.7 Prove (8.36).

8.8 Prove (8.39).

8.9 When $\eta_0 \notin \mathcal{H}$, substituting η_0^* of §8.3.5 for η_0 in §§8.3.1–8.3.4, show that the convergence rates remain valid under Conditions 8.3.0—8.3.5.

Section 8.4

8.10 Specify the definition $V(g) = \int_{\mathcal{X}} g^2(x) v_{\eta_0}(x) f(x) dx$ for the families of §5.4 and §7.5.

8.11 Show that the minimizer $\tilde{\eta}$ of (8.43) has Fourier coefficients

$$\tilde{\eta}_\nu = (\beta_\nu + \eta_{\nu,0})/(1 + \lambda \rho_\nu),$$

where

$$\beta_\nu = -\frac{1}{n} \sum_{i=1}^{n} u(\eta_0(x_i); Y_i) \phi_\nu(x_i)$$

satisfy $E[\beta_\nu] = 0$ and $E[\beta_\nu^2] = \sigma^2/n$.

8.12 Specify Conditions 8.4.3 and 8.4.4 for the families of §5.4 and §7.5.

8.13 For the families of §5.4 and §7.5, specify the extra conditions needed to extend Theorems 8.16 and 8.18 to the case $\eta_0 \notin \mathcal{H}$.

References

Abramowitz, M. and I. A. Stegun (1964). *Handbook of Mathematical Functions with Formulas, Graphs, and Mathematical Tables.* Washington, DC: National Bureau of Standards.

Akhiezer, N. I. and I. M. Glazman (1961). *Theory of Linear Operators in Hilbert Space.* New York: Ulgar.

Anderson, J. A. (1972). Separate sampling logistic regression. *Biometrika 59*, 19–35.

Anderson, J. A. and V. Blair (1982). Penalized maximum likelihood estimation in logistic regression and discrimination. *Biometrika 69*, 123–136.

Anderson, J. A. and A. Senthilselvan (1980). Smooth estimates for the hazard function. *J. Roy. Statist. Soc. Ser. B 42*, 322–327.

Ansley, C. F., R. Kohn, and C.-M. Wong (1993). Nonparametric spline regression with prior information. *Biometrika 80*, 75–88.

Antoniadis, A. (1989). A penalty method for nonparametric estimation of the intensity function of a counting process. *Ann. Inst. Statist. Math. 41*, 781–807.

Aronszajn, N. (1950). Theory of reproducing kernels. *Trans. Amer. Math. Soc. 68*, 337–404.

Azzalini, A. and A. W. Bowman (1990). A look at some data on the Old Faithful geyser. *Appl. Statist. 39*, 357–365.

Barry, D. (1986). Nonparametric Bayesian regression. *Ann. Statist. 14*, 934–953.

Bartoszyński, R., B. W. Brown, C. M. McBride, and J. R. Thompson (1981). Some nonparametric techniques for estimating the intensity function of a cancer related nonstationary Poisson process. *Ann. Statist. 9*, 1050–1060.

Bates, D. M., M. Lindstrom, G. Wahba, and B. Yandell (1987). GCVPACK – routines for generalized cross validation. *Commun. Statist.-Simulat. Comput. 16*, 263–297.

Becker, R. A., J. M. Chambers, and A. R. Wilks (1988). *The New S Language.* Pacific Grove, CA: Wadsworth & Brooks/Cole.

Berger, J. O. (1985). *Statistical Decision Theory and Bayesian Analysis* (2nd ed.). New York: Springer-Verlag.

Breiman, L. (1991). The $\prod$ method for estimating multivariate functions from noisy data. *Technometrics 33*, 125–160 (with discussions).

Breiman, L. and J. H. Friedman (1985). Estimating optimal transformations for multiple regression and correlation. *J. Amer. Statist. Assoc. 80*, 580–598 (with discussions).

Brinkman, N. D. (1981). Ethanol fuel – a single-cylinder engine study of efficiency and exhaust emissions. *SAE Trans. 90*, 1410–1424.

Brockwell, P. J. and R. A. Davis (1991). *Time Series: Theory and Methods* (2nd ed.). New York: Springer-Verlag.

Buja, A., T. Hastie, and R. Tibshirani (1989). Linear smoothers and additive models. *Ann. Statist. 17*, 453–555 (with discussions).

Chambers, J. M. and T. J. Hastie (Eds.) (1992). *Statistical Models in S.* New York: Chapman & Hall.

Chen, Z. (1991). Interaction spline models and their convergence rates. *Ann. Statist. 19*, 1855–1868.

Chhikara, R. S. and J. L. Folks (1989). *The Inverse Gaussian Distribution: Theory, Methodology, and Applications.* New York: Marcel Dekker.

Cleveland, W. S. (1979). Robust locally weighted regression and smoothing scatterplots. *J. Amer. Statist. Assoc. 83*, 829–836.

Cleveland, W. S. and S. J. Devlin (1988). Locally weighted regression: An approach to regression analysis by local fitting. *J. Amer. Statist. Assoc. 83*, 596–610.

Cogburn, R. and H. T. Davis (1974). Periodic splines and spectral estimation. *Ann. Statist. 2*, 1108–1126.

Cole, T. J. (1988). Fitting smoothed centile curves to reference data. *J. Roy. Statist. Soc. Ser. A 151*, 385–418 (with discussions).

Cole, T. J. and P. J. Green (1992). Smoothing reference centile curves: The LMS method and penalized likelihood. *Statist. Med. 11*, 1305–1319.

Cosslett, S. R. (1981). Maximum likelihood estimator for choice-based samples. *Econometrika 49*, 1289–1316.

Cox, D. D. (1984). Multivariate smoothing spline functions. *SIAM J. Numer. Anal. 21*, 789–813.

Cox, D. D. (1988). Approximation of method of regularization estimators. *Ann. Statist. 16*, 694–712.

Cox, D. D. and Y. Chang (1990). Iterated state space algorithms and cross validation for generalized smoothing splines. Technical Report 49, Department of Statistics, University of Illinois, Champion, IL.

Cox, D. D. and F. O'Sullivan (1990). Asymptotic analysis of penalized likelihood and related estimators. *Ann. Statist. 18*, 124–145.

Cox, D. R. (1969). Some sampling problems in technology. In N. L. Johnson and H. Smith Jr. (Eds.), *New Developments in Survey Sampling*. New York: Wiley, pp. 506–527.

Cox, D. R. (1972). Regression models and life tables. *J. Roy. Statist. Soc. Ser. B 34*, 187–220 (with discussions).

Craven, P. and G. Wahba (1979). Smoothing noisy data with spline functions: Estimating the correct degree of smoothing by the method of generalized cross-validation. *Numer. Math. 31*, 377–403.

Crowley, J. and M. Hu (1977). Covariance analysis of heart transplant survival data. *J. Amer. Statist. Assoc. 72*, 27–36.

Dalzell, C. J. and J. O. Ramsay (1993). Computing reproducing kernels with arbitrary boundary constraints. *SIAM J. Sci. Comput. 14*, 511–518.

de Boor, C. (1978). *A Practical Guide to Splines*. New York: Springer-Verlag.

de Boor, C. and R. Lynch (1966). On splines and their minimum properties. *J. Math. Mach. 15*, 953–969.

Dennis, J. E. and R. B. Schnabel (1996). *Numerical Methods for Unconstrained Optimization and Nonlinear Equations*. Philadelphia: SIAM; corrected reprint of the 1983 original.

Dongarra, J. J., C. B. Moler, J. R. Bunch, and G. W. Stewart (1979). *LINPACK User's Guide*. Philadelphia: SIAM.

Douglas, A. and M. Delampady (1990). Eastern Lake Survey – Phase I: Documentation for the data base and the derived data sets. Technical Report 160 (SIMS), Department of Statistics, University of British Columbia, Vancouver, BC.

Duchon, J. (1977). Splines minimizing rotation-invariant semi-norms in sobolev spaces. In W. Schemp and K. Zeller (Eds.), *Constructive Theory of Functions of Several Variables*. Berlin: Springer-Verlag, pp. 85–100.

Elden, L. (1984). A note on the computation of the generalized cross validation function for ill-conditioned least square problems. *BIT 24*, 467–472.

Fan, J. and I. Gijbels (1996). *Local Polynomial Modelling and its Applications.* London: Chapman & Hall.

Fleming, T. R. and D. P. Harrington (1991). *Counting Processes and Survival Analysis.* New York: Wiley.

Gao, F. (1999). *Penalized Multivariate Logistic Regression with a Large Data Set.* Ph.D. thesis, University of Wisconsin, Madison, WI.

Gao, F., G. Wahba, R. Klein, and B. E. K. Klein (2001). Smoothing spline ANOVA for multivariate Bernoulli observations, with application to ophthalmology data. *J. Amer. Statist. Assoc. 96*, 127–160 (with discussions).

Gill, P. E., W. Murray, and M. H. Wright (1981). *Practical Optimization.* New York: Academic Press.

Gill, R. D. (1984). Understanding Cox's regression model: A martingale approach. *J. Amer. Statist. Assoc. 79*, 441–447.

Gill, R. D., Y. Vardi, and J. A. Wellner (1988). Large sample theory of empirical distributions in biased sampling models. *Ann. Statist. 16*, 1069–1112.

Girard, D. A. (1989). A fast "Monte-Carlo cross validation" procedure for large least squares problems with noisy data. *Numer. Math. 56*, 1–23.

Girard, D. A. (1991). Asymptotic optimality of the fast randomized versions of GCV and C_L in ridge regression and regularization. *Ann. Statist. 19*, 1950–1963.

Golub, G. and C. Van Loan (1989). *Matrix Computations* (2nd ed.). Baltimore, MD: The Johns Hopkins University Press.

Good, I. J. and R. A. Gaskins (1971). Nonparametric roughness penalties for probability densities. *Biometrika 58*, 255–277.

Gray, R. J. (1992). Flexible methods for analyzing survival data using splines, with applications to breast cancer prognosis. *J. Amer. Statist. Assoc. 87*, 942–951.

Green, P. J. and B. W. Silverman (1994). *Nonparametric Regression and Generalized Linear Models.* London: Chapman & Hall.

Green, P. J. and B. Yandell (1985). Semi-parametric generalized linear models. In R. Gilchrist, B. Francis, and J. Whittaker (Eds.), *Proceedings of the GLIM85 Conference.* Berlin: Springer-Verlag, pp. 44–55.

Gu, C. (1989). RKPACK and its applications: Fitting smoothing spline models. In *ASA Proceedings of Statistical Computing Section*, pp. 42–51.

Gu, C. (1990). Adaptive spline smoothing in non Gaussian regression models. *J. Amer. Statist. Assoc. 85*, 801–807.

Gu, C. (1992a). Cross-validating non-Gaussian data. *J. Comput. Graph. Statist. 1*, 169–179.

Gu, C. (1992b). Diagnostics for nonparametric regression models with additive term. *J. Amer. Statist. Assoc. 87*, 1051–1058.

Gu, C. (1992c). Penalized likelihood regression: A Bayesian analysis. *Statist. Sin. 2*, 255–264.

Gu, C. (1992d). Smoothing spline density estimation: Biased sampling and random truncation. Technical Report 92-03, Department of Statistics, Purdue University, West Lafayette, IN.

Gu, C. (1993a). Interaction splines with regular data: Automatically smoothing digital images. *SIAM J. Sci. Comput. 14*, 218–230.

Gu, C. (1993b). Smoothing spline density estimation: A dimensionless automatic algorithm. *J. Amer. Statist. Assoc. 88*, 495–504.

Gu, C. (1994). Penalized likelihood hazard estimation: Algorithm and examples. In J. O. Berger and S. S. Gupta (Eds.), *Statistical Decision Theory and Related Topics, V.* New York: Springer-Verlag, pp. 61–72.

Gu, C. (1995a). Smoothing spline density estimation: Conditional distribution. *Statist. Sin. 5*, 709–726.

Gu, C. (1995b). Smoothing spline density estimation: Response-based sampling. Technical Report 267, Department of Statistics, University of Michigan, Ann Arbor, MI.

Gu, C. (1996). Penalized likelihood hazard estimation: A general procedure. *Statist. Sin. 6*, 861–876.

Gu, C. (1998a). Penalized likelihood estimation: Convergence under incorrect model. *Statist. Prob. Lett. 36*, 359–364.

Gu, C. (1998b). Structural multivariate function estimation: Some automatic density and hazard estimates. *Statist. Sin. 8*, 317–335.

Gu, C. and C. Qiu (1993). Smoothing spline density estimation: Theory. *Ann. Statist. 21*, 217–234.

Gu, C. and C. Qiu (1994). Penalized likelihood regression: A simple asymptotic analysis. *Statist. Sin. 4*, 297–304.

Gu, C. and G. Wahba (1991a). Discussion of "Multivariate adaptive regression splines" by J. Friedman. *Ann. Statist. 19*, 115–123.

Gu, C. and G. Wahba (1991b). Minimizing GCV/GML scores with multiple smoothing parameters via the Newton method. *SIAM J. Sci. Statist. Comput. 12*, 383–398.

Gu, C. and G. Wahba (1993a). Semiparametric analysis of variance with tensor product thin plate splines. *J. Roy. Statist. Soc. Ser. B 55*, 353–368.

Gu, C. and G. Wahba (1993b). Smoothing spline ANOVA with componentwise Bayesian "confidence intervals." *J. Comput. Graph. Statist. 2*, 97–117.

Gu, C. and J. Wang (2001). Penalized likelihood density estimation: Direct cross validation. Technical report, Department of Statistics, Purdue University, West Lafayette, IN.

Gu, C. and D. Xiang (2001). Cross-validating non-Gaussian data: Generalized approximate cross-validation revisited. *J. Comput. Graph. Statist. 10*, 581–591.

Gu, C., N. Heckman, and G. Wahba (1992). A note on generalized cross-validation with replicates. *Statist. Prob. Lett. 14*, 283–287.

Gu, C., D. M. Bates, Z. Chen, and G. Wahba (1989). The computation of GCV functions through Householder tridiagonalization with application to the fitting of interaction spline models. *SIAM J. Matrix Anal. Applic. 10*, 457–480.

Härdle, W. (1991). *Smoothing Techniques with Implementation in S*. New York: Springer-Verlag.

Harville, D. A. (1977). Maximum likelihood approaches to variance component estimation and to related problems. *J. Amer. Statist. Assoc. 72*, 320–340 (with discussions).

Hastie, T. and R. Tibshirani (1986). Generalized additive models. *Statist. Sci. 1*, 297–318 (with discussions).

Hastie, T. and R. Tibshirani (1990). *Generalized Additive Models*. London: Chapman & Hall.

Heckman, N. E. and J. O. Ramsay (2000). Penalized regression with model-based penalties. *Can. J. Statist. 28*, 241–258.

Hutchinson, M. F. (1989). A stochastic estimator of the trace of the influence matrix for Laplacian smoothing splines. *Commun. Statist.–Simulat. Comput. 18*, 1059–1076.

Hutchinson, M. F. and F. R. de Hoog (1985). Smoothing noisy data with spline functions. *Numer. Math. 47*, 99–106.

Ihaka, R. and R. Gentleman (1996). R: A language for data analysis and graphics. *J. Comput. Graph. Statist. 5*, 299–314.

Johnson, R. A. and D. W. Wichern (1992). *Applied Multivariate Statistical Analysis* (3rd ed.). Englewood Cliffs, NJ: Prentice Hall.

Jones, M. C. (1991). Kernel density estimation for length biased data. *Biometrika 78*, 511–520.

Kalbfleisch, J. D. and J. F. Lawless (1989). Inference based on retrospective ascertainment: An analysis of the data on transfusion-related AIDS. *J. Amer. Statist. Assoc. 84*, 360–372.

Kalbfleisch, J. D. and R. L. Prentice (1980). *The Statistical Analysis of Failure Time Data.* New York: Wiley.

Kaplan, E. L. and P. Meier (1958). Nonparametric estimator from incomplete observations. *J. Amer. Statist. Assoc. 53*, 457–481.

Karlin, S. and W. J. Studden (1966). *Tchebycheff Systems: With Applications in Analysis and Statistics.* New York: Interscience (Wiley).

Keiding, N. and R. D. Gill (1990). Random truncation models and markov processes. *Ann. Statist. 18*, 582–602.

Kernighan, B. W. (1975). Ratfor – A preprocessor for a rational Fortran. In *UNIX Programmer's Manual.* Murray Hill: Bell Laboratories.

Kimeldorf, G. and G. Wahba (1970a). A correspondence between Bayesian estimation of stochastic processes and smoothing by splines. *Ann. Math. Statist. 41*, 495–502.

Kimeldorf, G. and G. Wahba (1970b). Spline functions and stochastic processes. *Sankhya Ser. A 32*, 173–180.

Kimeldorf, G. and G. Wahba (1971). Some results on Tchebycheffian spline functions. *J. Math. Anal. Applic. 33*, 82–85.

Klein, R., B. E. K. Klein, S. E. Moss, M. D. Davis, and D. L. DeMets (1988). Glycosylated hemoglobin predicts the incidence and progression of diabetic retinopathy. *J. Amer. Med. Assoc. 260*, 2864–2871.

Klein, R., B. E. K. Klein, S. E. Moss, M. D. Davis, and D. L. DeMets (1989). The Wisconsin Epidemiologic Study of Diabetic Retinopathy. X. Four incidence and progression of diabetic retinopathy when age at diagnosis is 30 or more years. *Arch. Ophthalmol. 107*, 244–249.

Koenker, R., P. Ng, and S. Portnoy (1994). Quantile smoothing splines. *Biometrika 81*, 673–680.

Kooperberg, C., S. Bose, and C. J. Stone (1997). Polychotomous regression. *J. Amer. Statist. Assoc. 92*, 117–127.

Lauritzen, S. L. (1996). *Graphical Models.* New York: Oxford University Press.

Lehmann, E. L. and G. Casella (1998). *Theory of Point Estimation* (2rd ed.). New York: Springer-Verlag.

Leonard, T. (1978). Density estimation, stochastic processes and prior information. *J. Roy. Statist. Soc. Ser. B 73*, 113–146 (with discussions).

Leonard, T., J. S. J. Hsu, and K.-W. Tsui (1989). Bayesian marginal inference. *J. Amer. Statist. Assoc. 84*, 1051–1058.

Li, K.-C. (1986). Asymptotic optimality of C_L and generalized cross-validation in the ridge regression with application to spline smoothing. *Ann. Statist. 14*, 1101–1112.

Lin, X. (1998). *Smoothing Spline Analysis of Variance for Polychotomous Response Data.* Ph.D. thesis, University of Wisconsin, Madison, WI.

Lin, Y. (2000). Tensor product space ANOVA models. *Ann. Statist. 28*, 734–755.

Lindsey, J. K. (1997). *Applying Generalized Linear Models.* New York: Springer-Verlag.

Mallows, C. L. (1973). Some comments on C_P. *Technometrics 15*, 661–675.

Manski, C. F. (1995). *Identification Problems in the Social Sciences.* Cambridge, MA: Harvard University Press.

McCullagh, P. and J. A. Nelder (1989). *Generalized Linear Models* (2nd ed.). London: Chapman & Hall.

Meinguet, J. (1979). Multivariate interpolation at arbitrary points made simple. *J. Appl. Math. Phys. (ZAMP) 30*, 292–304.

Miller, R. G. (1976). Least squares regression with censored data. *Biometrika 63*, 449–464.

Miller, R. G. and J. Halpern (1982). Regression with censored data. *Biometrika 69*, 521–531.

Moreau, T., J. O'Quigley, and M. Mesbah (1985). A global goodness-of-fit statistic for the proportional hazards model. *Appl. Statist. 34*, 212–218.

Morgenthaler, S. and Y. Vardi (1986). Choice-based samples: A nonparametric approach. *J. Econometrics 32*, 109–125.

Nychka, D. (1988). Bayesian confidence intervals for smoothing splines. *J. Amer. Statist. Assoc. 83*, 1134–1143.

O'Sullivan, F. (1985). Discussion of "Some aspects of the spline smoothing approach to nonparametric regression curve fitting" by B. Silverman. *J. Roy. Statist. Soc. Ser. B 47*, 39–40.

O'Sullivan, F. (1988a). Fast computation of fully automated log-density and log-hazard estimators. *SIAM J. Sci. Statist. Comput. 9*, 363–379.

O'Sullivan, F. (1988b). Nonparametric estimation of relative risk using splines and cross-validation. *SIAM J. Sci. Statist. Comput. 9*, 531–542.

O'Sullivan, F., B. Yandell, and W. Raynor (1986). Automatic smoothing of regression functions in generalized linear models. *J. Amer. Statist. Assoc. 81*, 96–103.

Parzen, E. (1979). Nonparametric statistical data modeling. *J. Amer. Statist. Assoc. 74*, 105–131 (with discussions).

Pawitan, Y. and F. O'Sullivan (1994). Nonparametric spectral density estimation using penalized Whittle likelihood. *J. Amer. Statist. Assoc. 89*, 600–610.

Prentice, R. L. and R. Pyke (1978). Logistic disease incidence models and case-control studies. *Biometrika 66*, 403–411.

Priestley, M. B. (1981). *Spectral Analysis and Times Series.* London: Academic Press.

Ramsay, J. O. and C. J. Dalzell (1991). Some tools for functional data analysis. *J. Roy. Statist. Soc. Ser. B 53*, 539–572 (with discussions).

Rao, C. R. (1973). *Linear Statistical Inference and Its Applications.* New York: Wiley.

Robinson, G. K. (1991). That BLUP is a good thing: The estimation of the random effects. *Statist. Sci. 6*, 15–51 (with discussions).

Scheffe, H. (1959). *The Analysis of Variance.* New York: Wiley.

Schoenberg, I. J. (1964). Spline functions and the problem of graduation. *Proc. Natl. Acad. Sci. USA 52*, 947–950.

Schumaker, L. (1981). *Spline Functions: Basic Theory.* New York: Wiley.

Scott, A. J. and C. J. Wild (1986). Fitting logistic models under case-control and choice-based sampling. *J. Roy. Statist. Soc. Ser. B 48*, 170–182.

Scott, D. W. (1985). Averaged shifted histograms: Effective nonparametric density estimators in several dimensions. *Ann. Statist. 13*, 1024–1040.

Scott, D. W. (1992). *Multivariate Density Estimation: Theory, Practice and Visualization.* New York: Wiley.

Seber, G. A. F. (1977). *Linear Regression Analysis.* New York: Wiley.

Silverman, B. W. (1978). Density ratios, empirical likelihood and cot death. *Appl. Statist. 27*, 26–33.

Silverman, B. W. (1982). On the estimation of a probability density function by the maximum penalized likelihood method. *Ann. Statist. 10*, 795–810.

Snyder, D. L. (1975). *Random Point Processes.* New York: Wiley.

Stein, M. L. (1993). Spline smoothing with an estimated order parameter. *Ann. Statist. 21*, 1522–1544.

Stewart, G. W. (1987). Collinearity and least square regression. *Statist. Sci. 2*, 68–100 (with discussions).

Stone, C. J. (1985). Additive regression and other nonparametric models. *Ann. Statist. 13*, 689–705.

Stone, C. J., M. H. Hansen, C. Kooperberg, and Y. K. Truong (1997). Polynomial splines and their tensor products in extended linear modeling. *Ann. Statist. 25*, 1371–1470 (with discussions).

Tapia, R. A. and J. R. Thompson (1978). *Nonparametric Probability Density Estimation.* Baltimore, MD: Johns Hopkins University Press.

Tierney, L. and J. B. Kadane (1986). Accurate approximations for posterior moments and marginal densities. *J. Amer. Statist. Assoc. 81*, 82–86.

Tong, H. (1990). *Nonlinear Time Series.* New York: Oxford University Press.

Turnbull, B. W., B. W. Brown, and M. Hu (1974). Survivorship analysis of heart transplant data. *J. Amer. Statist. Assoc. 69*, 74–80.

Utreras, F. (1981). Optimal smoothing of noisy data using spline functions. *SIAM J. Sci. Statist. Comput. 2*, 349–362.

Utreras, F. (1983). Natural spline functions: Their associated eigenvalue problem. *Numer. Math. 42*, 107–117.

Utreras, F. (1988). Convergence rates for multivariate smoothing spline functions. *J. Approx. Theory 52*, 1–27.

Vardi, Y. (1982). Nonparametric estimation in the presence of length bias. *Ann. Statist. 10*, 616–620.

Vardi, Y. (1985). Empirical distributions in selection bias models. *Ann. Statist. 13*, 178–203.

Venables, W. N. and B. D. Ripley (1999). *Modern Applied Statistics with S-PLUS* (3rd ed.). New York: Springer-Verlag.

Wahba, G. (1978). Improper priors, spline smoothing and the problem of guarding against model errors in regression. *J. Roy. Statist. Soc. Ser. B 40*, 364–372.

Wahba, G. (1980). Automatic smoothing of the log periodogram. *J. Amer. Statist. Assoc. 75*, 122–132.

Wahba, G. (1983). Bayesian "confidence intervals" for the cross-validated smoothing spline. *J. Roy. Statist. Soc. Ser. B 45*, 133–150.

Wahba, G. (1985). A comparison of GCV and GML for choosing the smoothing parameter in the generalized spline smoothing problem. *Ann. Statist. 13*, 1378–1402.

Wahba, G. (1986). Partial and interaction spline models for the semiparametric estimation of functions of several variables. In *Computer Science and Statistics: Proceedings of the 18th Symposium on the Interface*, pp. 75–80.

Wahba, G. (1990). *Spline Models for Observational Data*, Volume 59 of CBMS-NSF Regional Conference Series in Applied Mathematics. Philadelphia: SIAM.

Wahba, G. and J. Wendelberger (1980). Some new mathematical methods for variational objective analysis using splines and cross validation. *Month. Weather Rev. 108*, 1122–1145.

Wahba, G., Y. Wang, C. Gu, R. Klein, and B. E. K. Klein (1995). Smoothing spline ANOVA for exponential families, with application to the Wisconsin Epidemiological Study of Diabetic Retinopathy. *Ann. Statist. 23*, 1865–1895.

Wang, M.-C. (1989). A semiparametric model for randomly truncated data. *J. Amer. Statist. Assoc. 84*, 742–748.

Wang, M.-C., N. P. Jewell, and W.-Y. Tsay (1986). Asymptotic properties of the product limit estimate under random truncation. *Ann. Statist. 14*, 1597–1605.

Wang, Y. (1997). GRKPACK: Fitting smoothing spline ANOVA models for exponential families. *Commun. Statist.–Simulat. Comput. 26*, 765–782.

Wang, Y. and M. B. Brown (1996). A flexible model for human circadian rhythms. *Biometrics 52*, 588–596.

Weinberger, H. F. (1974). *Variational Methods for Eigenvalue Approximation*, Volume 15 of CBMS-NSF Regional Conference Series in Applied Mathematics. Philadelphia: SIAM.

Whittaker, E. T. (1923). On a new method of graduation. *Proc. Edinburgh Math. Soc. 41*, 63–75.

Whittaker, J. (1990). *Graphical Models in Applied Multivariate Statistics*. Chichester: Wiley.

Woodroofe, M. (1985). Estimating a distribution function with truncated data. *Ann. Statist. 13*, 163–177.

Xiang, D. and G. Wahba (1996). A generalized approximate cross validation for smoothing splines with non-Gaussian data. *Statist. Sin. 6*, 675–692.

Zucker, D. M. and A. F. Karr (1990). Nonparametric survival analysis with time-dependent covariate effects: A penalized partial likelihood approach. *Ann. Statist. 18*, 329–353.

Author Index

Abramowitz, M., 35, 53
Akhiezer, N. I., 53
Anderson, J. A., 18, 204, 208, 228
Ansley, C. F., 144
Antoniadis, A., 228, 258
Aronszajn, N., 53, 54
Azzalini, A., 174

Barry, D., 54
Bartoszyński, R., 228
Bates, D. M., 103, 104
Becker, R. A., 82
Berger, J. O., 54, 103
Blair, V., 208
Bose, S., 208
Bowman, A. W., 174
Breiman, L., 93, 94, 105
Brinkman, N. D., 93
Brockwell, P. J., 170
Brown, B. W., 228
Brown, M. B., 144
Buja, A., 91, 92, 94, 104, 105
Bunch, J. R., 104

Casella, G., 53
Chambers, J. M., 82
Chang, Y., 174
Chen, Z., 54, 103, 104, 259
Chhikara, R. S., 174
Cleveland, W. S., 93, 94, 105, 208
Cogburn, R., 174
Cole, T. J., 208
Cosslett, S. R., 205, 208
Cox, D. D., 174, 234, 258, 259
Cox, D. R., 19, 207, 221, 229
Craven, P., 18, 53, 62, 66, 103, 143, 144
Crowley, J., 19

Dalzell, C. J., 144
Davis, H. T., 174
Davis, M. D., 175
Davis, R. A., 170
de Boor, C., 3, 18, 100, 101, 105
de Hoog, F. R., 105
Delampady, M., 12
DeMets, D. L., 175

Dennis, J. E., 131, 189
Devlin, S. J., 93, 94, 105
Dongarra, J. J., 78, 81, 104, 189
Douglas, A., 12
Duchon, J., 136, 144

Elden, L., 79, 104

Fan, J., 208
Fleming, T. R., 19, 215, 216, 223, 229, 246
Folks, J. L., 174
Friedman, J. H., 94

Gao, F., 208
Gaskins, R. A., vii, 18, 206, 208
Gentleman, R., 82
Gijbels, I., 208
Gill, P. E., 54, 80
Gill, R. D., 208, 229, 246
Girard, D. A., 102, 105
Glazman, I. M., 53
Golub, G., 61, 76, 78, 100, 102, 104, 207
Good, I. J., vii, 18, 206, 208
Gray, R. J., 229
Green, P. J., vii, 143, 173, 208
Gu, C., 18, 19, 54, 75, 78, 80, 81, 103, 104, 144, 154, 156, 173–175, 207, 208, 228, 229, 258

Härdle, W., 168, 174
Halpern, J., 15
Hansen, M. H., 207, 208
Harrington, D. P., 19, 215, 216, 223, 229, 246
Harville, D. A., 103
Hastie, T., 19, 82, 104, 105, 229
Heckman, N. E., 104, 134, 144
Hornik, K., 82
Hu, M., 19
Hutchinson, M. F., 105

Ihaka, R., 82

Jewell, N. P., 208
Johnson, R. A., 48, 72
Jones, M. C., 208

Kadane, J., 161
Kalbfleisch, J. D., 19, 229
Kaplan, E. L., 212
Karlin, S., 123
Karr, A. F., 18, 229
Keiding, N., 208
Kernighan, B. W., 81
Kimeldorf, G., vii, 18, 53, 54, 144
Klein, B. E. K., 54, 173–175, 208
Klein, R., 54, 173–175, 208
Koenker, R., 208
Kohn, R., 144
Kooperberg, C., 207, 208

Lauritzen, S. L., 19
Lawless, J. F., 19
Lehmann, E. L., 53
Leonard, T., 18, 161, 206
Li, K.-C., 66, 103
Lin, X., 208
Lin, Y., 259
Lindsey, J. K., 174
Lindstrom, M., 104
Lumley, T., 230
Lynch, R., 18

Mallows, C. L., 103
Manski, C. F., 208
McBride, C. M., 228
McCullagh, P., 151, 173, 174
Meier, P., 212
Meinguet, J., 136, 137, 139, 144
Mesbah, M., 229
Miller, R. G., 15, 19
Moler, C. B., 104
Moreau, T., 218, 229
Morgenthaler, S., 208
Moss, S. E., 175
Murray, W., 54

Nelder, J. A., 151, 173, 174
Ng, P., 208

Nychka, D., 75, 104

O'Quigley, J., 229
O'Sullivan, F., 18, 101, 105, 173, 174, 206, 228, 229, 258, 259

Parzen, E., 207
Pawitan, Y., 174
Portnoy, S., 208
Prentice, R. L., 19, 208, 229
Priestley, M. P., 143, 170
Pyke, R., 208

Qiu, C., 18, 54, 207, 258

Ramsay, J. O., 134, 144
Rao, C. R., 53
Raynor, W., 18, 173
Ripley, B. D., 82, 131, 174
Robinson, G. K., 103

Scheffe, H., 19
Schnabel, R. B., 131, 189
Schoenberg, I. J., 3, 18, 54
Schumaker, L., 3, 18, 53, 100, 101, 105, 123, 125–127, 144
Schur, I., 54
Scott, A. J., 208
Scott, D. W., 174, 190, 200, 207
Seber, G. A. F., 19
Senthilselvan, A., 18, 228
Silverman, B. W., vii, 18, 54, 143, 173, 183, 206, 258
Snyder, D. L., 182
Stegun, I. A., 35, 53
Stein, M. L., 144
Stewart, G. W., 87, 104
Stone, C. J., 19, 207, 208
Studden, W. J., 123

Tapia, R. A., 51, 54
Therneau, T. M., 230
Thompson, J. R., 51, 54, 228
Tibshirani, R., 19, 104, 105, 229
Tierney, L., 161
Tong, H., 170
Truong, Y. K., 207, 208
Tsay, W.-Y., 208
Turnbull, B. W., 19

Utreras, F., 233, 234, 258, 259

Van Loan, C., 61, 76, 78, 100, 102, 104, 207
Vardi, Y., 208
Venables, W. N., 82, 131, 174

Wahba, G., vii, 18, 19, 53, 54, 59, 62, 66, 68, 72, 75, 80, 103, 104, 136, 137, 139, 143, 144, 152, 155, 156, 173–175, 208, 234
Wang, J., 207
Wang, M.-C., 14, 19, 208
Wang, Y., 54, 144, 171, 173–175
Weinberger, H. F., 232, 258
Wellner, J. A., 208
Wendelberger, J., 136, 137, 139, 144
Whittaker, E. T., 18
Whittaker, J., 12, 19
Wichern, D. W., 48, 72
Wild, C. J., 208
Woltring, H. J., 101
Wong, C.-M., 144
Woodroofe, M., 208
Wright, M. H., 54

Xiang, D., 152, 155, 156, 174

Yan, L., 54
Yandell, B., 18, 104, 173

Zucker, D. M., 18, 229

Subject Index

Accelerated life models, 222–228, 230
ACE (alternating conditional expectation), 94, 105
Additive models, 6, 10, 12, 14, 19, 46, 47, 57, 94, 105, 109, 142, 172, 180, 192, 213, 242
AIDS incubation, 14, 191, 198
Algorithm
  back-fitting, 92, 94, 105
  convergence of, *see* Convergence
  for density estimation, 187–190
  for hazard estimation, 216
  for L-splines, 134–135
  for penalized least squares, 76–81, 99–103
  for penalized likelihood regression, 151–159
ANOVA decomposition, 6–12, 20, 31, 32, 34, 38, 39, 41, 85–87, 114, 140, 141, 178, 180, 199, 201, 202
ANOVA models, 6, 7, 9, 10, 18, 31, 46, 47
At-risk
  probability, 215
  process, 215, 221
Averaging operator, 6–9, 11, 20, 31, 34, 38–41, 56, 85, 114, 140, 178, 179, 199

B-splines, 18, 101, 105, 206
  basis, *see* Basis
Base hazard, *see* Hazard
Basis
  B-spline, 101, 206
  local-support, 99–101
  of null space, 35, 60, 73, 85, 105, 123, 125–127, 138, 140, 148, 175, 202
  orthonormal, 35, 56, 123, 125–127, 138, 140, 148, 234
  trigonometric, 114

Bayes model, 46–49, 54, 68–69, 72–75, 159–161
  empirical, 46, 54
Bayesian confidence intervals, 19, 72–75, 80, 83, 93, 99, 104, 132, 133, 136, 138, 170, 171, 173, 174, 227
  approximate, 159–161
Bernoulli
  polynomials, 35, 37, 53
  trials, 166
Biased sampling, *see* Sampling
Biasing function, 193, 194, 196, 198
Bilinear form, 23, 26, 60, 150
  non-negative definite, 23, 25
  positive definite, 23
  symmetric, 23
Binomial
  distribution, *see* Distribution
  family, *see* Family
BLAS, 81
Blood transfusion data, *see* Data sets
Box-Cox transformation, 208
Buffalo snowfall data, *see* Data sets

Cauchy sequence, 24, 29, 55
CDC blood transfusion data, *see* Data sets
Censored data, 5, 15, 20, 212, 218, 219, 224, 226, 227, 245
  regression, *see* Regression
Censoring, 15, 218, 224, 227
  rate, 217
  time, 5, 15, 212, 219
Chebyshev
  space, 127, 134, 135
  spline, 53, 120, 122–126, 144
  system, 122, 123, 127
Cholesky decomposition, 76, 189, 190
  band, 78, 100
  modified, 80
Closed
  set, 24
  space, 24, 26, 27, 30, 34, 55
Closedness, 24, 27
$\mathcal{C}^{(m)}[0,1]$ space, 32–38, 53, 122, 127
Collinearity, 86, 91, 92, 104
  indices, 87
Complete continuity, 232–235, 247, 254
Complete space, 24, 25, 29
Concurvity, 89–92, 104
Conditional
  density, 198–204, 208, 244, 245
  distribution, 11, 205
  independence, 10–12, 180, 182
  mean, 208
  median, 201
  non-negative definiteness, 136, 137
  probability, 202, 215
  quantiles, 201, 208
  variance, 72
Conditional density estimation, *see* Density estimation
Confidence intervals, 14
  Bayesian, *see* Bayesian confidence intervals
  parametric, 72, 75
Consistency, 63, 66, 67, 69
Constrained
  least squares, 3
  minimization, 137
  optimization, 50, 52–54, 206
Continuity, 22–24, 27, 31, 35, 50, 51, 56, 60, 103, 105, 136, 151, 175, 179, 213
  complete, *see* Complete continuity

Continuous
derivative, 100, 109
domain, *see* Domain
function, *see* Function
functional, *see* Functional
Contrast, 7, 32, 34, 38, 41, 47, 68, 69, 114, 202, 213, 219
Convergence, 23, 24, 29, 51, 57
of algorithms, 76, 80, 103, 154, 158, 159, 165, 166, 189
rates, 63, 180, 195, 200, 206, 213, 231, 234–259
Convex
combination, 152, 153, 256
function, 58
functional, *see* Functional
set, 237, 241, 244, 248, 252, 256
Convexity, 50–52, 58, 60, 105, 151, 175, 178, 205, 210, 212, 242
Covariance
function, 46, 47, 49, 68, 72, 73, 160, 170
matrix, 48, 107
posterior, 74
Cross-validation, 65–66, 103, 189, 190, 192, 196, 200, 206, 209, 230
approximate, 156, 174
for density estimation, 184–187
direct, 155–158, 183, 207, 229
generalized, 62, 66–72, 131, 144
generalized approximate, 152, 155–158
for hazard estimation, 214–217
indirect, 152–155
for logistic regression, 203
modified, 185–186, 190–192, 196, 200, 201, 218, 219, 221
Monte Carlo, 101–103
for non-Gaussian regression, 151–159
Cubic spline, 2–5, 32, 34, 37, 70, 82–84, 89–91, 93, 109, 125, 128–133, 137, 140, 162–164, 166, 172, 179, 187, 190, 192, 196, 198, 213, 217, 221, 224
tensor product, 41, 84–85, 88, 93, 96, 199, 213, 219, 233
with jump, 112
CV, *see* Cross-validation

Data sets
Buffalo snowfall, 190–191, 207
CDC blood transfusion, 14, 19, 191–192, 198, 207
EPA lake acidity, 12–14, 19, 141–143
gastric cancer, 218–219, 229
Los Angeles ozone, 94–98, 105
nitrogen oxides (NO_x), 93–94, 105
Old Faithful eruption, 168–170, 174, 191
Stanford heart transplant, 15–16, 19, 219–221, 227–229
weight loss, 130–134
WESDR, 171–173, 175
yearly sunspots, 170–171, 174
Decomposition
ANOVA, *see* ANOVA decomposition
Cholesky, *see* Cholesky decomposition
eigenvalue, *see* Eigenvalue

of kernel, *see* Reproducing kernel
QR, *see* QR-Decomposition
of reproducing kernel, *see* Reproducing kernel
spectral, *see* Spectral
tensor sum, *see* Tensor sum decomposition
Density, 4, 10, 11, 14, 20, 150, 151, 162, 164–166, 170, 176, 178, 182–184, 191–196, 199, 204, 210, 215, 221, 222
joint, 11, 20, 194, 198, 204, 206
limiting, 243, 254
marginal, 14, 192, 199, 205, 244
occurrence, 182
spectral, *see* Spectral
unimodal, 191
Density estimation, 170, 174, 207, 221
conditional, 198–204, 208, 244
convergence rates for, 234–245
penalized likelihood, 4, 14, 18, 178–210
Poisson, 182
Density-quantile autoregressive estimation, 207
Dimension of linear space, 22, 23
Discrete Fourier transform, 115, 143, 170, 171
inverse, 115
Dispersion, 150, 163–166, 171, 174, 225, 254
Distance, 23, 24
Euclidean, *see* Euclidean
Kullback-Leibler, *see* Kullback-Leibler distance
Distribution
binomial, 162, 174
conditional, 11, 205
empirical, 19, 208
exponential, 164, 171, 217
exponential family, 150, 152, 254, 259
extreme value, 222
gamma, 164, 167, 174
inverse Gaussian, 165, 166, 174
log logistic, 223, 230
log normal, 223, 230
logistic, 223
negative binomial, 166
normal, 48, 72, 223
Poisson, 163, 167, 174, 182
posterior, 72–74
sampling, 72, 93
survival, 222
uniform, 10, 195
Weibull, 222, 230
Domain
binary, 208
bounded, 4, 178, 195, 234
continuous, 7, 20, 37, 178
discrete, 6, 7, 18, 20, 30–32, 39, 40, 46–47, 53, 221
marginal, 10, 38, 40, 41, 85, 138, 192
multidimensional, 135, 140, 259
product, 5–7, 11, 12, 38–47, 198, 199, 204, 212
restricted, 91

Eigenfunction, 232
Eigenvalue, 31, 61, 64, 67, 68, 106, 115, 232–235, 247, 254, 258
decomposition, 31, 64
Eigenvector, 31
EPA lake acidity data, *see* Data sets
Estimate
Bayes, 46–49

existence of, *see* Existence of estimates
interval, 72–75, 159–161
Kaplan-Meier, 212
kernel, 208
martingale moment, 229
maximum likelihood, 4, 50, 167, 178, 180, 212
penalized least squares, 259
penalized likelihood, 50, 54, 178, 182, 223, 231, 258
shrinkage, 32, 46–47, 53
unbiased, 62, 63, 69
variance, 67, 68, 72, 75, 86, 103

Estimation
density, *see* Density estimation
hazard, *see* Hazard estimation
parametric, 12, 208, 229
penalized likelihood, 4, 5, 8, 46, 194–196, 199–200, 206, 208, 230
of relative risk, *see* Relative risk
spectral, *see* Spectral

Euclidean
distance, 23, 136
inner product, 26
norm, 23, 87
space, 23, 24, 26–28, 31, 55, 56, 58

Evaluation, 27, 28, 31, 150, 179, 180, 213
functional, 27, 136, 150, 179, 180, 213
representer of, 27, 31, 33, 36, 53

Existence of estimates, 50–51, 54, 60, 136, 151, 178, 180, 212

Exponential distribution, *see* Distribution

Exponential family distribution, *see* Distribution

Extreme value distribution, *see* Distribution

Family
binomial, 162–163
exponential, *see* Exponential family distribution
gamma, 164–165, 176, 254
inverse Gaussian, 165–166, 176, 254
location-scale, 222
log logistic, 222, 226–227, 230
log normal, 222, 225–226, 230
negative binomial, 166–168, 174, 176, 254
Poisson, 163–164
Weibull, 222, 224–225, 230

Fixed effects, 46, 47, 49, 73, 74, 108, 112, 121

Fourier series expansion, 113, 143, 232, 235, 254

Fréchet differentiability, 52, 105, 175, 184

Function
biasing, *see* Biasing function
continuous, 58, 113, 216, 246
convex, 58
covariance, *see* Covariance
cumulative distribution, 222, 223
delta, 178
on discrete domains, 6, 26, 31, 39
even, 170
Green's, *see* Green's function
hazard, *see* Hazard
inverse, *see* Inverse

non-negative definite, 22, 28, 39
periodic, 35, 113, 120, 232
predictable, 216
survival, *see* Survival
symmetric, 27
Functional, 23, 24, 52, 136, 147, 151
continuous, 23, 27, 50, 52
convex, 50, 52, 58, 183
evaluation, *see* Evaluation
least squares, 54, 60–62, 69, 105
linear, 23, 27, 31, 56, 232
log likelihood, 60, 63, 175, 178, 183
penalized least squares, 60, 151
penalized likelihood, 27, 50, 150, 195, 212, 247, 254
penalized partial likelihood, 230
quadratic, 4, 23, 50, 58, 231, 232, 245, 247, 253, 254

GACV (generalized approximate cross-validation), *see* Cross-validation
Gamma
distribution, *see* Distribution
family, *see* Family
regression, *see* Regression
Gastric cancer data, *see* Data sets
Gaussian
approximation, 161
inverse, *see* Inverse Gaussian
likelihood, 161
prior, *see* Prior
process, 49, 68, 72, 73, 160
regression, *see* Regression
GCV (generalized cross-validation), *see* Cross-validation
GML (generalized maximum likelihood), 68, 69, 84
Graphical models, 10–12, 19
Green's function, 123, 126, 127
GRKPACK, 171, 174

Hazard, 5, 12, 212, 213, 217–220, 228
base, 12, 15, 16, 212, 220, 221, 228
empirical, 217
function, 5, 212, 222, 223, 228
model, *see* Proportional hazard model
rate, 212, 220
Hazard estimation, 20
convergence rates for, 245–253
penalized likelihood, 5, 15–16, 18, 212–220, 228
Heart transplant data, *see* Data sets
Hilbert space, 22–28, 50, 52, 53, 55, 56, 58
finite-dimensional, 27, 55
reproducing kernel, *see* Reproducing kernel Hilbert space

Independence, 5, 11, 14, 47, 49, 69, 72, 73, 102, 107, 110, 155, 160, 170, 180, 191, 212, 215, 216, 246
conditional, *see* Conditional
linear, 22, 23
Inequality
Cauchy-Schwarz, 23, 29, 55, 106, 118, 119, 238, 249, 257
Hölder's, 178, 213
triangle, 23, 24, 55

Inner product, 23–26, 29–33, 35, 36, 40–45, 49, 52, 53, 55, 56, 60, 114, 120–124, 127–130, 139, 140, 146, 148, 150, 210
  semi, 25, 26, 35, 43, 45, 52, 139
Interaction, 6, 7, 9–11, 13, 40, 41, 43, 57, 84, 85, 88, 89, 95–97, 105, 109, 141, 142, 172, 205, 210, 219, 220, 242
Invariance, 66
  isotropic, 12
  rotation, 136, 141, 147
Inverse, 31, 56
  function, 125, 146
  matrix, 128–130
  Moore-Penrose, 31, 40, 156, 160
Inverse Gaussian
  distribution, *see* Distribution
  family, *see* Family
  regression, *see* Regression
Isomorphism, 27, 55, 56, 113, 180, 221, 229

Kernel
  method, 208
  reproducing, *see* Reproducing kernel
  semi-, 136–138, 140
Knot, 3, 100, 109
Kronecker
  delta, 35, 56, 117, 123, 127, 138, 232
  product, 40
Kullback-Leibler distance, 152, 155, 183, 187, 195, 196, 200, 203, 206, 215, 217, 245
  relative, 155, 184, 185, 196, 200, 203, 206, 215, 216, 242–245, 253, 257
  symmetrized, 152, 234, 238, 242, 244, 245, 247, 249, 250, 253
Kullback-Leibler projection, 243

L-splines, 70, 119–135, 144
$\mathcal{L}_2[0,1]$ space, 25–27, 32, 36, 41, 114, 127
Lagrange
  method, 2, 3
  multiplier, 2–4, 52
Lake acidity data, *see* Data sets
Least squares
  constrained, 3
  functional, *see* Functional
  parametric, 131, 136
  penalized, *see* Penalized least squares
  weighted, 61, 69, 76, 151, 159, 172, 173, 223
Likelihood, 4, 5, 20, 22, 27, 50, 52, 60, 62, 68, 106, 155, 160–164, 166, 167, 176, 178, 182, 184, 186, 205, 210, 212, 216, 230, 254
  under biased sampling, 194, 221, 229
  of lifetime, 20, 223–226
  maximum, *see* Maximum likelihood
  partial, 212, 221, 229
  penalized, *see* Penalized likelihood
  profile, 68
Limit point, 23, 24, 29
Linear
  approximation, 235, 247, 254
  combination, 74, 75
  functional, *see* Functional
  independence, 22, 23
  regression, *see* Regression
  system, 61, 76, 92, 99, 101–103, 105, 115, 137, 190, 209

Linear models, 2, 18, 19, 23, 82, 86, 87, 92, 230
  extended, 207
  generalized, 173, 174
  log, 19
  retrospective, 87, 108
Linear space, 22–26, 28–30, 34, 50, 52, 55, 100, 113, 127, 180
  dimension of, 22, 23
Linear spline, 33, 37, 85, 89
  tensor product, 40, 88–90, 95, 109, 172
Linearity, 2, 3, 52, 65, 173, 219, 235, 255
  non-, 132
LINPACK, 78, 81, 189
Log logistic
  distribution, *see* Distribution
  family, *see* Family
  regression, *see* Regression
Log normal
  distribution, *see* Distribution
  family, *see* Family
  regression, *see* Regression
Logistic
  distribution, *see* Distribution
  regression, *see* Regression
  spline, 130
Logistic density transform, 10, 178, 206, 221
  conditional, 11, 198, 199
Los Angeles ozone data, *see* Data sets
Loss, 63, 69, 71, 152, 158, 183, 188, 203, 215, 217, 230
  relative, 63, 66, 69

Main effect, 6, 7, 9–11, 40, 41, 43, 57, 95–98, 109
  multivariate, 141
Matrix
  banded, 99, 100, 134
  column space of, 30, 31, 47
  computation, 6, 76–81, 99–103, 134–135, 187–190
  conditionally non-negative definite, 137
  covariance, *see* Covariance
  Fourier, 115, 144
  of full column rank, 48, 73, 105, 151, 175
  Gram, 138
  identity, 28, 70
  inverse, 128–130
  non-negative definite, 23, 31, 39, 105
  orthogonal, 61, 64, 70, 77, 79, 107, 147, 156, 175
  positive definite, 23
  projection, 31, 108
  scatter plot, 95, 105
  smoothing, *see* Smoothing matrix
  sparse, 101
  symmetric, 23, 48, 61, 73, 106
  Wronskian, 127–130
Maximum likelihood, 2, 178
  estimate, *see* Estimate
  generalized, 68, 69, 84
  nonparametric, 212
  partial, 229
  restricted, 68–69, 103
  semiparametric, 19
  type-II, 103
Mean
  (as opposed to contrast), 34, 38, 47, 114, 213, 219
  conditional, 208
  geometric, 218
  posterior, 46–49, 72, 74, 83, 84, 160, 161
  sample, 184, 216
  treatment, 6

zero, 46–49, 63, 66, 68, 69, 72, 73, 102, 153, 160
Mean value theorem, 152, 237, 249, 250, 252, 253, 256
Mixed effect models, 46–49, 68, 72–75, 103
Model
accelerated life, 222–228, 230
additive, *see* Additive models
ANOVA, *see* ANOVA models
Bayes, *see* Bayes model
graphical, 10–12, 19
linear, *see* Linear models
mixed effect, *see* Mixed effect models
nonlinear, 131
parametric, 2, 75, 92, 120, 131, 212, 222
proportional hazard, *see* Proportional hazard model

Natural spline, 3, 54, 100, 101, 105, 109
Negative binomial
distribution, *see* Distribution
family, *see* Family
Nitrogen oxides (NO_x) data, *see* Data sets
Non-negative definite
bilinear form, *see* Bilinear form
function, *see* Function
matrix, *see* Matrix
Non-negative definiteness, 23, 28, 30, 33, 38, 39, 137, 139, 189
conditional, 136, 137
Norm, 23, 25, 29, 35, 36, 46, 50, 52, 139, 178, 189, 202, 234
Euclidean, *see* Euclidean
semi, 25, 46, 50, 52, 105, 136, 139, 175, 178
Normal distribution, *see* Distribution
Normality, 6, 62, 64, 102, 153
Null space, 2, 4, 25–27, 32, 35, 45, 46, 50, 60, 85, 112, 120–122, 125–127, 131, 136, 138, 140, 150, 178–180, 199, 201, 212, 213, 219, 234
augmented, 112
basis, *see* Basis
Numerical linear algebra, 76, 104
software, 54, 81

Old Faithful eruption data, *see* Data sets
Operator
averaging, *see* Averaging operator
differential, 119–123, 126, 128, 131, 146
identity, 7, 138
projection, 138
Optimization
constrained, 50, 52–54, 206
penalized, 50, 52–54
Orthogonal complement, 24, 26, 33, 46, 55
Orthogonality, 29, 30, 34, 47, 78, 87, 93, 114, 117, 145, 234
Orthonormal basis, *see* Basis
Ozone data, *see* Data sets

Partial spline, 112–113, 121, 143, 198, 201, 213
Penalized least squares, 2–4, 18, 60–110, 150, 155, 175, 223, 259
functional, *see* Functional
Penalized likelihood, 2, 4–5, 18, 208, 228, 233

density estimation, *see* Density estimation
estimate, *see* Estimate
estimation, *see* Estimation
functional, *see* Functional
hazard estimation, *see* Hazard estimation
partial, 221, 229, 230
regression, *see* Regression
Penalized optimization, 50, 52–54
Periodic spline, 113–119
Periodicity, 36, 57
Periodogram, 170, 171
Poisson
distribution, *see* Distribution
family, *see* Family
process, 182, 207
regression, *see* Regression
Polynomial spline, 32–38, 41, 47, 49, 53, 54, 99, 100, 104, 120, 124, 125, 233
periodic, 113, 115
Polynomials, 2, 33, 35, 100, 120, 136, 138, 140, 147, 148
Bernoulli, 35, 37, 53
piecewise, 3, 10, 100, 109
Positive definite
bilinear form, *see* Bilinear form
matrix, *see* Matrix
Positive definiteness, 23, 25, 80
Prior, 46, 47
diffuse, 46, 47, 49, 73
Gaussian, 46, 47, 49, 72, 73, 160
proper, 107
Probability, 204, 245
at-risk, 215
conditional, 202, 215
density, *see* Density
measure, 195
success, 166, 215
survival, 20
Process
at-risk, 215, 221
counting, 182, 229
event, 215
Gaussian, *see* Gaussian
Poisson, 182, 207
Projection, 24, 30, 39, 43, 46, 49, 52, 55, 87, 108, 138, 235, 240, 251
Kullback-Leibler, 243
matrix, *see* Matrix
operator, 138
Proportional hazard model, 12, 15, 16, 19, 212, 213, 220, 221, 224, 228, 229
generalization of, 229

QR-Decomposition, 61, 76, 77, 79, 140, 148, 175
Quadratic
approximation, 151, 161, 184, 216, 230, 235
form, 23
functional, *see* Functional

R, 82–87, 93, 94, 104, 112, 130, 131, 141, 143, 161–168, 170, 172, 174, 222, 224, 227, 230
R package
`MASS`, 130
`gss`, 82–87, 104, 112, 141, 161–168, 174, 222
`survival`, 227, 230
`ts`, 170
Radon-Nikodym derivative, 195
Random effects, 46, 47, 49, 73, 74, 108
Regression, 2, 4, 31, 174, 198
censored data, 224, 226, 227
convergence rates for, 253–258
gamma, 164–165, 170–171, 174
Gaussian, 4, 150

inverse Gaussian, 165–166
linear, 2, 173
local weighted, 105
log logistic, 226–227
log normal, 225–226
logistic, 19, 150, 162–163, 166–168, 171–173, 198, 202–203
multinomial logistic, 202–203
non-Gaussian, 174, 183, 212, 222
nonlinear, 130
parametric, 2, 92, 173
penalized least squares, *see* Penalized least squares
penalized likelihood, 4, 18, 19, 150–176, 203
Poisson, 19, 151, 163–164, 168–170, 174, 191
ridge, 101, 103
Weibull, 224–225
Relative risk, 12, 16, 212, 220, 221, 224, 228
estimation of, 221, 229, 230
REML (restricted maximum likelihood), 68–69, 103
Representer, 27
of evaluation, *see* Evaluation
Reproducing kernel, 22, 27–46, 49, 53, 56, 60, 85, 113, 114, 120–124, 127–130, 135–141, 150, 199, 201, 202, 210
decomposition of, 31, 34, 38, 140
marginal, 38–40, 53
product, 40
Reproducing kernel Hilbert space, 27–32, 38–46, 49, 53, 56, 60, 105, 139, 175, 178, 182, 212
periodic, 114
tensor product, 38–46, 54, 57, 199
tensor sum, 201, 202
Riesz representation theorem, 27, 53
RKPACK, 78, 81, 82, 85, 104, 161
G-, 171, 174

Sampling
biased, 192–198, 208, 221, 229, 243
case-control, 204
choice-based, 204, 205
distribution, 72, 93
length-biased, 194, 207, 208
response-based, 204–206, 208, 245
separate, 204, 205
two-stage, 194
Sampling points, 3, 34, 63, 65, 66, 72, 74, 75, 87, 91, 98, 100, 152, 161, 172, 173
marginal, 141
Side condition, 4, 6, 7, 11, 14, 32, 39, 178–180, 182, 183, 198, 199, 205, 206, 221, 228, 243
Singularity, 61, 70, 189
non-, 48, 73, 157, 160, 161, 175, 190
numerical, 78, 189, 207
Smoothing, 41, 182, 191, 208
under-, 72, 118, 141, 186, 187, 190, 196
Smoothing matrix, 61, 70, 74, 102, 103, 106, 107, 156
Smoothing parameter, 2, 4, 5, 60, 62, 63, 75–77, 79, 83, 84, 86, 88, 89, 99, 101, 102, 115, 151–155, 181–184, 186, 189–191, 207

selection of, 5, 92, 94, 102–104, 114, 116, 141, 151–159, 183–187, 192, 193, 198, 200, 201, 207, 218, 219, 221, 229
Smoothing spline, 2–4, 18, 32–38, 41, 45–47, 49, 53, 54, 70, 75, 82, 100, 258
computation of, 34–35, 37, 45, 54
on discrete domain, 30–32
interpolation, 18
on product domain, 38–46
tensor product, 38–46, 53, 54, 57, 58
Space
Chebyshev, *see* Chebyshev
closed, *see* Closed
$\mathcal{C}^{(m)}[0,1]$, *see* $\mathcal{C}^{(m)}[0,1]$ space
complete, *see* Complete space
Euclidean, *see* Euclidean
finite-dimensional, 4, 24, 25, 27, 31, 34, 40, 49–51, 58, 60, 100, 127, 135, 180, 202, 213, 235
Hilbert, *see* Hilbert space
$\mathcal{L}_2[0,1]$, *see* $\mathcal{L}_2[0,1]$ space
linear, *see* Linear space
marginal, 41, 43
null, *see* Null space
tensor product, 38–46, 202, 210
vector, *see* Vector
Spectral
decomposition, 114–116, 118, 144
density, 170, 171
estimation, 170–171, 174
Spline, 18, 85, 112, 162
B, *see* B-splines
Chebyshev, *see* Chebyshev
on the circle, 113–119
cubic, *see* Cubic spline
exponential, 125, 128
hyperbolic, 126
L, *see* L-splines
linear, *see* Linear spline
logistic, 130
natural, *see* Natural spline
partial, *see* Partial spline
periodic, *see* Periodic spline
polynomial, *see* Polynomial spline
regression, 207
smoothing, *see* Smoothing spline
tensor product, 136, 138, 180
thin-plate, *see* Thin-plate spline
trigonometric, 120–122, 129
Splus, 82, 230
Stanford heart transplant data, *see* Data sets
Survival, 15, 219, 220
analysis, 19, 229, 230
distribution, 212, 222
function, 5, 212, 217, 222, 223
probability, 20
time, 15, 218, 219
Symmetric
bilinear form, *see* Bilinear form
function, 27
matrix, *see* Matrix
Symmetry, 23, 37, 60, 100

Taylor expansion, 32, 37, 53, 155, 184, 237
generalized, 37, 124, 125, 128
Tensor product
cubic spline, *see* Cubic spline
linear spline, *see* Linear spline

smoothing spline, *see* Smoothing spline
space, *see* Space
thin-plate spline, 12, 19, 140, 141, 144
Tensor sum decomposition, 24, 25, 30, 31, 33, 34, 36–41, 43, 45
Thin-plate spline, 85, 135–144, 234
tensor product, 12, 19, 140, 141, 144
Transform
cubic root, 94
discrete Fourier, *see* Discrete Fourier transform
log, 13, 93–95
logistic density, *see* Logistic density transform
monotone, 68
multinomial logistic, 202
orthogonal, 66, 147
square root, 16, 206, 219
Trigonometric
basis, 114
spline, 120–122, 129
Truncation, 14, 19, 187, 191, 192, 194, 208, 217
left-, 5, 212

Variance
components, 103
conditional, 72
estimate, *see* Estimate
inflation factor, 87
posterior, 72, 73, 80, 86, 99, 134, 138, 160, 161
Vector
response, 70, 106
space, 22, 23, 25, 26, 30, 31, 40, 53
unit, 28, 56, 74, 78, 101, 115

Weibull
distribution, *see* Distribution
family, *see* Family
regression, *see* Regression
Weight loss data, *see* Data sets
WESDR data (Wisconsin Epidemiological Study of Diabetic Retinopathy), *see* Data sets

Yearly sunspots data, *see* Data sets

Springer Series in Statistics

(continued from p. ii)

Ibrahim/Chen/Sinha: Bayesian Survival Analysis.
Kolen/Brennan: Test Equating: Methods and Practices.
Kotz/Johnson (Eds.): Breakthroughs in Statistics Volume I.
Kotz/Johnson (Eds.): Breakthroughs in Statistics Volume II.
Kotz/Johnson (Eds.): Breakthroughs in Statistics Volume III.
Küchler/Sørensen: Exponential Families of Stochastic Processes.
Le Cam: Asymptotic Methods in Statistical Decision Theory.
Le Cam/Yang: Asymptotics in Statistics: Some Basic Concepts, 2nd edition.
Liu: Monte Carlo Strategies in Scientific Computing.
Longford: Models for Uncertainty in Educational Testing.
Mielke, Jr./Berry: Permutation Methods: A Distance Function Approach.
Miller, Jr.: Simultaneous Statistical Inference, 2nd edition.
Mosteller/Wallace: Applied Bayesian and Classical Inference: The Case of the Federalist Papers.
Pan/Fang: Growth Curve Models and Statistical Diagnostics.
Parzen/Tanabe/Kitagawa: Selected Papers of Hirotugu Akaike.
Politis/Romano/Wolf: Subsampling.
Ramsay/Silverman: Functional Data Analysis.
Rao/Toutenburg: Linear Models: Least Squares and Alternatives.
Read/Cressie: Goodness-of-Fit Statistics for Discrete Multivariate Data.
Reinsel: Elements of Multivariate Time Series Analysis, 2nd edition.
Reiss: A Course on Point Processes.
Reiss: Approximate Distributions of Order Statistics: With Applications to Non-parametric Statistics.
Rieder: Robust Asymptotic Statistics.
Rosenbaum: Observational Studies, 2nd edition.
Rosenblatt: Gaussian and Non-Gaussian Linear Time Series and Random Fields.
Särndal/Swensson/Wretman: Model Assisted Survey Sampling.
Schervish: Theory of Statistics.
Shao/Tu: The Jackknife and Bootstrap.
Siegmund: Sequential Analysis: Tests and Confidence Intervals.
Simonoff: Smoothing Methods in Statistics.
Singpurwalla and Wilson: Statistical Methods in Software Engineering: Reliability and Risk.
Small: The Statistical Theory of Shape.
Sprott: Statistical Inference in Science.
Stein: Interpolation of Spatial Data: Some Theory for Kriging.
Taniguchi/Kakizawa: Asymptotic Theory of Statistical Inference for Time Series.
Tanner: Tools for Statistical Inference: Methods for the Exploration of Posterior Distributions and Likelihood Functions, 3rd edition.
Tong: The Multivariate Normal Distribution.
van der Vaart/Wellner: Weak Convergence and Empirical Processes: With Applications to Statistics.
Verbeke/Molenberghs: Linear Mixed Models for Longitudinal Data.
Weerahandi: Exact Statistical Methods for Data Analysis.
West/Harrison: Bayesian Forecasting and Dynamic Models, 2nd edition.